Pollen Biology
and
Biotechnology

Pollen Biology
and
Biotechnology

K.R. Shivanna

CRC Press
Taylor & Francis Group
Boca Raton London New York

CRC Press is an imprint of the
Taylor & Francis Group, an **informa** business
A SCIENCE PUBLISHERS BOOK

First published 2003 by Science Publishers, Inc.

Published 2018 by CRC Press
Taylor & Francis Group
6000 Broken Sound Parkway NW, Suite 300
Boca Raton, FL 33487-2742

© 2003 by copyright reserved
CRC Press is an imprint of Taylor & Francis Group, an Informa business

No claim to original U.S. Government works

ISBN-13: 978-1-57808-241-4 (pbk)
ISBN-13: 978-1-57808-242-1 (hbk)

Visit the Taylor & Francis Web site at
http://www.taylorandfrancis.com

and the CRC Press Web site at
http://www.crcpress.com

Library of Congress Cataloging-in-Publication Data

Shivanna, K.R.
  Pollen biology and biotechnology/K.R. Shivanna.
    p. cm.
  Includes bibliographical references (p.).
  ISBN 1-57808-241-2
   1, Pollen. 2. Pollen-Biotechnology. I. Title.

  QK658.S5353 2003
  571.8'45--dc21

Dedicated to my Guru and Mentor,
Professor NS Rangaswamy

Preface

The last three decades have been the most exciting period in the history of pollen biology. Integrated studies using diverse techniques particularly of advanced microscopy, cell and molecular biology, and genetics have revolutionized our knowledge of the structural and functional aspects of pollen, leading to the development of a 'new biology' of pollen. This remarkable progress on pollen biology has resulted in the development of a number of effective and viable pollen-based technologies, broadly termed pollen biotechnology, for practical benefit, especially for crop improvement.

The vast knowledge generated on pollen biology and biotechnology is scattered in Journals devoted to a number of disciplines such as morphology, physiology, biochemistry, cell and molecular biology, plant tissue culture, agriculture, horticulture, forestry and plant breeding. The books available at present are confined to the coverage of a few selected areas of pollen biology/biotechnology. It has thus become difficult for teachers, students and researchers to get an overview of recent developments covering the whole spectrum of pollen biology and biotechnology.

The primary objective of the present volume is to give a coherent and concise account of pollen biology and biotechnology with an emphasis on recent developments. The discussion is largely confined to pollen grains of flowering plants. The number of papers published on pollen is so large that the over 1500 references cited had necessarily to be selective and subjective. The reader is able to access most of the literature in the field by referring to the cited papers and reviews. The topics included in the book are interdisciplinary and cater not only to pollen biologists and biotechnologists but also to reproductive biologists, pollination biologists, aerobiologists, plant breeders, horticulturists and foresters. I hope that the book, apart from providing an overview of pollen biology and biotechnology to students, teachers and researchers will facilitate integration of pollen biotechnology into the traditional methods of crop production and improvement.

My involvement with pollen goes back to 1964 when I started my Ph.D. programme under the supervision of Professor NS Rangaswamy. Another dimension was added to my career on pollen when I began, in 1974, collaborative research with late Prof. J Heslop-Harrison FRS, and Dr (Mrs) Y Heslop-Harrison, first at the Royal Botanic Gardens, Kew and then at the Welsh Plant Breeding Station, UK, which has continued over the years. My long association with Professor Rangaswamy and Heslop-Harrisons has immensely benefited my comprehension of pollen; I owe all those a great debt. I also had the privilege of discussing various aspects of pollen during my visits abroad with many renowned pollen biologists and biotechnologists, in particular Late Prof. RB Knox, Prof. HF Linskens, Prof. M Cresti and Prof. VK Sawhney. I thank them most sincerely for their time, interest and encouragement.

It is a pleasure to acknowledge the help and encouragement received over the years from many of my senior colleagues in the Department in particular Prof. BM Johri, Prof. HY Mohan Ram, Prof. PS Ganapathy and Prof CR Babu. I have greatly benefited in my career from the researches of and discussions with my students on pollen biology and biotechnology for over 25 years. Also, Dr Madhu Bajaj and Dr Rajesh Tandon, my former students, read the entire text of this book and gave constructive suggestions. I convey my deepest feelings of gratitude to all my students. I thank Mr Bhaskar Bhandari for his very able help in preparation of illustrations.

I cannot adequately thank my wife Pramila for her understanding and encouragement throughout my career.

April 2002

KR SHIVANNA
Department of Botany
University of Delhi

Contents

Part I
POLLEN BIOLOGY

The role of pollen grain as the male partner in sexual reproduction of seed plants was established by the end of the 19th century (Maheshwari 1950, 1963). Pollen grains develop in anthers as a result of reduction division and represent highly reduced male gametophytes. In gymnosperms the microspore undergoes 3–5 mitotic divisions to produce male gametes. At the time of shedding, the gymnosperm pollen grains contain a variable number of prothallial cells, a tube cell and an antheridial cell; following pollen germination the antheridial cell divides to form a stalk cell and a body cell. The body cell eventually divides and produces two male gametes (Pacini 2000). In flowering plants the microspore undergoes only two mitotic divisions to produce male gametes. At the time of shedding pollen grains of flowering plants contain either 2 cells (vegetative and generative) or 3 cells (vegetative cell and two male gametes (formed by division of the generative cell)) at the time of dispersal (Fig. 1A-C).

Following anther dehiscence, pollen grains are released into the atmosphere. At the time of release, they are desiccated and their moisture level is generally reduced to less than 15%. Pollen grains survive for varying periods after release as independent functional units. This phase of pollen, termed the free dispersed phase, is an important aspect of pollen biology as it facilitates gene flow.

In gymnosperms, pollen grains land at the micropylar part of the ovule that houses the female gametes inside the archegonia (Fig. 1D). A pollination drop is produced at the micropyle in most of the gymnosperms and aids in pollen passage into the ovule and in pollen germination. The male gametes are released near the archegonia, pass through the neck of the archegonia, reach the egg cell and effect fertilization (Friedman 1993). In flowering plants, pollen grains are deposited on the stigma (Fig. 1E) where they germinate and the resultant pollen tubes grow through the tissues of the stigma and style, enter the ovule and eventually the embryo sac (female gametophyte) present inside the

ovule, where the two male gametes are discharged. This process of conveying the male gametes through directional growth of the pollen tube is termed siphonogamy. Apart from flowering plants, siphonogamy is also present in advanced gymnosperms such as *Pinus* and *Gnetum*.

This phase of pollen biology, from pollination until the pollen tube reaches the embryo sac, is generally referred to as pollen-pistil interaction. Pollen-pistil interaction plays an important role in sexual reproduction. The pollen grain is screened during pollen-pistil interaction; the pistil facilitates germination of compatible pollen and growth of the resultant pollen tube until it reaches the embryo sac while incompatible pollen is inhibited before germination or during the growth of pollen tube before its entry into the embryo sac. One of the two sperms released in the embryo sac fuses with the egg to form the zygote and the other with the secondary nucleus of the central cell to form the primary endosperm nucleus, thus completing the process of double fertilization, characteristic of flowering plants. The zygote gives rise to the embryo and the primary endosperm nucleus to the endosperm which nourishes the embryo. The ovule develops into the seed and the ovary into the fruit. Thus, development of functional pollen, their transfer to the stigma and successful completion of pollen-pistil interaction and fertilization are the prerequisites for fruit- and seed-set. Pollen biology involves a comprehensive understanding of the structural and functional details associated with the aforesaid pollen events.

Seeds and fruits are the economic products of more than 90% of our crop plants. A thorough knowledge of pollen biology and its manipulation are required for any rational approach to increase crop productivity. This realization, in recent years, has greatly stimulated researches on pollen biology. Early studies on pollen biology used only conventional techniques such as microtomy and acetolysis. Since the 1950s, a number of physiological and biochemical techniques were introduced particularly in studies

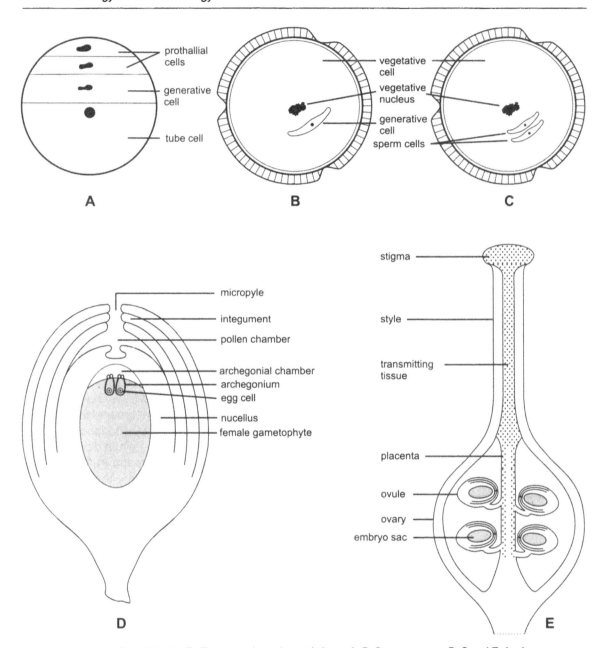

Fig 1. Male (A–C) and female (D, E) gametophytes in seed plants. A, D. Gymnosperms, B, C and E. Angiosperms.

on pollen viability, storage and germination. Invention of electron microscopy in the 1960s added another dimension to the study of pollen. Detailed fine structural studies have since been carried out on almost all facets of pollen biology.

In recent years pollen biologists have been using better fixatives, more diverse and versatile microscopes, and the techniques of cell and tissue culture, and immunofluorescence. During the last 15 years techniques of molecular biology and genetics have been integrated effectively in studies on pollen biology. These integrated approaches have revolutionized our knowledge of the structural and functional aspects of pollen.

The demonstration that pollen grains of many species present in air are responsible for allergy led to comprehensive studies on the distribution of pollen in air and the details of allergy. This area of pollen biology has now developed into an independent branch, aeropalynology (Knox 1979, Mohapatra and Knox 1996).

Since pollen morphology is a conserved character and often specific to a given species/genus, it has been used widely in understanding taxonomic and phylogenetic relationships of species. Studies on fossil pollen grains have added greatly to our understanding of the origin, distribution and evolution of flowering plants. Also, because of the correlation that exists between pollen flora in geologic ages and the presence of petroleum fuel, studies on fossil pollen have become an integral part of the petroleum industry (Faegri and Iversen 1989).

Pollination is another area that has long attracted the attention of pollen biologists (McGregor 1976, Faegri and van der Pijl 1979, Real 1983, Free 1993). Initial studies were largely confined to identifying pollinating agents and floral adaptations. Gradually the pollinator-flower interaction and co-evolution of the flower and the pollinator have become the main focus. In-depth studies are being carried out on pollinator attractants and rewards, specialization of the flower for pollinators and the breeding system (Richards 1986). Two manuals on the techniques of pollination biology were recently published (Dafni 1992, Kearns and Inouye 1993).

Pollen-pistil interaction was largely ignored until the 1960s. Following the realization that the main barrier to fertilization occurs during pollen-pistil interaction, interest in studies on pollen-pistil interaction has steadily increased over the years and this aspect has become one of the front-line areas of research during the last two decades (Raghavan 1999, de Nettancourt 2001).

Fertilization in flowering plants has been a difficult area for research as it takes place deep inside the ovule and imposes technical problems for basic studies as well as effective experimentation. One of the important goals of embryologists has been to achieve in-vitro fertilization using isolated egg and sperm cells so as to facilitate basic as well as experimental studies on fertilization in flowering plants. Considerable advances have been made in recent years on the structural details of fertilization at the light as well as electron microscopic levels. In-vitro fertilization using isolated egg and sperm cells has now become a reality and progress in understanding fertilization is expected to be rapid in the coming years (Kranz and Dresselhaus 1996, Shivanna and Rangaswamy 2000).

Studies on pollen biology have also acquired significance in the light of recent progress in developing transgenic plants in several crop species and their release. Since engineered genes from crop species can escape and spread to other species or cultivars through pollen, generation of detailed information on the extent of pollen flow, pollen viability and its compatibility with related species has become a requirement for the release of transgenics (Bhatia and Mitra 1998).

Extensive literature accumulated on various aspects of pollen biology has been reviewed in a number of books over the years (Maheshwari 1950, 1963, Stanley and Linskens 1974, de Nettancourt 1977, 2001, Knox 1979, Johri 1984, Shivanna and Johri 1985, Raghavan 1999). Many edited volumes have also appeared on different areas of pollen biology/biotechnology (Mulcahy et al. 1986, Linskens 1974, Mulcahy and Ottaviano 1983, Cresti et al. 1988, 1992, Ottaviano et al. 1992, Russel and Dumas 1992, Williams et al. 1994, Mohapatra and Knox 1996, Shivanna and Sawhney 1997, Clement et al. 1999, Dafni et al. 2000).

1

Pollen Development

In flowering plants pollen grains develop in the anther. A young anther consists of a homogeneous mass of meristematic cells surrounded by an epidermis. Further development of the anther involves a major histodifferentiation resulting in several highly specialized cells and tissues. Some of these cell types continue to differentiate while others degenerate. The differentiation and degeneration events take place in a precise spatial and temporal order in the anther and ultimately result in the development and dispersal of pollen grains (Goldberg et al. 1993). Details of pollen development are quite uniform in different species and have been reviewed from time to time (Heslop-Harrison 1971, 1972, Mascarenhas 1975, 1989, Shivanna and Johri 1985, Raghavan 1999). The following are the major events associated with pollen development.

DIFFERENTIATION OF ANTHER LAYERS

The mature anther is usually a bithecous and tetrasporangiate structure. Different cell and tissue layers demarcated in the mature anther can be traced to the three primary germ layers—L_1, L_2 and L_3—present in the floral meristems (see Goldberg et al. 1993). L_1 gives rise to the epidermis, L_2 (hypodermal layer) to the archesporium and L_3 to the connective tissue and the central vascular bundle (Fig 1.1). The epidermis undergoes no further elaboration except for differentiation of the stomium, along the line of anther dehiscence in the mature anther. The connective tissue also shows only a limited differentiation; it gives rise to the inner tapetum which lines the inner portion of the anther locules and a small cluster of cells separating the two pollen sacs of each half anther (intersporangial septum/circular cell cluster) along the stomium. Histodifferentiation events of the anther are largely confined to the derivatives of the hypodermal layer (L_2). As the anther develops, it assumes a four-lobed appearance and four groups of archesporial cells, corresponding to the four microsporangia (one in each anther lobe), differentiate in the hypodermal position. The archesporial cells are easily distinguishable from other cells by their dense cytoplasm and large nuclei. Archesporial cells divide periclinally to form an outer primary parietal layer and an inner primary sporogenous layer. The former undergoes a few more periclinal divisions and gives rise to the endothecium, 2 or 3 middle layers and the outer tapetum lining the outer face of the anther locules. The primary sporogenous layer undergoes a few mitotic divisions and gives rise to the microspore mother cells. The microspore mother cells and the tapetum are the sites of intense metabolic activity during meiosis.

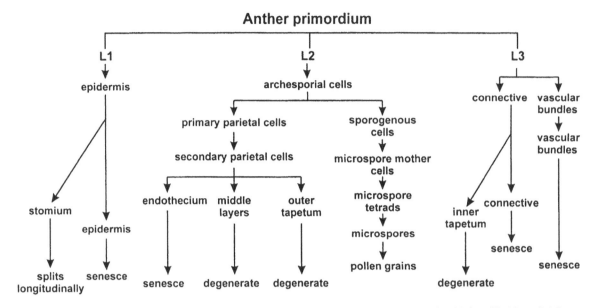

Fig. 1.1 Schematic representation of cell lineages from the three floral meristem layers—L1, L2, L3 (modified from Goldberg et al. 1993)

MICROSPOROGENESIS

Meiosis is one of the most critical events during microsporogenesis and thus in the development of pollen. Apart from reduction in number of chromosomes, it facilitates genetic recombination, the most significant feature of sexual reproduction. Anthers provide an excellent system for studying meiosis because: (a) they are easily available and accessible for experimental manipulation, (b) each anther contains a large number of meiocytes, more or less in synchrony, and (c) different anthers of a flower bud often show good synchrony so that one of them can be used for identifying the stage of meiosis while others can be used for experimentation. Anthers of *Lilium* and *Trillium* are the most widely used in these studies because of their large size and good synchrony.

The duration of meiosis is highly variable in different species; it is <24 h in *Petunia* and extends up to 3 months in some members of Liliaceae (Bennett 1971, Bennett and Smith

1976). In many conifers also meiosis, in particular the prophase, lasts for several weeks or even months (Owens and Molder 1974, Pennell and Bell 1987).

Initiation of Meiosis

Before the onset of meiosis, the mitotic divisions in the archesporial cells are non-synchronous. Irreversible commitment to meiotic division occurs late in the premeiotic S-phase (Takegami et al. 1981, Ito and Takegami 1982). Initiation of meiosis is synchronous. In spite of extensive studies, the causative factor(s) which trigger(s) meiosis is/are not clear. Culture of anthers before the initiation of meiosis has been largely unsuccessful. In such anthers the sporogenous cells continue to divide mitotically and eventually differentiate into vacuolated parenchymatous cells. Meiosis is not initiated even in cultures of entire flower buds (Tepfer et al. 1963). Meiosis appears to proceed normally in flowers developed on cultured shoot tips, as revealed by the formation of seeds in such cultures (Raghavan and Jacobs 1961). These studies have been interpreted to indicate that the stimulus for initiation of meiosis originates in some other part of the plant body and is transmitted to the anther just before initiation of meiosis (see Sauter 1971).

Meiosis continues in anthers cultured after the initiation of meiosis. In anthers cultured at the leptotene stage, meiocytes show several abnormalities, such as lack of chromosome pairing, desynapsis, absence of chiasma formation and failure of callose wall formation around meiocytes (Reznickova and Bogdanov 1972). Meiosis proceeds normally even on a simple nutrient medium in anthers or isolated meiocytes cultured at zygotene or later stages (Ito and Stern 1967). Following the reports on induction of pollen embryos in cultured anthers, this powerful technique of anther/meiocyte culture has hardly been used to understand meiosis; studies in this area have been diverted largely to induction of androgenic embryos in cultured anthers/microspores (see Chapter 14).

Synthesis of Macromolecules

Biochemical studies on the synthesis of DNA, RNA and proteins during meiosis have been conducted in a limited number of species (see Stern 1966, Hotta and Stern 1974, Stern and Hotta 1974). Although most of the DNA in microspore mother cells is synthesized during the premeiotic S-phase, a small amount of DNA (0.3%) is synthesized during zygotene and pachytene stages of meiosis (Stern and Hotta 1974, Bouchard and Stern 1980, Porter et al. 1982). This is essential for the orderly progress of meiosis. Inhibition of prophase DNA synthesis results either in zygotene arrest or in fragmentation of the chromosomes.

The DNA synthesized during zygotene (when chromosome pairing occurs) is typically semiconservative, and represents a delayed replication of portions of chromosomes which were not replicated in the S-phase. The DNA synthesized during pachytene (when chiasma formation occurs) has the characteristics of the repair type of replication and is likely to be involved in repair events associated with the crossing over (Hotta and Stern 1971).

Synthesis of RNA and proteins has been studied in only a few species such as *Lilium, Trillium* (Heslop-Harrison 1972, Porter et al. 1982) and *Cosmos* (Knox et al. 1970, 1971). Marked changes in RNA and protein synthesis have been recorded during meiosis. Synthesis of RNA is prominent during the premeiotic S-phase. There is a continuous drop in RNA synthesis during the prophase. From metaphase I onwards, until the completion of meiosis, there is hardly any synthesis of RNA (see Sauter 1971, Porter et al. 1982). The drop in RNA synthesis during the prophase is correlated with cytochemical and ultrastructural studies reporting a continuous drop in RNA content from the preleptotene stage onwards and a steady depletion of the ribosome population (Fig. 1.2A) in the cytoplasm (Moss and Heslop-Harrison 1967, Mackenzie et al. 1967).

Protein synthesis also follows a pattern similar to that of RNA synthesis. Protein synthesis decreases as the meiocytes approach leptotene and remains low during subsequent stages of meiosis (Sauter 1971, Albertini and Souvre 1974). During premeiotic S-phase period many meiosis-specific proteins, especially histones are synthesized (Sheridan and Stern 1967, Riggs and Hasenkampf 1991, Ueda and Tanaka 1994, 1995). One of these, np47 (apparent mol. wt. 47 kD) is synthesized abundantly during premeiotic interphase, reaches peak concentration during the leptotene/early zygotene interval and disappears after meiotic division (Riggs and Hasenkampf 1991). Immunological studies using np47 antiserum showed that this protein is present only in meiotic cells and absent in vegetative tissues. This protein has been termed meiotin 1 (Riggs and Hasenkampf 1991). Since the appearance of meiotin 1 roughly coincides with the timing of irreversible commitment to meiosis, it has been suggested that meiotin 1 is related to meiotic commitment of archesporial cells (Riggs and Hasenkampf 1991).

Cytoplasmic Reorganization

Many cytochemical and ultrastructural studies have shown marked changes in the cytoplasmic organization of meiocytes during meiosis (Heslop-Harrison 1971, Dickinson and Heslop-Harrison 1977). Initiation of meiosis is associated with a significant fall in the cytoplasmic RNA and elimination of a major part of the ribosome population (Knox et al. 1971, Dickinson and Heslop-Harrison 1977, see also Sangwan et al. 1989). The ribosome population is restored only after diakinesis (Fig.1.2A). Restoration of cytoplasmic ribosomes takes place through the production of 'nucleoids' packed with ribosomes or ribosome subunits, which are extruded from the nucleus into the cytoplasm (Dickinson and Heslop-Harrison 1970). Just before the initiation of ribosome elimination from the cytoplasm, a part of the cytoplasm is enclosed within double or multiple membrane units (Fig. 1.2B). The cytoplasm enclosed by these membrane units forms 10–15% of the cytoplasmic area and does not undergo the ribosome degradation observed in other parts of the cytoplasm. However, in many conifers, such as *Pinus* (Willemse 1971) and *Taxus* (Pennell and Bell 1987), there is no substantial fall in ribosome population during the meiotic prophase.

Both plastids and mitochondria undergo structural reorganization in the form of dedifferentiation and redifferentiation (Marumaya 1968, Dickinson and Heslop-Harrison 1977). As the prophase advances, both these organelles round off and become isodiametric. They lose most of the internal membrane system and reach a period of maximum simplification during zygotene-pachytene stages. Redifferentiation of plastids and mitochondria is initiated after metaphase I; both organelles have regained normal profiles by the time microspores are released from the tetrads. Similar reorganization of organelles during meiosis is also reported in a pteridophyte, *Lycopodium* (Pettitt 1978); a proportion of plastids in this taxon seem to be degraded during this reorganization.

Syncytium and Isolation

An important structural feature of meiosis is the formation of a syncytium of microspore mother cells in each sporangium during the prophase, and subsequent isolation of individual microspore mother cells and microspores by a callose wall (Fig 1.3A-D) (Heslop-Harrison 1966, 1971, Whelan 1974). In the young anther, before initiation of meiosis, all cell layers of the anther (wall layers, tapetum and sporogenous tissue) show plasmodesmatal connections between the cells of the same layer, as well as with those of adjacent layers. Cell divisions in any of the layers are not synchronous. Following initiation of

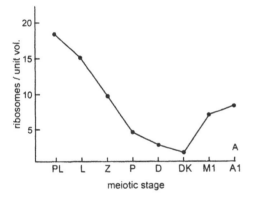

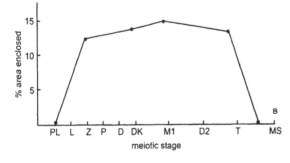

Fig. 1.2 Changes in ribosome population (A), and proportion of cytoplasm encapsulated in double- and multimembrane inclusions (B) in meiocytes of *Lilium* during meiosis. Meiotic stages: PL—pre-leptotene, L—leptotene, Z—zygotene, P—pachytene, D—diplotene, DK—diakinesis, M1—metaphase I, A1—anaphase I, D2—division II, T—tetrads, MS microspores (A— after Dickinson and Heslop-Harrison 1977, B—Dickinson and Andrews 1977).

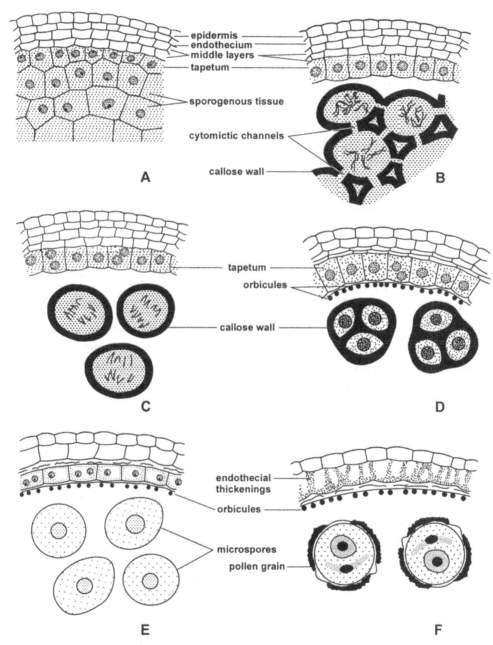

Fig 1.3 Diagrammatic representation of pollen development. Only a part of the microsporangium is shown. A. Differentiation of wall layers. B. Formation of a syncytium by the development of cytoplasmic channels between neighbouring microspore mother cells through incomplete callose wall (shown in black). C. Isolation of individual microspore mother cells by callose wall. D. Tetrads of microspores; individual microspores are also enclosed by a callose wall. Middle layers have started to degenerate. E. Release of microspores by dissolution of callose wall. Middle layers have degenerated and the cells of the endothecium have enlarged. Tapetum has become vacuolated. F. Mature 2-celled pollen. Tapetum is degenerated and endothecium has developed thickenings (after Shivanna et al. 1997).

meiosis, the plasmodesmatal connections between the wall layers and tapetum and between the tapetum and sporogenous tissue are progressively severed, but plasmodesmatal connections are retained between cells of the same layer (Clement and Audran 1995).

Onset of meiosis is characterized by initiation of callose deposition around individual microspore mother cells, followed by dissolution of the primary cell wall. Initially the callose wall is incomplete, leaving many gaps in the sites of some of the plasmodesmata; these gaps enlarge to form massive cytoplasmic channels (cytomictic channels) connecting neighbouring microspore mother cells (Fig.1.3B). Cytoplasmic channels reach maximum development during the zygotene-pachytene period, covering up to 20% of the interface. At this stage the discontinuities in callose wall measure up to 2.4 μm in some species. The cytoplasm and organelles can easily pass through these channels from one meiocyte to the other. Thus, all microspore mother cells of a sporangium form a single cytoplasmic entity, the syncytium. Eventually the callose wall around the microspore mother cells becomes continuous, severing the cytoplasmic channels (Fig.1.3C). This takes place at the end of metaphase I in species in which the cleavage of microspore mother cells is successive, and at the end of metaphase or anaphase II in species with simultaneous cell cleavage. Completion of meiotic division results in the formation of tetrads of microspores. The callose extends to the walls separating individual microspores also (Fig. 1.3D). Depending on the orientation of the second meiotic spindle the tetrads may be of different orientation—tetrahedral, isobilateral, linear, decussate or T-shaped (Davis 1966); tetrahedral is the most common type.

The syncytium acts as an efficient system for the distribution of nutrients between the microspore mother cells of each sporangium. As meiotic synchrony is established before t formation of cytoplasmic channels, the syncytium does not play a role in the initiation of synchrony. However, it seems to have a role in the maintenance of synchrony (Heslop-Harrison 1968). Synchrony between meiocytes is progressively lost after the breakdown of cytoplasmic channels. In some orchids (Heslop-Harrison 1968) and several members of Winteraceae (Sampson 1981), the cytoplasmic channels between microspores persist until pollen mitosis and the synchrony in these species is maintained during pollen mitosis also. The sporogenous tissue is interrupted in some members of Leguminosae and Orchidaceae by plates of somatic tissue and each isolated group of sporogenous tissue forms a syncytium. Synchrony is maintained between the cells of a syncytium but not between cells of different syncytia. In *Podocarpus* no cytoplasmic channels are formed during early meiosis and this is correlated with absence of synchrony in meiotic stages between meiocytes of the same microsporangium (Vasil and Aldrich 1970).

Significance of Cytoplasmic Reorganization and Isolation

Cytoplasmic reorganization in meiocytes and isolation of meiocytes and microspores with a callose wall seem to be a general feature of meiosis. Both these events have been interpreted as requirements for the transition from the diplophase to the haplophase (Heslop-Harrison 1972). Elimination of sporophytic information seems to be necessary to provide a suitable milieu for expression of the gametophytic genome. Apparent lack of RNA elimination in a few gymnosperms (Willemse 1971, Pennell and Bell 1987) is probably related to the prolonged prophase in these species. Sporophytic informational macromolecules in these species are likely to be eliminated during their normal turnover and gradually replaced by informational macromolecules produced by the gametophytic genome.

Isolation of meiocytes by a callose wall seems to be essential for normal development of pollen (Scott et al. 1991). Lack of callose deposition or its early breakdown results in pollen sterility (see Chapter 5). Many functions have been attributed to the isolation of meiocytes and microspores by a callose wall.

1. Isolation seems to be necessary for the meiocytes to embark on (a) transition from the sporophytic phase to the gametophytic phase and (b) expression of gametophytic genome without interference either from other spores or from the parent sporophytic tissue. The callose wall provides a degree of isolation to the meiocytes/microspores by selectively preventing entry of some classes of molecules. For example, thymidine (Heslop-Harrison and Mackenzie 1967), phenylalanine (Southworth 1971) and fluorescein diacetate (Knox and Heslop-Harrison 1970a) do not enter meiocytes when they are enclosed by a callose wall, while sucrose readily penetrate the callose wall (Southworth 1971).

2. Much evidence indicates that isolation of young microspores by a callose wall seems to be essential for orderly deposition of pollen exine (Waterkeyn and Bienfait 1970). In members of Cyperaceae, in which three of the microspores of a tetrad degenerate, a callose wall does not develop around individual microspores although it develops around the microspore mother cell (Dunbar 1973). Exine is laid down around the microspore mother cell wall but not along the walls separating individual microspores. SEM studies on exine formation in pollinium of an orchid, *Dendrobium*, using freeze-fracture and freeze-substitution techniques (Fitzgerald et al. 1994) have shown a clear relationship between callose wall and exine deposition. Exine is confined to the outer surfaces of the outer microspores of the pollinium. Some sea-grasses such as *Amphibolis antarctica* (Ducker et al. 1978) and *Halophila stipulacea* (Pettitt 1980) exhibit no detectable callose around microspore tetrads and pollen grains show no exine deposition. Similarly, *Arabidopsis* mutants that produce no callose wall show no ornamented exine either (see Fitzgerald et al. 1994).

3. The callose wall serves as a source of nutrients to the developing microspores by providing soluble carbohydrates following its breakdown (Barskaya and Balina 1971).

The importance of physical or physiological isolation for a marked change in morphogenetic pathway has also been highlighted in other groups of plants. According to Desikachary and Swamy (1976), any cell or group of cells embarking on a new morphogenetic pathway in all major groups of plants, from algae to angiosperms, generally becomes isolated. In apomictic species the nucellar cells which embark on an embryogenic pathway also show isolation from the surrounding nucellar cells with a callose wall (Gupta et al. 1996).

MICROGAMETOGENESIS

The callose wall around microspores is degraded (Fig.1.3E, F) by the activation of callase (β-1,3 glucanase) and the microspores are released into the anther locule. Although dissolution of the callose wall is a prerequisite for the release of microspores, its breakdown may not invariably lead to their release, as happens in species which release pollen as permanent tetrads/dyads. In *Arabidopsis* pollen grains are released as monads. However, two of the mutants termed quartet 1 and quartet 2 release pollen grains as tetrads (Preuss et al. 1994). This is due to fusion of exine in the region of microspore contact. All the pollen grains of the tetrad are viable in both mutants.

Soon after their release from the tetrads, the microspores contain a large nucleus at the centre and normal cytoplasmic organelles. The spores expand rapidly after release often up to 2.5-fold in volume (Fitzgerald et al. 1993a). The foot layer of the exine is still sufficiently flexible to permit this expansion (see Chapter 2). The carbohydrates released into the anther locule as a result of breakdown of the callose wall are likely to provide nutrients during this phase of microspore growth.

Division of the microspore is a major morphogenetic event after its release from the tetrads (Fig. 1.4). A large vacuole develops at the centre of the microspore, accompanied by the migration of the nucleus to the periphery. Both vacuolation and cytoskeleton elements (Brown and Lemmon 1991) seem to be involved in nuclear migration. Following nuclear migration, a major reorganization takes place in the cytoplasm; most of the plastids and mitochondria move away from the region of the nucleus, resulting in polarity of organelle distribution. Cytoskeleton elements have been implicated in polarized distribution of organelles also (van Lammeren et al. 1985). The nucleus undergoes mitotic division followed by formation of a cell plate between the daughter nuclei. A cell wall appears between the plasma membranes of the two nuclei and fuses with the intine at the margin (Angold 1968). This asymmetric division of the microspore results in a larger vegetative cell (VC) which receives most of the plastids and mitochondria of the microspore and a smaller generative cell (GC) with fewer or no organelles (Fig. 1.4 A-D).

Vegetative and Generative Cells

To begin with, the GC is attached to the intine of the pollen grain (Angold 1968, Pandolfi et al. 1993). The vacuole is resorbed and the GC becomes detached from the intine by inward extension of the newly formed wall (Fig. 1.4 D, E). After its release, the GC lies in the cytoplasm of the VC (Fig. 1.4 F-H). The GC is surrounded by a transitory callose wall (Gorska-Brylass 1970, Mepham and Lane 1970, Echlin 1972). Eventually the callose wall breaks down and the VC and GC are separated by the membranes of the respective cells (Shivanna and Johri 1985). A few studies, however, report the presence of a non-fibrillar wall material (Tiwari 1994) between the two plasma membranes. Similarly, a number of studies have shown lack of plasmodesmata between the GC and the VC (Cresti et al. 1987) while a few investigations have reported their presence (Burgess 1970, Lancelle and Hepler 1992). The GC is spheroidal initially but soon becomes spindle-shaped (Fig. 1.4 G, H) by orientation of the microtubules parallel to its

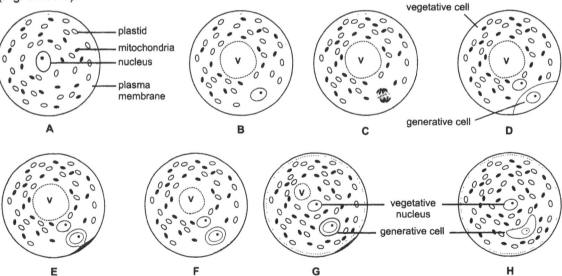

Fig. 1.4 Diagrammatic representation of asymmetric division of the microspore. A. Microspore soon after release. B-D. Formation of vacuole (v), migration of the nucleus, polarized distribution of mitochondria and plastids away from the nucleus, and division of microspore. E-H. Separation of generative cell from the microspore wall and resorption of vacuole.

long axis (Cresti et al. 1984, Derksen et al. 1985, Pierson and Cresti 1992). The GC shows normal profiles of endoplasmic reticulum (ER), ribosomes, dictyosomes and mitochondria.

The cytoplasm of the VC contains all the normal organelles (Sanger and Jackson 1971a, b). The number of organelles increases steadily until pollen maturation. The nucleus of the VC is generally lobed in mature pollen. The plastids, which are free from starch grains at the vacuolate microspore stage, develop starch grains soon after the division. Reserve food material in the form of lipids or starch grains accumulates in the VC; this is generally associated with the breakdown of the tapetum (Echlin 1972, Christensen and Horner 1974), indicating that the degradation products of the tapetum are taken up by the developing pollen. In many taxa, such as tobacco, stacked rough endoplasmic reticulum (RER) appear in the vegetative cell (Jensen et al. 1974, Cresti et al. 1975).

Although both the VC and the GC are products of mitotic division of the microspore, the two cells are structurally and functionally very different. Many differences have been recorded in the chromatin of the two nuclei (Table 1.1). These differences are considered responsible for the differences observed in the transcriptional and translational activities of the two cells (Reynolds and Raghavan 1982)

Protocols for isolation of a large number of generative nuclei free from vegetative nuclei

Table 1.1 Differences between vegetative cell and generative cell

Feature	Vegetative cell	Generative cell
Metabolism	Active with high levels of RNA and protein synthesis	Almost dormant except for a mitotic division in 3-celled pollen
DNA	Remains at 1C level	Increases to 2C level
Nucleus	Low amount of DNA-associated lysine-rich histone	Higher amount of DNA-associated lysine-rich histone
Chromatin	Diffuse	Highly condensed

have been developed (Ueda and Tanaka 1994). Isolated generative nuclei have been used to characterize basic protein components (Ueda and Tanaka 1994, 1995). These studies have shown the presence of five basic proteins specific to or highly concentrated in the generative nuclei with apparent molecular masses ranging from 18.5 to 33 kD. Further analyses of two of these proteins (18.5 and 22.5 kD) revealed that although they resembled somatic histones H2B and H3 respectively, histones from the generative nucleus showed unique peptide fragments and their antibodies did not cross-react with any somatic histones. One of these proteins (22.5 kD) is lysine-rich and the other (18.5 kD) arginine-rich. Immunofluorescence studies showed that both proteins are present only in the generative nucleus and not in the vegetative.

The technique of genetic ablation which involves targeted expression of a cytotoxic protein under the control of cell-specific regulatory sequences has been used to understand the expression of some genes in the two cells and their functional interaction. In tobacco, the promoter of the *LAT 52* gene, which is activated specifically in the VC following pollen mitosis, was used to express cytotoxic diphtheria toxin A chain (DTA) (Twell 1994). Analysis of pollen of such transgenic plants showed that *LAT 52-DTA* resulted in ablation of VC soon after microspore division (as revealed through FDA test). The GC retained its viability for several days following VC ablation but progressively lost viability. Interestingly, absence of functional VC prevented migration of the GC away from the pollen grain wall into the VC cytoplasm. These studies elegantly demonstrate the dependence of GC on the VC for its functioning.

Several studies have shown that the vegetative nucleus forms an intimate association with the GC in mature pollen or during pollen tube growth (Ciampolini et al. 1988, Hu and Yu 1988, Wagner and Mogensen 1988). In *Medicago sativa* (Shi et al. 1991) the vegetative nucleus in mature pollen surrounds a major part of the GC. The nuclear pore density on the surface of the vegetative nucleus facing the GC is 69% higher than that on the surface away from the GC. As the macromolecular traffic into and out of the nucleus occurs through the pores, these observations suggest a functional relationship between the vegetative nucleus and the GC. This is in agreement with the report (Tang 1988) on the increased level of ATPase activity at the vegetative nuclear envelope in the area of its association with the GC in *Amaryllis* and *Clivia*.

An interesting variation in the disposition of the GC has been reported in *Rhododendron* (Kaul at al. 1987) and *Acacia* (McCoy and Knox 1988), both characterized by polyads. In *Rhododendron* the spindle-shaped generative cell possess two extensions, one at either end. One of these extensions is connected to the intine wall ingrowth situated near the external aperture (Theunis et al. 1985). Following in-vitro germination, this connection breaks off and the generative cell enters the pollen tube where it becomes closely associated with the vegetative

nucleus (Kaul et al. 1987). In *Acacia* (McCoy and Knox 1988) the spindle-shaped GC lies across the width of the pollen grain and is attached to both ends of the intine by means of membrane labyrinths.

Sperm Cells

The GC divides mitotically to form two sperm cells (Fig. 1.5 A, B). In species in which pollen grains are shed at the 2-celled stage, the GC generally enters dormancy at the prophase stage. In species in which pollen grains are shed at the 3-celled stage, the GC completes division before dormancy. Early electron microscopic studies on the division of the GC in barley (Cass and Karas 1975) indicated that the nuclear division of the GC takes place while it is still attached to the pollen wall. An incipient wall appears between the two sperm cells and the two-celled unit detaches from the pollen wall. Eventually the partition wall as well as the surrounding walls of the sperm becomes degraded, releasing the

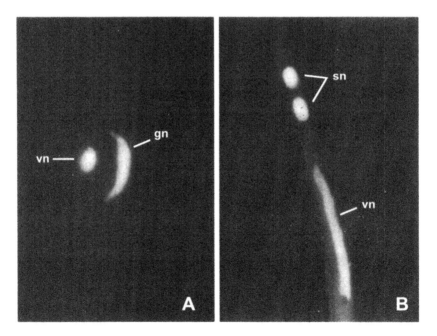

Fig. 1.5 Fluorescence micrographs of squash preparations of pollen grain (A), and pollen tube (B) of tobacco stained with DAPI, a DNA fluorochrome, to show the vegetative nucleus (vn), the generative nucleus (gn) and the sperm nuclei (sn). In B pollen tube tip is toward the top (after Shivanna and Rangaswamy 1992).

sperms. Subsequent studies on barley (Charzynska et al. 1988) and a few other species (Charzynska et al. 1989, Murgia et al. 1991) have shown that the GC detaches from the pollen wall before its division and a regular cell plate is formed between the two nuclei during division. In several species the two sperm cells remain associated with each other even after division as part of the male germ unit (see below)

In nearly 70% of the flowering plants, pollen grains are shed at the 2-celled stage (Fig 1.5A). In the remaining species they are shed at the 3-celled stage (Brewbaker 1959). In 2-celled pollen the generative cell divides and gives rise to the two sperm cells in the pollen tube after germination (Fig.1.5B). The cytological status of the pollen (2 or 3 cells) at the time of shedding is correlated with a number of physiological and genetical features (Table 1.2).

Table 1.2 Correlation between pollen cytology and some other traits

Character	2-celled pollen	3-celled pollen
Viability	prolonged	short
Storage	store well	difficult to store
In-vitro germination	comparatively easy	comparatively difficult
Respiratory rate	low	high
Self-incompatibility	gametophytic	sporophytic

Dimorphism of sperm cells and organization of male germ unit

The classical concept assumes that the two sperm cells formed by the division of the GC are isomorphic. Recent studies in several species have clearly shown distinct morphological differences between the two sperm cells of the pollen. Such studies became possible only after the technical advances made to produce computer-assisted three-dimensional reconstruction of serial transmission electron micrographs. In *Plumbago* (Fig. 1.6 A) (Russell and Cass 1981, Russell 1984) the sperm cells are linked together by a common transverse wall traversed by plasmodesmata. One of the sperm cells (Svn) is

associated with the vegetative nucleus through a long extension. The two sperm cells also show differences in nuclear size and number of cytoplasmic organelles (Table 1.3) (Russell 1984). The smaller sperm, which is not associated with the vegetative nucleus (Sua), contains an average of 24 plastids and 40 mitochondria, while the larger sperm cell (Svn) usually contains no plastids or very few plastids and has an average of 256 mitochondria. Such a physical association between the two sperm cells and the vegetative nucleus (linking all nuclear and cytoplasmic DNA of male heredity units) has been termed the 'male germ unit' (MGU). The MGU in *Plumbago* is formed in mature pollen and is maintained throughout pollen tube growth.

Table1.3 Quantitative details of the two sperm cells in *Plumbago zeylanica* studied through three-dimensional reconstruction (based on data presented in Russell 1984)

Feature	Svn	Sua
Mean cell volume (μm^3)	69.5	48.9
Mean cell surface (μm^2)	147.9	84.7
Nuclear volume (μm^3)	19.9	12.1
Nuclear surface (μm^2)	32.4	20.4
Cytoplasmic volume (μm^3)	45.2	33.4
No. of mitochondria	256.2	39.8
No. of plastids	0.45	24.3

Sperm dimorphism and MGU have subsequently been demonstrated in many other 3-celled pollen species such as *Beta vulgaris* (Hoefert 1969), *Catananche caerulea* (Barnes and Blackmore 1987) and a few other taxa (Knox and Singh 1987, Mogensen 1992). In *Brassica* (Fig 1.6B) the two sperm cells are linked by a common cell junction and one of the sperm cells is associated with the vegetative nucleus. Plastids are absent in both sperm cells; however, the sperm cell associated with the vegetative nucleus (Svn) has a larger number of mitochondria than the other sperm cell (Sua) (McConchie et al. 1985, 1987). Although differences in the contents have not been documented in other species, the difference in size of the two sperm cells has been reported in several species: e.g.

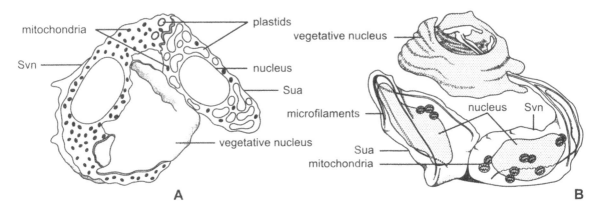

Fig. 1.6 Diagrammatic representation of the male germ unit in *Plumbago zeylanica* (A) and *Brassica campestris* (B). Long projection of one of the sperms, Svn (sperm physically associated with vegetative nucleus) wraps around and lies within embayments of the vegetative nucleus in A. In B, Svn projection enters the enclaves within the vegetative nucleus. In *Plumbago* the Svn contains a majority of mitochondria and only two plastids whereas the Sua (sperm physically unassociated with vegetative nucleus) contains most of the plastids and fewer mitochondria. In *Brassica* the plastids are absent in both the sperm cells but the Svn is mitochondria-rich compared to Sua (A—after Russell 1984, B—Knox and Singh 1987).

Gladiolus and *Rhododendron* (Shivanna et al. 1987), *Euphorbia* and *Gerbera* (Mogensen 1992).

The concept of MGU has now been extended to 2-celled pollen species in which the vegetative nucleus and the generative cell are in association (Knox and Singh 1987). Some of the 2-celled pollen systems which show MGU are: *Gossypium* (Jensen and Fischer 1968), *Hippeastrum vittatum* (Mogensen 1992), *Petunia hybrida* (Wagner and Mogensen 1988), *Nicotiana tabacum* (Yu et al. 1989), *Rhododendron* sp. (Shivanna et al. 1987), *Acacia retinoids* (McCoy and Knox 1988), *Medicago sativa* (Zhu et al. 1992), *Aloe ciliaris* (Ciampolini et al. 1988) and *Gladiolus gandavensis* (Shivanna et al. 1987).

Mature pollen grains of Poaceae, which are 3-celled, show neither sperm dimorphism nor MGU (Mogensen and Rusche 1985, Mogensen 1992). However, a close association is formed between the vegetative nucleus and the sperm cells soon after pollen germination, thus forming the MGU (Mogensen and Wagner 1987). The association between vegetative nucleus and sperm cells is transitory and short; by the time sperm cells enter the pollen tube, the vegeta-

tive nucleus gets separated but the two sperm cells remain connected during pollen tube growth until the pollen tube approaches the ovule. A similar situation exists in *Zea* also (Rusche and Mogensen 1988). However, in *Zea* the vegetative nucleus remains associated with the sperm cells during their passage in the pollen tube.

The significance of MGU in fertilization is not clearly understood. MGU, in particular the association of the two sperm cells in the pollen tube, seems to be common. So far, analysis of sperm cells during pollen tube growth and its entry into the ovule has been confined to a limited number of taxa. MGU is likely to play an important role in the transfer and discharge of the two sperm cells together into the synergid (Russell 1992).

TAPETUM

The tapetum is a transitory layer that surrounds the sporogenous tissue and plays an important role in pollen development (Pacini 1997). In most species the origin of the tapetum is dual. The outer tapetum (facing the outer surface of the anther) is derived from the secondary parietal layer, while the inner tapetum (facing the inner side of the anther locule) is derived from the cells of the connective tissue (Goldberg et al. 1993). In general, the tapetum is single layered. In some species, especially of marine angiosperms, the tapetum is 2–4 layered (Ducker et al. 1978).

The tapetal cells are metabolically very active during both the premeiotic and meiotic period; they show active synthesis of RNA and proteins. This is correlated with the presence of well-developed mitochondria, abundant RER cisternae and active dictyosomes in the cytoplasm of tapetal cells (Steer 1977). A unique cytological feature of tapetal cells is an increase in their DNA content initiated soon after the onset of meiosis in the sporogenous cells. As this increase in DNA is not followed by regular cell division, it results in one or more cytological abnormalities: multinucleate cells, polyploid nuclei (formed by incomplete mitosis or nuclear fusion) and polyteny. Often the extent of DNA increase in the tapetal cells is up to 16 times that present in the sporogenous cells (for a general review of the tapetum see Echlin 1971, Heslop-Harrison 1972, Bhandari 1984, Pacini et al. 1985, Chapman 1987, Pacini 1990, 1997). In members of Mimosoideae and Papilionoideae (Leguminosae), the tapetal cells remain uninucleate, a condition considered to be derived from bi- and multinucleate conditions (Buss and Lersten 1975, Albersten and Palmer 1979).

Two main types of tapetum are distinguished: secretory/parietal tapetum and plasmodial/invasive/amoeboid tapetum. The distinction is largely based on the nature of the tapetal cells at the peak of their activity. In the secretory tapetum the cells of the tapetum maintain their position and identity and eventually undergo degeneration in situ towards the end of pollen development. In the plasmodial tapetum the inner tangential wall breaks down and the protoplasts of the tapetal cells enter the anther locule (Fig. 1.7A, B). The secretory tapetum is more common in dicots than in monocots. According to one survey (Pacini et al. 1985), the secretory tapetum occurred in 154 dicot and 21 monocot families, while a plasmodial tapetum was present in 14 dicot and 18 monocot families. Both types of tapeta have been reported in 11 dicot and one monocot family.

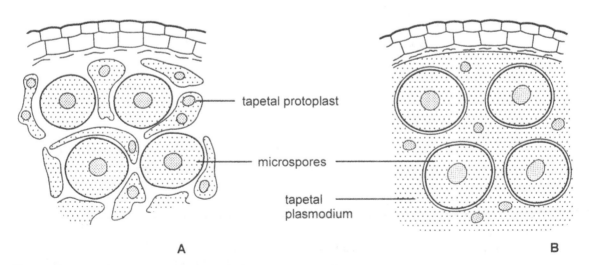

tapetal protoplast

microspores

tapetal plasmodium

A

B

Fig 1.7 Diagrammatic representation of plasmodial tapetum at microspore stage. Protoplasts of tapetal cells have entered anther locule between the microspores. The identity of individual protoplasts is maintained in A, whereas the protoplasts have fused to form tapetal plasmodium in B (after Shivanna et al. 1997).

Secretory Tapetum

Cells of the secretory tapetum retain plasmodesmatal connections between the neighbouring cells. In some species they develop cytomictic channels similar to those of sporogenous cells and thus form a tapetal syncytium (Heslop-Harrison 1972, Clement and Audran 1995). In several species, the inner tangential wall and later the radial walls break down, but the protoplasts remain in situ. Transfer of products from the tapetum is an integral part of pollen development. Transport of substances from the tapetal cells into the locule may occur through exocytosis (Ciampolini et al. 1993) or through transport across the inner plasma membrane. In *Brassica oleracea* the inner tapetal membrane forms tubular evaginations that increase the surface area and might be involved in translocation of solutes (Murgia et al. 1991).

The time of tapetum degeneration varies greatly from species to species. It is generally towards the end of pollen development but in *Pterostylis*, an orchid, tapetal degeneration occurs before microspore mitosis (Fitzgerald et al. 1993a). Before their breakdown the populations of ribosomes, dictyosomes and RER cisternae diminish in the tapetal cells, and lipid bodies and plastids become conspicuous features of the cytoplasm. Eventually the tapetal cell membrane degenerates and the contents enter the locule and are deposited on the pollen surface as pollenkitt/tryphine.

In many taxa characterized by secretory tapetum, sporopollenin granules, termed 'Ubisch bodies' or 'orbicules' are deposited (Fig. 1.3) on the inner tangential surface of the tapetal cells (Pacini and Franchi 1992, 1993). Orbicules originate in the cytoplasm of the tapetal cells as lipoidal pro-orbicular bodies covered with a membrane. Pro-orbicular bodies accumulate below the membrane and eventually extrude to the cell surface (facing the locule) where they acquire a sporopollenin coating. Orbicules are absent in taxa characterized by plasmodial tapetum and also in a few taxa with a secretory tapetum (Murgia et al. 1991, Pandolfi et al. 1993).

Plasmodial Tapetum

The inner tangential and radial walls of the tapetal cells break down and the tapetal protoplasts intrude into the thecal cavity amid the microspores. In advanced monocots the intrusion generally occurs earlier, while in primitive monocots and dicots the intrusion occurs at the microspore stage (Pacini et al. 1985). The identity of individual tapetal protoplasts may be maintained or the protoplasts may fuse in the locules to give a periplasmodium (Fig. 1.7A, B) (Tiwari and Gunning 1986a-c, Pacini and Keijzer 1989). Ultrastructural studies of the plasmodial tapetum indicate that the protoplasts/periplasmodium are/is not a degeneration product but an organized and functional unit with normal organelle distribution (Pacini and Juniper 1983, Tiwari and Gunning 1986a-c, Fernando and Cass 1994, Galati 1996). An increase in number of mitochondria and elaboration of ER membrane, reported in tapetal protoplasts/periplasmodium, is indicative of their/its high metabolic activity (Fernando and Cass 1994, Galati 1996). The plasmodial tapetum comes into close contact with the microspores/young pollen grains, facilitating transport of substances from the tapetum to the microspores. Towards the end of pollen development the tapetal protoplasts/periplasmodium degenerate(s) and adhere(s) to the pollen surface as pollenkitt.

Sporopollenin bodies (orbicules) are generally absent in the plasmodial tapetum. However, sporopollenin bodies have been reported in a few species with plasmodial tapetum such as *Tradescantia* (Tiwari and Gunning 1986a, b) and *Butomus* (Fernando and Cass 1994). In *Butomus* vesicles produced by extensive arrays of RER in tapetal cells/protoplasts have been implicated in the formation of sporopollenin-like bodies and sporopollenin precursors.

Tapetal Membrane

Development of tapetum is associated with formation of an acetolysis resistant membrane, termed the tapetal membrane in many species of angiosperms as well as gymnosperms (Heslop-Harrison 1972). This appears to be a general feature of seed plants. In species characterized by a secretory tapetum, the tapetal membrane is generally formed on the inner surface of the tapetal cells (towards the locule) while in species characterized by a plasmodial tapetum, the membrane is formed on the outer surface of the tapetum (towards the endothecium) (see also Parkinson and Pacini 1995). The membrane is largely made up of sporopollenin; insoluble polysaccharides such as cellulose, callose and pectin also seem to be present in small amounts (Gupta and Nanda 1972).

The tapetal membrane forms a culture sac around the pollen at a later phase of its development. The function of the tapetal membrane is not clearly understood (Shivanna and Johri 1985). As it is largely made up of sporopollenin, it may restrict the free passage of materials into and out of the pollen mass.

Role of Tapetum

Much direct and indirect evidence clearly shows that the tapetum plays a crucial role in pollen development. Pollen sterility (nuclear/cytoplasmic/environmental) is invariably associated with tapetal abnormality (see Chapter 5). It has also been possible to induce pollen sterility by targeting the tapetum through recombinant DNA technology (see Chapter 5). The role of the tapetum in pollen development seems to be manifested largely after completion of meiosis. The major functions attributed to the tapetum are described below.

Supply of nutrients to developing pollen

The tapetum has traditionally been considered to be the nurse tissue for the developing pollen. As the tapetum encloses the sporogenous tissue all around, any nutrients entering the sporogenous cells have to pass through the tapetum. In the secretory tapetum, metabolites in the form of soluble carbohydrates, amino acids and peptides are released into the locular fluid through exocytosis or secretion, from where they are taken up by the developing pollen (Pacini 1994). In the plasmodial tapetum, the plasma membrane of the tapetal protoplasts are closely adpressed to the developing pollen making the passage of nutrients more efficient than in the secretory tapetum. The build-up of nutritional reserves in the pollen is generally associated with breakdown of the tapetum. In *Sorghum* and many other members of Poaceae the pore end of the pollen grains lies adjacent to the tapetal cells. Starch build-up in the pollen is initiated at the germ pore end of the pollen, indicating absorption of tapetal products through the pore (Christensen and Horner 1974). In many species the intine in the poral region produces wall ingrowths similar to transfer cells (Charzynska et al. 1990). The wall ingrowths have been suggested to facilitate nutrient and water uptake initially from the tapetum and subsequently from the anther locule. In *Acacia* also (McCoy and Knox 1988) membrane labyrinths present in the apertural intine have been implicated in the transport of metabolites.

Breakdown of callose wall around microspore tetrads

The enzyme callase, required for breakdown of the callose wall around the microspore tetrads, is released by the tapetum. The microspores themselves are incapable of callase synthesis; culture of isolated microspore tetrads does not result in breakdown of callose. The tapetal plasmodium, however, is able to break down the callose of microspore tetrads as well as sieve tubes (Mepham and Lane 1969). Biochemical analysis of callase in the developing anther has shown that the dissociation of tetrads is correlated with a sharp increase in callase activity; most of the activity is localized in the wall of the anther and not in the meiocytes (Stieglitz and

Stern 1973, Scott et al. 1991). In *Pterostylis* a proteinaceous layer is deposited in the inner tangential wall of tapetal cells at the tetrad stage (Fitzgerald et al. 1993a). However, the wall becomes uniformly thin following microspore release. It is suggested that the material deposited in the tapetal wall is callase and is released before callose breakdown (Fitzgerald et al. 1993b). Tapetal activation of callase at the right stage is very important for normal development of pollen. Mistiming of callase activation results in pollen sterility (see Chapter 5).

Supply of sporopollenin precursors to pollen exine

Tapetum is considered to provide sporopollenin precursors for exine formation. Although the blueprint of exine is laid down before the microspores are released, in most taxa the bulk of the exine is deposited after release of the microspores (Echlin 1971, Heslop-Harrison 1971, Owens and Dickinson 1983). The orbicules produced in the secretory tapetum seem to be the end products of metabolism; so far no enzyme capable of degrading sporopollenin has been identified. Orbicules, therefore, do not seem to take part in the formation of exine. The precursors secreted by the secretory as well as plasmodial tapetum are likely to be involved in exine formation (Heslop-Harrison 1971, Horner and Pearson 1978, Owens and Dickinson 1983, Fernando and Cass 1994). Available evidence suggests that the development of ektexine proceeds centrifugally through the deposition of tapetally derived sporopollenin; the endexine develops centripetally by the deposition of sporopollenin derived from pollen protoplast, deposited on membrane-like strands. In *Butomus* (Fernando and Cass 1994) the tapetal cells, before formation of the periplasmodium, extrude particles into the locule that condense to form electron dense aggregates. These aggregates are similar to the primexine and are incorporated into the developing ektexine. Sporopollenin produced from the tapetum is laid down on the developing ektexine even after the tapetum becomes plasmodial and completely encloses the microspores.

Supply of pollen coat substances and exine proteins

The tapetal origin of pollen coat substances generally known as pollenkitt or tryphine is well established (see Chapter 2). The proteins are located in exine cavities (tectate grain) or in the surface depressions (non-tectate grain). During meiosis, proteins and lipids accumulate in tapetal cells, the former in single membrane-bound vesicles apparently derived from RER and the latter in plastids. Following the breakdown of tapetal cells, these proteins and lipids are released into the thecal cavity and get deposited in the exine. The lipoidal components generally remain on the surface of the pollen (Heslop-Harrison et al. 1973, 1974). The pollenkitt is predominantly made up of lipids, flavonoids and carotenoids, and other degenerated products of the tapetum (Wiermann and Vieth 1983; see Chapter 2), and plays an important role in pollen dispersal and pollen function. Tryphine is often distinguished from pollenkitt; the former is a complex mixture of hydrophilic substances derived from the breakdown of tapetal cells, while the latter is principally the hydrophobic lipids containing species-specific carotenoids (Echlin 1971, Dickinson 1973). In *Pterostylis* pollenkitt acts as a glue to bind pollen grains into a pollinium; in this species there are no exine bridges between pollen grains (see Chapter 2 for details).

GENE EXPRESSION DURING ANTHER DEVELOPMENT

Gene expression during anther and pollen development has been an active area of research in many laboratories and considerable information has accumulated during the last 15 years. An excellent account of the available data is given by Mascarenhas (1990, 1992), Goldberg et al. (1993), McCormick (1993), Hamilton and

Mascarenhas (1997) and Raghavan (1999). The following is a brief summary:

Over 24,000 different RNAs have been reported in mature pollen. About 65% of these genes are also expressed in the sporophytic tissues and the remaining (ca. 10,000) seem to be expressed exclusively or predominantly in the anther and pollen. The expression of these genes is regulated both temporally and spatially. A number of anther-specific genes and/or their cDNAs have been isolated. Several genes (such as *Ta 13, Ta 26, Ta 29*) are expressed only in the tapetum, while some are expressed in the connective tissue, endothecium and middle layers (*Ta 20, Ta 55*). Anther-specific mRNAs have been shown to encode a range of proteins such as lipid transfer proteins, protease inhibitors, thiol endopeptidase, glycine-rich and proline-rich polypeptidases, pectate lyases, and chalcone synthase.

Comprehensive studies have been carried out on the temporal expression of genes during pollen development. The genes expressed in pollen have been grouped as 'early' and 'late' genes. The mRNAs of early genes are first detectable soon after meiosis, reaching a maximum at late pollen interphase, and decreasing thereafter. Expression of late genes starts at or after microspore mitosis and reaches maximum at the mature pollen stage. Late genes form the bulk of the pollen-specific genes identified to date. Early genes are presumed to be associated with pollen development, while late genes are likely to play a role in pollen maturation, germination and pollen tube growth (Mascarenhas 1990). Numerous cDNAs have been characterized, in particular of 'late' genes. Some of these are *BP 10* (*Brassica napus*), *Bcp 1* (*B. campestris*), *NTP 303* (tobacco), *LAT 51* (tomato), *Zm 13* (maize), *Org cl* (rice) and *Bet v 1* (white birch). *LAT 51* sequences seem to be highly conserved as similar sequences are found in diverse plant species. Most of the putative proteins of different genes exhibit sequence homologies to known proteins such as wall-degrading enzymes, cytoskeleton proteins and

allergens (Hamilton and Mascarenhas 1997). One of the genes, *Bcp 1*, expressed in the tapetum as well as microspores (Theerakulpisut et al. 1991), is essential for pollen fertility in *Arabidopsis* (Muschietti et al. 1994, Xu et al. 1995a). *Pex 1* (pollen extensin-like) gene in maize, expressed exclusively in pollen, has an extensin-like domain (Rubinstein et al. 1995). A pollen-specific gene in rice (*PS 1*) shows significant levels of homology to 'late' expressed maize *Zm 13* and tomato *LAT 52* (Zou et al. 1994).

TRANSMISSION OF PATHOGENS THROUGH POLLEN

Only limited studies have been carried out to determine whether pathogens can pass through pollen grains to the progeny. There are some reports of transmission of viruses through pollen (Mandahar 1981, Mink 1993). Electron microscopic studies of pollen grains from infected plants have shown the presence of viruses (Carroll 1974, Hamilton et al. 1977). Transmission of barley stripe mosaic virus via pollen has been demonstrated through ultrastructural studies of pistils of healthy plants pollinated with virus-infected pollen (Brlansky et al. 1986). Virus particles were detected not only in the pollen tubes but also in the zygote and the resultant embryo and endosperm. Although some early studies (Ark 1944) reported pollen as a source of some bacterial infection, there are no conclusive reports on the transmission of bacterial diseases through pollen.

POLLEN MATURATION AND ANTHER DEHISCENCE

Towards the end of its development, pollen builds up reserve materials (starch/lipids). The plastids differentiate into amyloplasts (Pacini et al. 1992, Franchi et al. 1996). In many species mature pollen grains contain starch grains (starchy pollen), while in others starch is hydrolyzed before pollen grains are shed (starchless pollen). On the basis of analyses of pollen of 124 families, Baker and Baker (1979) found that primitive families tend to have starchy pollen and advanced families starchless pollen. Also, entomophilous species pollinated by Hymenoptera and Diptera generally produce starchless pollen and those pollinated by Lepidoptera as well as birds produce starchy pollen. Starchy pollen grains tend to be larger than starchless pollen (Endress 1994).

Irrespective of the nature of reserve material present in mature pollen, plastids invariably develop starch either during microspore development (after microspore mitosis) and/or during maturation (Franchi et al. 1996, Speranza et al. 1997). In some species the starch present in amyloplasts is the only carbohydrate source in mature pollen and the cytoplasm has no PAS-positive material (eg. *Lolium, Cucurbita*). In others amyloplast starch is partially hydrolyzed and the cytoplasm contains low molecular weight insoluble PAS-positive material (eg. *Dactylis, Cucumis*). In several species starch is completely hydrolyzed during maturation and mature pollen is starchless; but the cytoplasm contains PAS-positive material and simple sugars such as glucose and fructose (e.g. *Borago, Lycopersicon*).

Recent survey of pollen grains of 901 species belonging to 104 dicot and 15 monocot families (Franchi et al. 1996, Pacini 1997) showed variations in physicochemical properties of starch in the intensity of colour development following I-KI staining and the presence or absence of birefringence under polarized light. The results also indicated that starchless pollen grains withstand desiccation better than starchy pollen. Desiccation tolerance has been attributed to low molecular weight carbohydrates in the cytoplasm of starchless grains (see Chapter 3).

Anther dehiscence is the result of precise co-ordination between differentiation and/or degeneration of cell layers of the anther and dehydration. Although extensive studies have been carried out on the details of pollen development, only limited information is available on the struc-

tural and physiological details associated with anther dehiscence (Keijzer 1987a, b, Bonner and Dickinson 1989, Goldberg et al. 1993). In one of several comprehensive studies on anther dehiscence, Bonner and Dickinson (1989) have presented structural and cytochemical aspects of anther dehiscence in *Lycopersicon esculentum*. The earliest event associated with anther dehiscence is observed in the cells of the intersporangial septum (ISS) (Bonner and Dickinson 1989) or circular cell cluster (Koltunow et al. 1990). In members of Solanaceae, a large number of calcium oxalate crystals accumulate in the cells of the ISS followed by their degeneration. However, in many other species such as *Gasteria*, degeneration of the cells of the ISS is not accompanied by crystal formation. ISS degeneration is associated with activation of many hydrolytic enzymes such as cellulases, pectinases and acid phosphatases. Degeneration of cells of ISS results in rupture of the septum leading to the fusion of the two locules. Degeneration of ISS is also associated with differentiation of the stomium in the epidermal layer along the line of dehiscence. Unlike the neighbouring epidermal cells which show enlargement with thickened cuticle, the stomium cells remain small and are covered with thin cuticle, thus creating a point of weakness in the anther wall. The protoplasts of cells of the stomium and epidermis degenerate. Simultaneous with degeneration of the epidermal layer, cells of the endothecium enlarge and develop characteristic thickenings. Eventually the cells of the anther wall desiccate and collapse, resulting in a narrow split along the stomium.

Using the strategy of ablation of specific cell types, Beals and Goldberg (1997) showed that anther dehiscence depends on the presence of a functional stomium. Ablation of cells in the stomium region leads to failure of anther dehiscence. Desiccation of anther tissue seems to result from a combination of active resorption of water through anther filament and cuticular evaporation (Heslop-Harrison et al. 1987, Pacini 1994). Removal of anthers from their filaments delayed anther dehiscence when compared to those still attached to the flowers (Schmid and Alpert 1977). In many members of Poaceae, anther filaments show remarkable elongation just before anther dehiscence and bring the anthers above the level of other floral organs. During this elongation water is retracted from anthers and moved to the filaments (Keijzer et al. 1987, Keijzer 1999); this results in synchronous dehydration of anthers and supply of water to the vacuoles of the filament cells facilitating their elongation.

Dehiscence requires activation of many genes, particularly those that encode hydrolytic enzymes. One of the genes *TA 56* (thiol peptidase) in tobacco seems to be involved in the dehiscence process (see Goldberg et al. 1993). *TA 56* mRNA accumulates first in the ISS prior to its destruction, subsequently in the stomium and finally in the connective.

2

Pollen Morphology and Aeropalynology

POLLEN MORPHOLOGY

Pollen morphology is of great significance in taxonomy, phylogeny, palaeobotany, aeropalynology and pollen allergy. Analysis of fossil pollen is the most important approach to reconstruction of past flora, vegetation and environment (Faegri and Iversen 1989). Pollen morphology is also important in understanding the functional aspects of pollen such as pollination biology and pollen-pistil interaction. Pollen identification, the basis of palynology, is based exclusively on pollen morphology. This chapter provides a brief description of technical terms used in pollen morphology. Several books (Erdtman 1966, 1969, Nair 1970, Moore and Webb 1978, Nilsson and Muller 1978, Thanikaimoni 1978, Faegri and Iversen 1989) are available for a comprehensive coverage of pollen morphology and pollen analysis.

Applications

Petroleum exploration

One of the most important applications of palynology is in the field of petroleum exploration. Until the first quarter of the 20th century, oil exploration was largely confined to the recovery of oil and gas in shallow deposits. Potential sources were identified on the basis of rock strata. However, these approaches proved in-sufficient for exploring oil in deeper wells. Analysis of fossil pollen present in rock samples collected from various depths in oil wells proved very useful in predicting oil and gas zones. Presently almost all oil exploration companies have well-established palynology laboratories to assist oil exploration work (Faegri and Iversen 1989, Bryant 1990).

Archaeology

Pollen analysis has important applications in understanding past climatic changes, the origin and spread of agriculture and prehistoric cultures (Bryant 1990). By comparing fossil pollen records of archaeological sites with well-dated pollen records, archaeologists are often able to date specific archaeological events. Analysis of pollen in archaeological human burials and samples of coprolites (preserved faeces) has given definitive information about ancient diets.

Criminology

Unlike many other macroevidences, which can be easily removed by criminals, pollen grains cannot be removed easily. Analysis of such pollen can be effectively used in solving crimes (Bryant 1990). For example, in one of the well-published murder trials in Austria in 1959 evidence from pollen grains was responsible for identifying the murderer. A man disappeared near Vienna while on a journey down the Danube

River, but his body was not found. The man who had a motive for killing was arrested and charged with murder. In the absence of confession or the body, prosecution could not prove the murder. Analysis of pollen in the mud found on the defendant's shoes revealed the pollen of spruce, willow and alder together with fossil hickory pollen (from exposed Miocene-age deposits). Only the soil from one small area 20 km north of Vienna along the Danube River valley contained this pollen mixture. When confronted with this information the shocked defendant confessed his crime and showed the authorities the site where the body was buried which turned out to be the region pinpointed on the basis of pollen analysis (Bryant 1990). Although pollen can provide definitive clues in solving crime, it is not yet being used widely, largely because of lack of sufficient basic pollen data of different regions and of experts to identify pollen grains.

Testing purity of honey

Honey-bees collect nectar and pollen from plants available around the hive and use them to make honey. Thus analysis of pollen present in the honey indicates not only the geographic region of honey production, but also the plant species which have been used.

Three Domains of Pollen Wall

The pollen wall is probably the most complex wall system found in higher plants. It includes three main domains—intine, exine and pollen coat—which differ in morphology, chemistry and function.

Intine

The intine envelopes the pollen protoplast and is comparable to the primary wall of other plant cells. It is primarily composed of cellulose, hemicellulose and pectic polymers. In some taxa such as grasses, a middle layer between the exine and intine rich in pectic polysaccharides, termed the Z-layer, is distinguishable. The Z-layer is thickened at the germ pore region and is termed

'Zwischenkorper' (Rowley 1964, Heslop-Harrison 1979c). The intine invariably contains protein in the form of radially elongated tubules generally concentrated in the germ pore region (see page 33).

Exine

The exine layer is highly sculptured and ornamented (Fig. 2.1). It is composed of sporopollenin, a highly resistant organic biopolymer. Viscin threads that bind together pollen grains and tetrads in some families such as Onagraceae and Ericaceae are also made up of sporopollenin (Hesse 1981, 1984). Because of the difficulties in purification of sporopollenin and its insolubility in most solvents, sporopollenin is not readily amenable to chemical analysis. Since the 1960s sporopollenin has been considered a biopolymer derived from carotenoids and carotenoid esters (Brooks and Shaw 1971). However, recent comprehensive studies involving tracer and degradation experiments and spectrometric analysis have indicated sporopollenin to be a mixed polymer with a large amount of aliphatics containing additional compounds such as phenols.

Pollen identification is largely based on the structure and surface sculpturing of the exine. Two sets of terminologies on exine stratification are in vogue (Faegri and Iverson 1989, Erdtman 1969) (Fig. 2.2). On the basis of chemical resistance, autofluorescence and staining capacity, two layers of exine, ektexine and endexine are distinguished (Kress and Stone 1982). The ektexine dissolves with 2- and 3-ethanolamine and stains deeply with alcoholic fuchsin and auramine O; the endexine resists hydrolysis with ethanolamine and stains weakly or not at all with alcoholic fuchsin and auramine O. The endexine forms a homogeneous layer and is continuous in the non-apertural region in most pollen. However, in some taxa, such as grasses, there are fine passages across the endexine (Rowley 1973, Dahl and Rowley 1974).

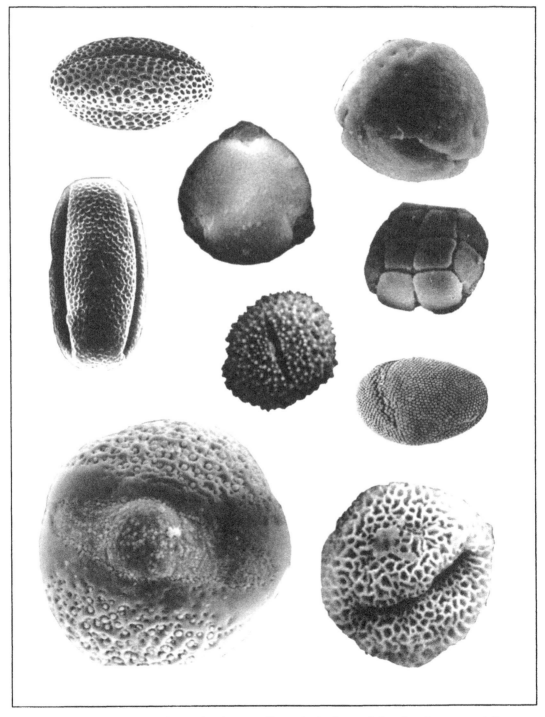

Fig. 2.1 Scanning electron micrographs of some pollen grains to show variations in exine ornamentation.

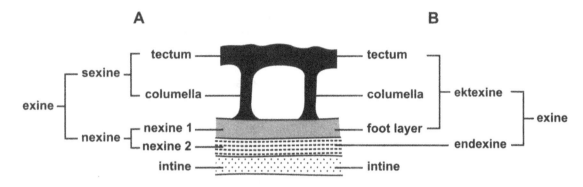

Fig. 2.2 Pollen wall architecture and two sets of terminologies (A, B) used to describe exine layers.

Pollen coat substances

Pollen grains, especially of entomophilous plants, are coated with an oily, sticky and often coloured material commonly termed 'pollenkitt' or 'tryphine'. In other species, it is less conspicuous. The pollen coat substances contain a range of chemicals; lipids, carbohydrates, proteins, glycoproteins, carotenoids, and flavonoids; these and other phenolics are the major components of the pollen coat (Wiermann and Gubatz 1992). Pollen coat substances are derived from the tapetum.

The carotenoids are synthesized in the tapetum and accumulate in osmiophilic globules (Heslop-Harrison and Dickinson 1969). After the breakdown of the tapetum, the carotenoids are released into the locule and eventually get deposited on the exine surface and/or cavities. Flavonoids are important phenolic compounds of the pollen coat substances. Kaempferol, quercetin and isorhamnetin are the principal flavonoids. Flavonoid biosynthesis occurs largely in the tapetum (Wiermann and Gubatz 1992). L-phenylammonia-lyase (PAL) and chalcone synthase (CHS) are the key enzymes in flavonoid biosynthesis; both these enzymes are predominantly distributed in the tapetum. It is suggested that the enzymes involved in flavonoid biosynthesis are released into the locule and are transported to the pollen where flavonoid synthesis takes place. In *Brassica napus* oleosin-

like proteins, major pollen coat components, are synthesized in the tapetum (Piffanelli et al. 1997, Murphy and Ross 1998).

Pollen Wall Morphogenesis

The wall of the pollen grain is rather uniform in architecture (see Fig. 2.2 for details). Pollen wall morphogenesis occurs in two stages. The first stage is completed when the microspores are still enclosed by a callose wall and the second after the release of microspores (Heslop-Harrison 1963). Details of pollen wall morphogenesis are diagrammatically represented in Figure 2.3. This scheme of exine deposition, generally referred to as the primexine scheme (Heslop-Harrison 1963), has been extensively adopted in studies on pollen wall development. The blueprint of exine, termed primexine, is laid down between the plasma membrane and the callose wall. The primexine has a matrix, presumably made up of cellulose, and radially directed rods, the probaculae. Probaculae become connected at their bases to form the foot layer. The probaculae may remain free above or form a roof over the primexine matrix. The foot layer represents the future nexine 1, and the probaculae and the roof layer (tectum) the future sexine (Erdtman 1966, 1969). The primexine and foot layer are acetolysis resistant. Since the electron opacity of the probaculae and foot layer differs from that of mature exine, the component of these layers has been termed

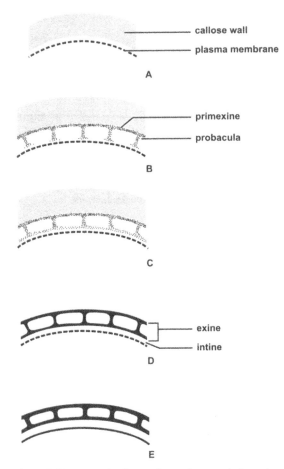

Fig. 2.3. Diagram of pollen wall morphogenesis based on primexine model.

'protosporopollenin'. In *Dendrobium* freeze fracture and freeze substitution studies have shown tubular SERs terminating at and fusing with the plasma membrane (Fitzgerald et al. 1994). It has been suggested that the enzymes for polymerization of sporopollenin precursors are transported via SER to the plasma membrane. According to the primexine scheme, the structural features of exine are determined by the distribution pattern of probaculae.

In many taxa the presumptive germinal apertures (pores/colpae) are demarcated during the formation of primexine by the presence of ER oriented parallel to the plasma membrane. These regions remain free of primexine deposition. It is suggested that ER physically prevents the movement of membranous structures coated with primexine material to the cell surface (Dickinson and Potter 1976, Sheldon and Dickinson 1983). In a few species, however, presumptive pore regions are not associated with ER localization (Shoup et al. 1980).

In a number of species the position of the probaculae and other features of the mature exine can also be correlated with specific features in the cytoplasm of young microspores. For example, the location of probaculae in *Silene* (Heslop-Harrison 1963) and *Berberis* (Gabara 1974) is marked by the presence of short membrane profiles below the plasma membrane, and of those of *Lilium* (Vazart 1970) by accumulation of ribosomes. In *Cosmos* (Dickinson and Potter 1976), large banks of ER and microtubules in the cytoplasm are associated with the development of spines of the exine. Microspores are released by the dissolution of callose wall soon after the differentiation of the primexine.

Recent studies on exine development in *Caesalpinia* and *Lilium* (Takahashi 1989, 1993, 1995), and a few other taxa (Takahashi and Skvarla 1991, Fitzgerald and Knox 1995) do not support the primexine scheme of exine deposition. These studies have shown that the exine pattern is initiated by invaginations of the microspore plasma membrane at an early tetrad stage (Fig. 2.4). In *Lilium* (Takahashi 1995), the invaginated plasma membrane takes the form of a reticulate pattern that corresponds to the pattern of mature exine. The invaginated regions correspond to the regions of future lumina and protuberants (raised areas) correspond to the muri of mature exine. At the regions of raised areas of the plasma membrane, fibrous threads of 10–20 nm diameter aggregate together with granules of 10 nm diameter. Gradually the aggregated fibrous threads and granules develop into 0.5–0.7 μm wide smooth protectum. Probaculae form subsequently below the protectum (between the protectum and the plasma membrane). At the late tetrad stage the protectum and probaculae become distinct

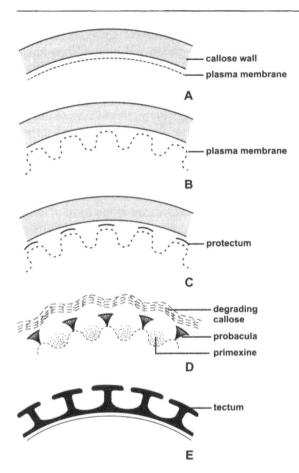

Fig. 2.4 Diagram of pollen wall morphogenesis based on plasma membrane undulation model.

below the callose wall and the plasma membrane becomes smooth. The fibrous primexine matrix becomes visible in the spaces between probaculae. The callose wall dissolves at this stage, releasing the microspores. At the time of release the surface of the protectum is smooth. Further differentiation of exine continues after microspore release. Thus the reticulate pattern of the invaginated plasma membrane forms the blueprint of mature exine. The protectum is the first exine layer to be deposited on the reticulate patterned plasma membrane. According to this scheme, plasma membrane plays a very important part in pollen wall morphogenesis. However, the mechanism by which plasma membrane de-

velops patterned invaginations is not known. It is possible that cytoskeleton elements play a role in this differentiation.

Studies of Fitzgerald and Knox (1995) on pollen wall initiation in *Brassica* microspores showed basically the same pattern of plasma membrane undulations as described above. However, before the undulations became visible continuous primexine deposition could be observed between the microspore plasma membrane and callose wall. Deposition of flakes of condensed material was observed in the depressions of the plasma membrane before deposition of the probaculae; this material represents interbaculate sites, referred to as spacers. This type of exine morphogenesis (based on undulations of plasma membrane) seems to be a feature specific only to reticulate exine. In species such as *Acacia* (Fitzgerald et al. 1993a) and *Dendrobium* (Fitzgerald et al. 1994) which do not have reticulate exine, plasma membrane undulations are not associated with exine deposition.

Control of exine pattern

Exine ornamentation is one of the unique features of pollen. Although there is great variation in exine ornamentation between species it is characteristic of a given species/genus. Initial observations that the blueprint of the exine is laid down when the microspores are still enclosed by the callose wall led to the suggestion that the exine pattern is under the control of the gametophytic genome (Godwin 1968). However, subsequent studies have clearly shown that the exine pattern is under the control of the sporophytic genome (Heslop-Harrison 1971, 1972). Some of the evidence is presented below.

In an interspecific hybrid of *Linum* (Rogers and Harris 1969), many chromosomes are left in the cytoplasm during meiotic division because of irregular chromosomal pairing. During cleavage of the microspore mother cell, small spores form around the scattered chromosomes. All these spores develop typical exine although they have an incomplete chromosome complement

(Fig. 2.5). If the exine pattern is to be under the control of the microspore genome, typical exine formation in all such spores cannot be expected, as it is not possible for all these spores to receive the genome complement necessary for control of the exine pattern. Development of typical exine in microspores with incomplete chromosome complement has also been reported in triploid clones of *Tradescantia* (Mepham and Lane 1970) and male sterile mutants of *Impatiens* (Tara and Namboodiri 1974).

It has been a common observation that there is no segregation of exine pattern in pollen grains of hybrids of parents showing differences in exine features. For example, in the hybrid *Lycopersicon esculentum* × *Solanum pennellii* (Quiros 1975), the genes of *S. pennellii* are dominant for pollen size and exine density, and those of *L. esculentum* for pollen shape and exine spine structure. There is no segregation of exine pattern in the pollen grains of the hybrid.

Although all the above evidence confirms sporophytic control of the exine pattern, there is no clear information on morphogenetical details of such control. Apparently the information is transcribed before meiosis and is carried in the cytoplasm of the meiocytes in the form of long-lived mRNAs or protein moiety (Heslop-Harrison 1972). It is possible that long-lived mRNAs and/or proteins are retained in the cytoplasm en-

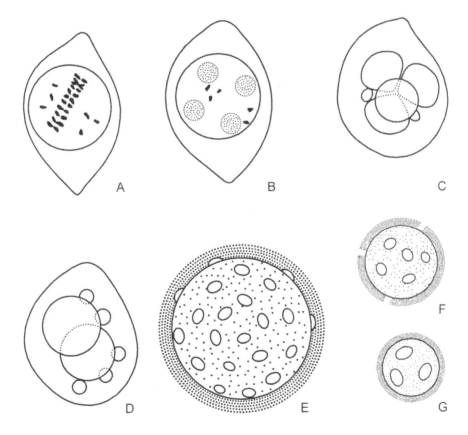

Fig. 2.5 Anomalous development of pollen grains in *Linum* hybrid. A. Metaphase I with several scattered laggards. B-D. Microspore formation resulting in a tetrad (C) or a dyad (D) accompanied by several smaller spores. E-G. Pollen grains of various sizes with fully developed exine (after Rogers and Harris 1969).

closed by double or multimembrane units (see Chapter 1), which are not affected by the cytoplasmic reorganization that takes place in other parts of the cytoplasm (Dickinson and Heslop-Harrison 1977).

Pollen wall proteins

It is now well established that both layers of the pollen wall, the exine and intine, contain a considerable amount of proteins (Figs 2.6, 2.7) (Heslop-Harrison 1975, Heslop-Harrison et al. 1975, Knox et al. 1975, Shivanna and Johri 1985). Exine proteins are located in the interporal regions in exine cavities (in tectate pollen) or surface depressions (in pilate pollen). Exine proteins originate from the surrounding tapetum and thus are sporophytic in origin (Fig. 2.8). The intine proteins are generally present in the form of radially oriented tubules and are concentrated in the germ-pore region. Intine proteins originate in the pollen cytoplasm (Fig. 2.8) and are therefore gametophytic in origin. Soon after the initiation of intine development, the plasma membrane of pollen produces radially oriented tubules into the developing intine. Eventually, these tubules with their protein inclusions are cut off from the plasma membrane and become incorporated in the intine.

Pollen wall proteins include many enzymes, particularly hydrolytic enzymes such as esterases, amylases and ribonucleases (Knox and Heslop-Harrison 1969, 1970b). Esterases occur predominantly in the exine and acid phosphatases in the intine. These enzymes can therefore be used as marker enzymes for exine and intine proteins. Proteins responsible for pollen allergy are also present in the pollen wall. Pollen wall proteins play an important role in pollen function (see Chapters 8, 9).

In many species belonging to Cannaceae, Zingiberaceae, and marine angiosperms in which exine is absent or highly reduced, tapetally derived proteins are absent. The intine is generally thick and often differentiated into many layers and the wall proteins are exclusively confined to the intine (Heslop-Harrison 1975, Knox 1984a, Kress et al. 1978, Kress and Stone 1982, Ducker et al. 1978).

Apart from the deposition of tapetally derived proteins in the interporal region of pollen, in many species such as *Olea* (Pacini and Juniper 1979, Pacini et al. 1981) and *Pterostylis* (Pandolfi et al. 1993) tapetal proteins are also deposited in the poral region (where exine is absent) outside the intine (Shivanna and Johri 1985). Unlike interporal exine proteins which are deposited following tapetal breakdown, poral proteins are deposited earlier, when the tapetal cells are still intact.

Pollen Analysis

Pollen size and shape

The size of the pollen is highly variable and ranges from 5 to 200 μm. Smallest grains are recorded in members of Boraginaceae and larger pollen grains are produced in members of Cucurbitaceae, Malvaceae and Nyctaginaceae. The pollen of *Cymbopetalum odoratissimum* (Annonaceae) measuring up to 350 μm is perhaps the largest reported (Walker 1971). In a majority of species pollen size varies between 15 and 50 μm. In marine angiosperms such as *Amphibolis* and *Zostera*, pollen grains are filiform and long measuring up

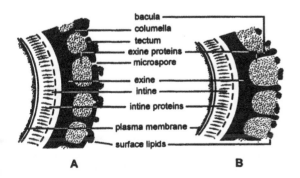

Fig. 2.6 Location of intine and exine proteins, and pollen coat lipids. A. Tectate exine. B. Pilate exine (modified from Shivanna 1977).

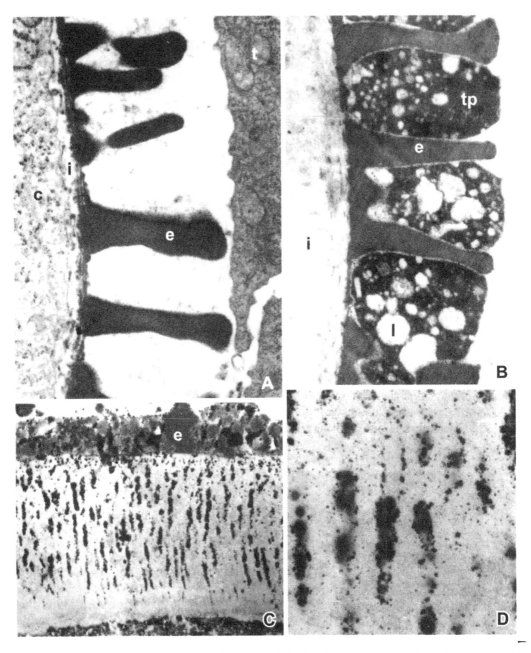

Fig. 2.7 Transmission electron micrographs of part of a pollen wall to show incorporation of pollen wall proteins. A, B. *Iberis*. A. Microspore before insertion of tapetal material in exine depressions. The tapetum (t) is still intact and the intine (i) is not fully developed. B. After insertion of the tapetum-derived components in the cavities of the exine. C, D. *Crocus* characterized by thick intine following localization of acid phosphatase activity. The enzyme is localized in radially oriented tubular inclusions in the intine. D. Part of the intine magnified to show details (e— exine, i—intine, c—cytoplasm, l—lipids, t—tapetum, tp—tapetal proteins) (after Heslop-Harrison 1975).

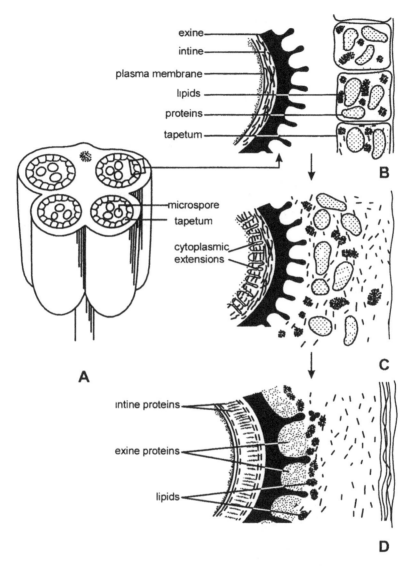

Fig. 2.8 Origin of intine and exine proteins. A. Transection of a young anther with microspores and intact tapetum. Only part of the microspore wall and tapetum are shown in B-D to indicate incorporation of proteins during pollen development. Exine proteins are derived from the tapetum and intine proteins from the pollen cytoplasm (after Shivanna 1977).

to 5 mm (Ducker et al. 1978). In *Crossandra*, a terrestrial species, pollen grains are long (>500 µm) (Brummitt et al. 1980).

The size of pollen grains varies to some extent, depending on the method of preparation and its hydration level. Therefore, pollen size can hardly serve as a diagnostic character (Faegri and Iversen 1989). However, with other morphological characters, size of the pollen may help in demarcating the taxa.

The shape of the pollen is also variable. Desiccated grains are frequently subprismatic because of the contraction of the exine, while hydrated grains tend to be ellipsoidal or spherical.

Also the shape of pollen grains varies depending on whether it is observed in polar view or equatorial view. Pollen grains of a large number of species in polar view are circular or elliptic. However, triangular, quadrangular and rectangular pollen are also present. In angular pollen, the angles may be acute or obtuse (Moore and Webb 1978, Faegri and Iversen 1989).

Compound pollen

Associations of two or more pollen grains as dispersal units are referred to as compound or composite pollen grains (Fig. 2.9) (Walker and Doyle 1975, Knox and McConchie 1986, Pacini 1997).

The number of pollen grains in a compound unit may be 2 (dyad), 4 (tetrad) or multiples of four (polyad). In several species, a large number of pollen grains associate to form massulae (some members of Orchidaceae). In a number of Asclepiadaceae and Orchidaceae all the microspores in a sporangium remain together to form a single mass called the pollinium. Compound pollen grains have been reported in more than 56 angiosperm families (Walker and Doyle 1975). Compound pollen may be formed by fusion of their exine layers (calymmate) or by connecting wall bridges (acalymmate). Wall bridges may be confined only to exine layers (sexine and

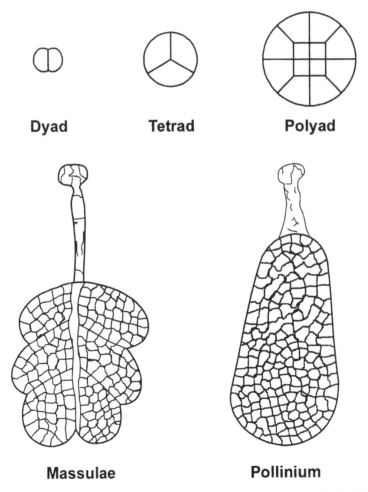

Dyad   **Tetrad**   **Polyad**

Massulae   **Pollinium**

Fig. 2.9 Diagram of compound pollen grains (modified and redrawn from Pacini 1997).

nexine) or include intine also. Some species show cytoplasmic bridges between individual units of the compound pollen.

Polarity

The morphological terms used to describe pollen are far too many and often confusing (Kremp 1965, Nilsson and Muller 1978, Thanikaimoni 1978). The following paragraphs provide only the basic terms used to describe pollen. Pollen morphology is largely described in relation to the polarity of the pollen grain. The polarity of the pollen is referenced to the arrangement of microspores in the tetrad (Fig. 2.10). The pole closest to the center of the tetrad is the proximal pole and that farthest from the center is the distal pole. A hypothetical axis connecting the two poles is the polar axis. The axis at a right angle to the polar axis is the equatorial axis. It demarcates the two halves of the pollen; if the two halves are similar, the pollen grain is described as isopolar, and if different, as heteropolar. Even after the release of microspores from the tetrads, their surface pattern is generally related to the orientation of the grain in the tetrad.

The classical method of studying the morphology of pollen has been through acetolysis. This method removes the protoplast, the intine and pollen coat substances, leaving only the exine. Studies of acetolyzed pollen provide information on the surface features. To obtain the best possible details, pollen grains should be mounted in an embedding medium of proper refractivity. The most commonly used mounting media have been glycerol, glycerol jelly and silicon oil.

It is possible to get information on the structure of the wall at different levels by studying acetolyzed pollen through optical sectioning (careful focusing at different levels from the surface towards the centre). Studies using microtome sections provide a better understanding of exine elements. In recent years, electron microscopy (transmission and scanning) has provided a powerful technique for studying the finer details of the exine.

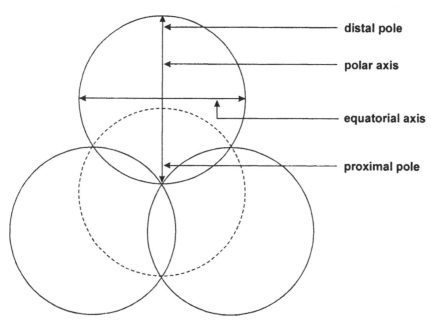

Fig. 2.10 Diagram of a tetrahedral tetrad of microspores to show the two poles and two axes (after Shivanna and Rangaswamy 1992).

Apertures

Apertures are the areas through which the pollen tube emerges during pollen germination (Rowley 1975). In the region of apertures, the exine is absent or very thin, thus facilitating tube emergence. Apertures are distinguished on the basis of shape, number and position (Fig. 2.11). They may be in the form of pores or in the form of furrows (colpi). The colpi are elongate boat-shaped structures with more or less acute ends. Pollen grains with pores are referred to as porate and those with colpi as colpate. Pollen grains are called colporate when a furrow and a pore are combined in the aperture. In some pollen, the pores occur only in some furrows (generally located in half or one-third of the pollen). Such pollen grains are heterocolporate.

The apertures, particularly the colpi, are also involved in accommodating changes in the volume of pollen as a result of hydration and desiccation (harmomegathic); the intine in desiccated pollen buckles inside in the apertural region bringing the two margins of the colpi together. It is suggested that this feature prevents excessive desiccation of pollen. In hydrated pollen the margins expand, exposing the colpal surface. Phylogenetically the colpi are considered primitive and the pores derived by contraction (Faegri and Iversen 1989).

In pollen grains of many species the aperture is covered with an isolated piece of exine called the operculum and such pollen grains are called operculate (Heslop-Harrison 1979c). In the porate type the operculum forms a disc (e.g. grasses) and in the colpate type a median strip (e.g. *Potentilla*). Edges of apertures may be similar to the rest of the exine (non-bordered) or may appear different to form a border (bordered). The border is generally in the form of a change in

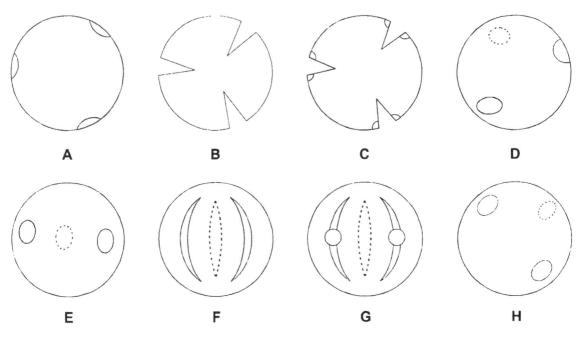

Fig. 2.11 Diagram of the four types of pollen grains showing the arrangement of apertures in polar view (top row) and equatorial view (bottom row). Dotted lines show the apertures as seen in different focal planes. A, E. Trizonoporate. B, F. Trizonocolpate. C, G. Trizonocolporate. D, H. Tripantoporate.

size and/or density of exine sculpturing elements.

The number (from one to forty or more) and arrangement of apertures varies between species. The prefixes mono-, bi-, tri-, tetra-, penta-, hexa- and poly- are used to describe porate or colpate pollen. In a large number of dicotyledonous species, three apertures (pores/colpi) corresponding to the three contact surfaces with other microspores of the tetrad are arranged equidistantly from each other along the equator. A great majority of monocotyledons show a single colpus or pore.

When the apertures are arranged equidistantly from each other along the equator or evenly distributed over the whole surface, the term zono- is used (e.g. zonopolyporate). When the pores are arranged irregularly, the term panto- is used (e.g. pantopolyporate). In some pollen grains, two or more colpi are combined into rings or a spiral surrounding the whole or a part of the grain (syncolpate). The colpi extend to different lengths towards the proximal and distal poles. The polar area is that part of the pollen which is above/below the level of colpi. The number, character and arrangement of apertures are very useful for identification of the pollen.

Exine sculpture

Exine sculpturing refers to the external surface features of the exine without reference to the internal structure of the exine (Faegri and Iversen 1989). Ornamentation of the exine is confined to the ektexine. The ektexine is basically a 3-layerd structure with radially oriented columellae/baculae separating the outer roof, the tectum and the lower foot layer (nexine 1). When the tectum covers all or most (ca 75%) of the surface, the pollen grain is called tectate (Fig. 2.12 A). When the tectum covers the pollen partially (ca <75%), the grains are semitectate. Pollen grains without tectum are called etectate/intectate. The tectum may have perforations (pits) of different shapes and sizes; when the pores are small (<1 μm) the tectum is called perforate and when they are >1 μm the tectum is foveo-

late. A distinction is often made between the columellae and baculae; the columellae support a tectum or a knob while baculae are free and cylindrical (Faegri and Iverson 1989).

Sculpturing is more constant than other pollen characters and provides an effective means of pollen identification. In intectate grains the pattern of arrangement of the columellae/baculae forms the surface pattern. In semitectate and tectate grains the sculpturing does not reveal the internal structure of the exine. In intectate grains the baculae, depending on their morphology, are referred to as gemmate (short and globular), baculate (rod-shaped), clavate (club-shaped), pilate (with swollen head) and echinate (pointed) (Fig. 2.12 B). When the surface of the pollen is smooth, it is referred to as psilate. In some pollen the surface may be covered with hemispherical warts (verrucae), or tiny flakes (scabrae) or small granules; the corresponding terms for such exines are verrucate, scabrate and granulate.

The baculae are often arranged regularly and fused with neighbouring ones, resulting in elevated wall-like structures (muri) and depressed areas (lumina), which together produce different patterns (Fig. 2.13). When the muri form a network, the pollen surface is described as reticulate. When the muri and lumina run parallel to one another it is striate, and when they are irregularly distributed, the surface is referred to as rugulate. Tectate and semitectate pollen may also bear, on their tectum, structures which may be baculate, clavate, echinate, gemmate, pilate, scabrate, verrucate, reticulate, striate or rugulate.

LO analysis

When a pollen grain is focused on its outermost surface, the raised areas appear lighted and the depressed areas/pits appear dark. On gradual downward focusing through the exine, the pattern reverses (raised areas appear dark and depressed areas lighted) due to a change in the diffraction images produced. One can obtain considerable information on the morphology of

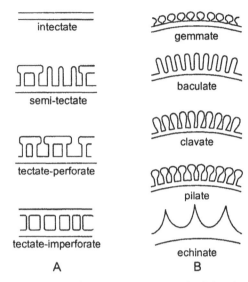

intectate

semi-tectate

tectate-perforate

tectate-imperforate

A

gemmate

baculate

clavate

pilate

echinate

B

Fig. 2.12 Diagram of intectate and tectate exine (A) and types of exine baculae as seen in sectional view in intectate grains (B).

various levels of the exine by careful studies of such optical sections of pollen by focusing at different levels.

Erdtman (1956) called this type of study LO and OL analysis (from Latin *Lux* = light and *obscuritas* = darkness). LO for the sequence: light islands and dark channels (high focus) followed by dark islands and light channels (low focus), and OL for the reverse sequence: dark islands and light channels followed by light islands and dark channels. In recent years this method is hardly used in pollen analysis because simpler and more effective techniques for studying exine structure are available.

AEROPALYNOLOGY

Aerobiology is the study of biological particles such as pollen, fungal spores, dust mites, insect

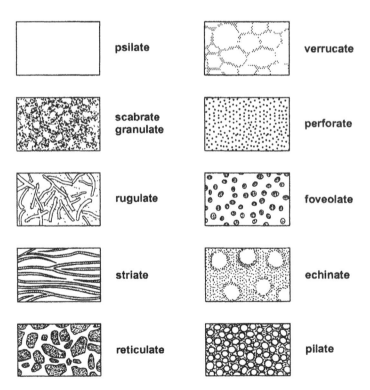

psilate

scabrate
granulate

rugulate

striate

reticulate

verrucate

perforate

foveolate

echinate

pilate

Fig. 2.13 Basic sculpturing patterns of exine in surface view.

debris and organic dusts present in the air (Hyde 1969, 1972). Aeropalynology is the study of the release, dissemination, deposition and allergic effects of pollen grains and spores present in the air. It has been well established for more than a century that pollen grains are responsible for many allergic diseases, such as hay fever, asthma, allergic rhinitis and atopic dermatitis (Hyde 1969, Stanley and Linskens 1974, Knox 1979, 1993, Leuschner 1993, Agashe 1994). Allergy, also termed immediate hypersensitivity, is defined as an altered and accelerated reaction of a person to a second or subsequent exposures to a substance to which he/she has been sensitized during the first exposure. About 10–20% of the population is known to suffer allergic disorders caused by bioparticles (Singh and Singh 1994).

Allergic Response

Pollen grains induce allergic responses in susceptible individuals. Allergic pollen grains belong to three broad categories: grasses, weeds and trees. Rye-grass, timothy-grass, Kentucky grass and Bermuda grass are the major grasses. Ragweed (*Ambrosia*) is one of the important allergenic weeds, and birch, oak and hazel are some examples of tree allergens. Some of the common allergenic species in India are given in Table 2.1 (Singh and Rawat 2000). A characteristic feature of pollen allergy is its seasonal occurrence associated with the prevalence of pollen of that particular species in the atmosphere. Airborne pollen grains generally travel short distances; however, when they are blown into the upper strata of the atmosphere, pollen grains travel long distances before they are deposited. Meteorological factors, in particular temperature, precipitation, humidity and wind speed strongly influence airborne pollen counts. Pollen dispersal is facilitated by dry weather and higher wind velocity. Pollen count is reduced after precipitation. Respiratory tracts, lungs, bronchi, skin and gastrointestinal tracts are commonly involved in allergic reaction.

Table 2.1 Some common plants in India in different seasons which cause pollen allergy (*Source:* Dr A.B. Singh, Centre for Biochemical Technology, Delhi)

Spring (February–April)	Autumn (September–October)	Winter (November–January)
Grasses		
Cynodon dactylon	Bothriochloa pertusa	Cynodon dactylon
Dichanthium annulatum	Cenchrus ciliaris	Eragrostris tenella
Imperata cylindrica	Heteropogon contortus	Phalaris minor
Paspalum disticum	Pennisetum typhoides	Poa annua
Poa annua	Sorghum vulgare	
Polygonum monspeliensis		
Weeds		
Cannabis sativa	Amaranthus spinosus	Ageratum conyzoides
Chenopodium murale	Artemisia scoparia	Argemone mexicana
Parthenium hysterophorus	Cassia occidentalis	Chenopodium album
Plantago major	Ricinus communis	Asphodelus tenuifolius
Suaeda fruticulosa	Xanthium strumanium	Ricinus communis
Trees		
Alnus nitida	Anogeissus pendula	Cassia siamea
Ailanthus excelsa	Eucalyptus sp.	Salvadora persica
Holoptelea integrifolia	Prosopis juliflora	Mallotus phillipensis
Prosopis juliflora	Cedrus deodara	Cedrus deodara
Putranjiva roxburghii		

In hay fever and asthma responses, pollen grain enters the nose and lands on the mucous membrane of the upper or lower respiratory tract. Pollen grain gets hydrated by the mucus secretion and releases pollen allergens which penetrate the mucous tissue. The allergens induce formation of antibodies (largely of IgE class) from B-lymphocytes (Fig. 2.14). The B-cells are stimulated to proliferate clonally by cytokines (interleukins) which are produced by T-cells.

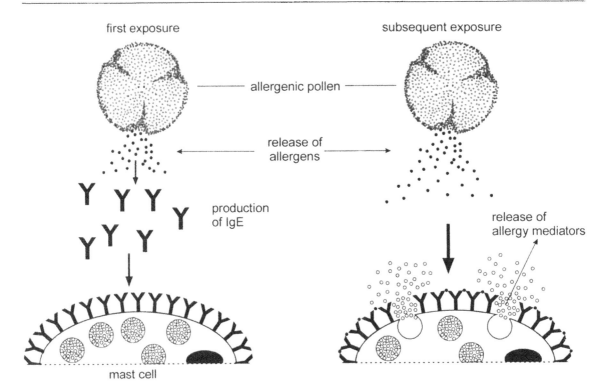

Fig. 2.14 Schematic details of allergic response to pollen grains. First exposure to the allergen leads to the production of IgE antibodies which bind to the corresponding receptors on the surface of mast cells. On subsequent exposure the allergens released from the pollen bind to adjacent IgE molecules located on mast cells; this binding triggers release of various mediators of allergy such as histamines which induce allergic symptoms.

T-cell epitope, a specific region of the allergen molecule, is necessary for activation of T-cells. The antibodies produced from B-cells circulate through the serum and bind to the corresponding receptors on the surface of the mast cells (Fig. 2.14) and basophils present in the connective tissue of the skin and other endothelial systems. There are about 10^{5-6} receptor sites on each mast cell. On subsequent exposure, the allergens released from the pollen bind and cross-link specific IgE antibodies present on the surface of mast cells. This binding activates enzymes which release mediators such as histamines, leukotrienes and prostaglandins from the mast cells (Fig. 2.14). These mediators induce allergic symptoms such as dilation of blood capillaries, contraction of nasal and bronchial muscles, oversecretion of watery nasal fluid,

constriction of the nasal or bronchial passages, sneezing, itching and edema of the mucous membrane (Stanley and Linskens 1974, Chanda 1994). Elevated levels of IgE are detectable in the blood serum of allergic patients. The serum containing specific IgE can therefore be used as a probe for identification of allergens.

Until recently, the relationship between pollen and asthma was not clearly understood as pollen grains are too large to reach the lower respiratory tract. Allergic asthma is the disease of airways and the allergens have to reach the lower respiratory tract to trigger allergic asthma. Recently allergen-containing microscopic particles released from bursting of pollen of grasses have been reported to occur in the atmosphere (Suphioglu et al. 1992). They provide a mechanism by which pollen allergens may reach the lower respiratory tract (bronchi and lungs) and cause asthma (Ong et al. 1995a, b, 1996, Knox and Suphioglu 1996). These microscopic particles are < 5 μm and reach the respiratory tract. In grass pollen, allergen-containing microscopic particles have been shown to be intracellular starch granules/p-particles (Singh et al. 1991). These granules are released from pollen during rainfall when pollen grains rupture osmotically. Starch granules are readily detected in the atmosphere during the grass pollen season. On days following rainfall, a 50-fold increase in granule number per cubic metre of air has been reported (Suphioglu et al. 1992).

Diagnostic Tests

Several diagnostic tests, in vivo as well as in vitro, are available to screen allergenicity of different antigenic extracts. Among in-vivo tests, the skin test (scratch test and prick test) is the most commonly used; it is simple, convenient and highly specific (Freeman and Noon 1911, Dreborg 1989, Lin et al. 1993). Bronchial provocation test has also been used for asthmatic patients.

Among the in-vitro assays, Paper Radio Immuno Sorbent Test (PRIST), Radio Allergo Sorbent Test (RAST) (Wide et al. 1967) and Enzyme Linked Immuno Sorbent Assay (ELISA) (Engvall and Pearlman 1971) are important. These tests estimate the amount of IgE in the serum. In the RAST test, allergens placed on filter paper discs are treated with human serum followed by radiolabelled antihuman IgE. The amount of radioactivity bound to the disc is estimated. In the ELISA test, the radioactive label is replaced with an enzyme label. Dot immunoblotting is another method in which small drops of antigen are applied on nitrocellulose paper and processed according to ELISA procedures. This is very convenient for routine diagnostic application since it permits analysis of many samples on a single strip of paper. Another method termed FAST (Fluorescent Allergo Sorbent Test), which uses a fluorescent

substrate, was recently developed (see Singh and Malik 1992).

Allergens

The pollen allergens are generally proteins or glycoproteins of molecular weight ranging from 5000 to 90,000 daltons. Using clinical and immunological approaches, a large number of allergens have been identified in pollen grains of different species. Crude pollen extract contains many other components besides allergens. The methods of pollen collection and processing also affect the amount of allergens recovered (Stanley and Linskens 1974, Singh and Singh 1994). Variations in allergens have also been reported between pollen samples of different populations and also those collected in different years. Standardization of allergenic extracts is necessary for effective clinical trials and for immunotherapy.

Recently the immunoprint/immunobiot technique has been extensively used to analyse allergenic components in different pollen samples (Towbin et al. 1979). The extract separated through immunoelectrofocusing (IEF) is transferred from the gel to the nitrocellulose paper and the allergens are detected through specific IgE antibodies. In rye-grass pollen, 14 IgE binding proteins have been reported using sera of allergic patients (Ford and Baldo 1986). In *Dactylis glomerata* SDS-PAGE showed 13 allergenic bands in the range of 14-17 kD (Ford et al. 1985). Similarly, of the 44 components of *Cynodon dactylon*, 17 are IgE binding proteins (Ford and Baldo 1986). Four allergenic components have been detected from pollen of *Parthenium hysterophorus* (Subba Rao 1984). Allergenic components have also been identified in pollen of *Artemisia* (Jaggi and Gangal 1986). In *Betula verrucosa* a polypeptide of 17 kD has been identified as the major allergen (Ipsen and Lowenstein 1983). In recent years, monoclonal antibodies have been prepared for several major allergens (Westphal et al. 1988, Walsh et al. 1990, Valenta et al. 1992, Perez et al. 1990, Roberts et al. 1992) and are being used for identifying/localizing allergens (Singh et al. 1991, Grote et al. 1994, Knox and Suphioglu 1996).

During the last 10 years several major allergens (for example, pollen of ragweed, many members of Poaceae and white birch) have been cloned and their amino acid sequence deduced (see Singh et al. 1991, Swoboda et al. 1996, Ong et al. 1996, Smith et al. 1966, Knox and Suphioglu 1996). This has led to a better understanding of their primary structures and their biological functions. *Lol p 1*, a major allergen of rye-grass is expressed by a pollen specific gene (Griffith et al. 1991, 1993) and is an acidic glycoprotein with about 5% carbohydrate moiety (mol wt 35 kD). Deduced amino acid sequences of many of the allergens show high homology to a new class of conserved pathogenesis-related proteins. Rye-grass pollen allergens have been localized immunocytochemically using monoclonal antibodies and immunogold labelling (Knox and Suphioglu 1996). Lol p 1 proteins are localized in pollen cytoplasm and to a lesser extent in exine cavities (Singh et al. 1991, Taylor et al. 1993, 1994) whereas Lol p 5 occurs largely in the starch granules and its molecular weight ranges from 28–32 kD.

Ragweed pollen is one of the most important allergens in the USA. Apart from the major allergenic proteins, Amb a 1 and Amb a 2, its pollen contains many other allergenic proteins (Rogers et al. 1996). Several allergens from ragweed pollen have been characterized by protein sequencing and/or c-DNA cloning (see also Chapter 15). The c-DNAs of some of the allergens have been expressed in *E. coli* and purified. Most of the recombinant allergens retain their IgE binding capacity. However, recombinant *Cyn d 1* allergen required expression in yeast, a eukaryote, in order to restore IgE binding (Suphioglu et al. 1996).

Biological Standardization of Allergens

This refers to the calibration of the potency of different allergenic preparations for their biological activity (Aas 1978, Turkeltaub 1989). Intradermal skin test and prick skin test have frequently been used for standardization of the potency. One of the standard units followed by European workers is the HEP unit. An HEP unit is defined as the concentration of the extract that gives, in the skin prick test, a wheal diameter equivalent to 1 mg l^{-1} of histamine hydrochloride (Dirksen et al. 1985). In the American method, intracutaneous skin tests are performed and comparison of extracts is made on their ability to induce a mean erythema diameter sum of 50 mm. (Turkeltaub 1989). This method is not related to histamine.

3

Pollen Viability and Vigour

Pollen grains, following their release, are exposed to the prevailing environmental conditions for varying periods before they land on the stigma. Depending on the conditions and the duration of exposure, the quality of pollen grains may be affected. The quality of pollen is assessed on the basis of viability and vigour.

POLLEN VIABILITY

Pollen viability refers to the ability of the pollen to perform its function of delivering male gametes to the embryo sac. The period for which pollen grains remain viable after they are shed varies greatly from species to species. On the basis of their longevity pollen grains of different species can be grouped into three categories (Harrington 1970, Barnabas and Kovacs 1997):

1. Short-lived pollen: Pollen grains belonging to members of Poaceae, Cyperaceae, Alismataceae, Juncaceae, Commelinaceae and Asteraceae, which lose viability within a few days. In some species viability is lost in less than an hour. For example, under dry and warm conditions pollen grains of sorghum (Fig 3.1), wheat and many other cereals lose viability within 30 min (Fritz and Lukaszewski 1989, Lansac et al. 1994), and those of Compositae in about 3 h (Hoekstra and Bruinsma 1975a).

2. Pollen with medium lifespan: Pollen of a majority of families, such as Solanaceae, Rutaceae, Cruciferae, Ranunculaceae, Liliaceae and Amaryllidaceae, fall within these extremes; they maintain viability for 1–3 months.

3. Long-lived pollen: Pollen of many Gymnosperms (Pinaceae and Gingkoaceae) and members of several angiosperm families, such as Leguminosae, Rosaceae, Anacardiaceae, Saxifragaceae and Arecaceae, maintain viability for over 6 months.

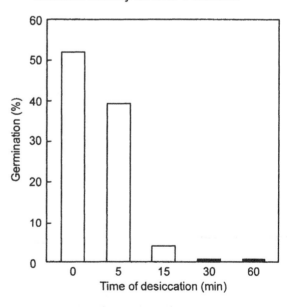

Fig. 3.1 Effect of desiccation on viability of pollen of *Sorghum* assessed on the basis of in-vitro germination. Viability is lost within 30 min. (after Lansac et al. 1994).

Comparative studies carried out on 2- and 3-celled pollen (Brewbaker 1959, Hoekstra and Bruinsma 1975b, Johri and Shivanna 1977, Hoekstra 1979) have demonstrated a broad correlation between the cytology of pollen and their viability (see Table 1.2). Two-celled pollen grains in general retain viability for a longer period than 3-celled pollen. Studies have shown that the rate of respiration in 3-celled pollen (maintained under high RH) is 2–3 times higher than 2-celled pollen (Hoekstra and Bruinsma 1975b). Although higher temperatures increased the rate of respiration in both 2- and 3-celled pollen, it was always higher in 3-celled pollen than that in 2-celled pollen (Fig. 3.2). These studies indicate that high respiratory activity may be a causative factor for rapid loss of viability of 3-celled pollen (Hoekstra and Bruinsma 1975b); however, there is no direct evidence for such a suggestion.

Rapid loss of viability in pollen grains of Poaceae is associated with their inability to withstand desiccation. Unlike pollen of other species which are shed under desiccated condition (moisture level < 20%), pollen grains of Poaceae are shed under hydrated conditions (moisture

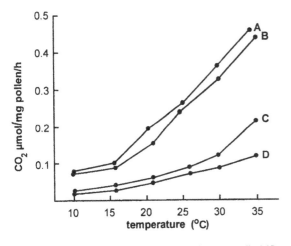

Fig. 3.2 Respiration rate in 3-celled (A, B) and 2-celled (C, D) pollen at 97% humidity and different temperatures. A—*Aster*, B—*Chrysanthemum*, C—*Nicotiana*, D—*Corylus* (after Hoekstra and Bruinsma 1975b).

level ca 50%); loss of moisture is detrimental to their viability.

The moisture content of the pollen is conventionally expressed as the percentage of fresh weight (based on loss with oven-drying to constant weight). Dumas and his associates (Dumas et al. 1985, Kerhoas et al. 1987) have used nuclear magnetic resonance (NMR) spectrometry, which is non-destructive, and seems to be more accurate in determining the moisture level of pollen. Most of the water present in desiccated pollen is in bound form.

Tests for Viability

For any experimental studies on pollen, assessment of their viability is a prerequisite. Consideration of pollen viability is also important in studies on pollen storage, reproductive biology and hybridization (Heslop-Harrison et al. 1984, Stone et al. 1995, Dafni and Firmage 2000). Standardization of a simple, rapid and dependable test to assess pollen viability is important. A number of tests have been developed over the years to test pollen viability. Detailed protocols for commonly used tests are given in Shivanna and Rangaswamy (1992).

Fruit- and seed-set

As viability refers to the ability of pollen to deliver functional gametes to the embryo sac, the most authentic test for viability would be to assess the fertilization capacity of the pollen sample as measured by fruit- and seed-set following controlled pollination (Heslop-Harrison et al. 1984, Shivanna and Rangaswamy 1992). However, this test has many limitations for use as a routine test: (i) it is laborious and time-consuming; it may take many days or weeks before seed-set is assessed; (ii) many other factors such as stigma receptivity and incompatibility have to be taken into consideration to perform this test; (iii) seed-set is not an inevitable outcome of fertilization as many post-fertilization factors associated with seed development may influence seed-set; (iv) this test cannot be used in apomictic species; (v) it can be used only during the flowering period of the species; (vi) it can be used more as a qualitative than a quantitative test, particularly in systems with fewer ovules, as germination of a limited number of pollen is enough to induce full seed-set. Therefore, assessment of pollen viability through fruit- and seed-set is not practicable as a routine test, although it can be used to confirm the results of other tests.

Pollen germination and pollen tube growth in the pistil

As an alternative to fruit- and seed-set, some attempts have been made to assess pollen viability by studying pollen germination and pollen tube growth in the pistil following controlled pollinations. In *Brassica oleracea*, for example, pollen samples which produce 70 pollen tubes in the style are considered fully viable (Ockendon 1974). Although this method markedly reduces the time taken compared to the fruit- and seed-set method, it has most of the other limitations associated with fruit- and seed-set. Also, it is not always feasible to quantify the number of pollen tubes growing in the style.

Non-vital stains and other tests of limited use

Because of the limitations of the above methods, many alternative methods which are simple, convenient and rapid have been developed over the years. Many of the staining tests using non-vital stains such as iodine in potassium iodide, aniline blue in lactophenol, acetocarmine, acid fuchsin and Alexander's stain (Alexander 1980), essentially assess the presence of contents in the pollen; they are satisfactory in assessing pollen sterility but are not dependable for testing viability (see Heslop-Harrison et al. 1984). Some non-permeating stains such as Evans blue and phenosafranin (Widholm 1972) which do not enter plasma membranes of living cells but stain the cytoplasm of dead cells are reportedly suitable for assessing pollen viability. However, studies using these tests have been limited so far. It would be worthwhile to extend them

to a larger number of species to assess their general applicability.

Inorganic acid tests (Koul and Paliwal 1961, see Shivanna and Johri 1985) based on bursting of viable pollen in inorganic acids and formation of instant pollen tubes (Linskens and Mulleneers 1967, Maheshwari and Mahadevan 1978) from hydrated pollen, although very simple and rapid, are largely not used because data for establishing their correlation with true viability are lacking. Similarly, estimation of ATP content (assayed by luciferase-luciferin method), shown to correlate with germinability in many conifers (Ching et al. 1975), has not been much used either, probably because of the time and expense involved in this test.

A cytochemical test, termed the benzidine test, which is based on the oxidation of benzidine by peroxidase in the presence of hydrogen peroxide, has been used in the past for assessing viability of pollen of many species (King 1960). In recent years, however, this test has not been used because of benzidine toxicity and also the availability of better and effective tests.

Tetrazolium test

The tetrazolium test is one of the most widely used, especially by early investigators (see Stanley and Linskens 1974). This test is based on reduction of soluble colourless tetrazolium salt to reddish insoluble formazan in the presence of dehydrogenase. Following incubation of pollen grains in tetrazolium solution for 30–60 min, pollen grains which take a reddish colour are scored as viable (Fig. 3.3A). Although 2,3,5-triphenyl tetrazolium chloride is the most commonly used salt, many other salts such as nitroblue tetrazolium (which is specific to succinic dehydrogenase) (Hauser and Morrison 1964) and tetrazolium red (Aslam et al. 1964) have also been used.

Some investigators have reported satisfactory results with this test in assessing pollen viability in several species (Hauser and Morrison 1964, Norton 1966, Collins et al. 1973).

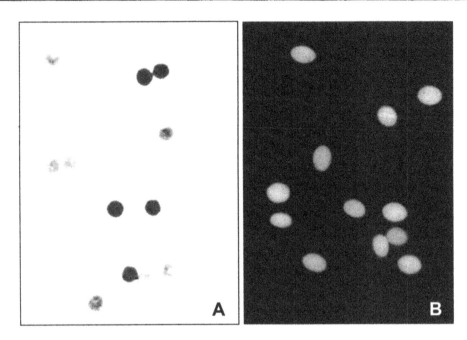

Fig. 3.3 Tests for pollen viability. A. Tetrazolium test; viable pollen grains are darkly stained. B. Fluorescein diacetate test (after Shivanna and Rangaswamy 1992).

However, many investigators have reported false positive responses with tetrazolium test (Oberle and Watson 1953). Often, results of the tetrazolium test did not correlate with seed-set data (Barrow 1983) or the in-vitro germination test (see Heslop-Harrison et al. 1984). For example, in *Simmondsia* over 95% of stored pollen developed colour reaction in tetrazolium solution even in samples which did not show in-vitro germination (Beasley and Yermanos 1976). Similarly, over 90% of the pollen of *Helleborus niger* subjected to DMSO or heat treatment, which failed to respond to germination or FDA tests, were positive for the tetrazolium test (Fig. 3.4) (Heslop-Harrison et al.1984). Another limitation of the tetrazolium test is that the colouration of responding pollen shows a gradation from very light to dark red with the result that the cut-off point for scoring viable pollen becomes subjective. Hence the tetrazolium test has not been popular in recent years.

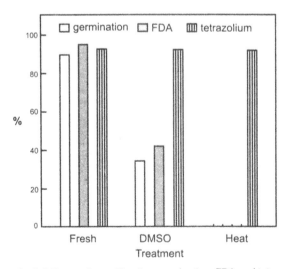

Fig. 3.4 Comparison of in-vitro germination, FDA and tetrazolium tests on fresh, DMSO-treated and heat-treated pollen of *Helleborus niger*. A good correlation was found between all the tests with fresh pollen. Tetrazolium test did not reflect viability in DMSO- and heat-treated pollen (after Heslop-Harrison et al. 1984).

In-vitro germination test

The most commonly used and acceptable test for assessing pollen viability is the in-vitro germination test. It is rapid, simple and the results of in-vitro germination generally correlate with seed-set data (Akihama et al. 1978, Janssen and Hermsen 1980). However, this correlation depends on optimization of the medium and other cultural conditions to induce germinability in most of the viable pollen. In suboptimal medium this test gives false negative results. Further, many of the stored pollen samples which fail to germinate in vitro are often found to be capable of inducing fruit- and seed-set following pollination (Visser 1955, King 1963, 1965, Ghatnekar and Kulkarni 1978). A major limitation of this test is lack of optimal germination medium for pollen of many species, particularly the 3-celled pollen species.

Fluorescein diacetate test

The fluorescein diacetate (FDA) test, often referred to as the fluorochromatic reaction (FCR) test, was introduced by Heslop-Harrison and Heslop-Harrison (1970) as a test for pollen viability. This has been the most commonly used test in recent years. The FDA test assesses two properties of the pollen: (i) integrity of the plasma membrane of the vegetative cell and (ii) presence of active esterases in the pollen cytoplasm. Non-polar, non-fluorescing FDA passes freely through the pollen membrane and enters the pollen cytoplasm. Hydrolysis of FDA by the activity of esterases results in fluorescein which is fluorescent. Since fluorescein (a polar substance) does not pass through the intact plasma membrane as readily as FDA, it accumulates in the pollen cytoplasm (Fig. 3.5). Such pollen grains show bright fluorescence when observed under a fluorescence microscope (Fig. 3.3B) (blue light excitation). Pollen grains that do not have intact plasma membrane allow fluorescein to move out readily (Fig. 3.5) and thus result in uniform background fluorescence. Likewise, if there are no active esterases in the pollen cytoplasm, fluorescein is not formed and hence pollen grains do not fluoresce.

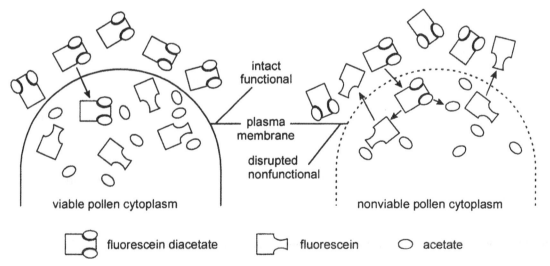

Fig. 3.5 Basis of fluorescein diacetate (FDA) test. FDA readily enters the pollen cytoplasm; in the presence of active esterases, FDA is hydrolyzed to release fluorescein, a fluorescent substance, and acetate. As fluorescein cannot readily pass through the intact membrane, it accumulates inside the cytoplasm in pollen grains with intact plasma membrane (left) and thus such pollen grains show bright fluorescence (as seen in Fig. 3.3 B). Fluorescein readily moves out of pollen cytoplasm with disrupted plasma membrane (right); such pollen preparations show a general background fluorescence and not the bright pollen fluorescence.

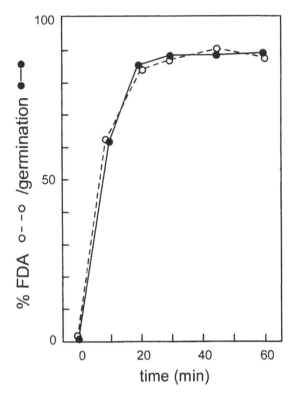

Fig. 3.6 Effect of controlled hydration on FDA and in-vitro germination tests of desiccated pollen of *Carex ovalis*. Desiccated pollen grains neither responded to FDA test nor showed in-vitro germination. Controlled hydration dramatically restored membrane integrity and germinability (after Shivanna and Heslop-Harrison 1981).

The FDA test has proven satisfactory in assessing pollen viability in a number of species (Shivanna and Heslop-Harrison 1981, Heslop-Harrison et al. 1984, Knox et al. 1986, Shivanna and Cresti 1989, Shivanna et al. 1991a, b, Rao et al. 1992, 1995, Sedgley and Harbard 1993). It reportedly has wider applicability and better resolution than other prevailing tests for assessing pollen viability of cotton (Gwyn and Stelly 1989).

In species in which the medium used for in-vitro germination of pollen is optimal, a close correlation exists between the FDA test and in-vitro germination test (Shivanna and Heslop-Harrison 1981, Heslop-Harrison et al. 1984). In

the absence of optimal medium, the FDA test gives a better index of viability than in-vitro germination (Shivanna et al. 1991a). The FDA test reflected the fertilizing ability of pollen grains exposed to high humidity (> 90% RH) and high temperature stresses (up to 60°C) better than the in-vitro germination test (Shivanna et al. 1991b).

To obtain valid results with the FDA test some important precautions have to be taken. For assessing viability of dry and desiccated pollen samples, they have to be exposed to controlled hydration (by maintaining them under high humidity for about 30 min) before testing for viability (Fig. 3.6); such pollen may not respond if tested directly (Shivanna and Heslop-Harrison 1981). The FDA test may not reflect fertilizing ability in pollen samples subjected to prolonged exposure to very high temperature (75°C) (Rao et al. 1995). In pollen samples of *Brassica* subjected to 75°C, the extent of retention of pollen fluorescence gave a better indication of pollen viability than initial fluorescence. Pollen samples which induced seed-set retained fluorescence even after 2 h, while those ineffective in inducing seed-set (treated at 75°C for 24 h) lost fluorescence within 60 min due to leakage of fluorescein.

Causes for Loss of Viability

Until recently there were hardly any studies to understand the causative factors for the loss of viability. It was suggested that deficiency of respiratory substrates and/or inactivation of enzymes and growth hormones are responsible for loss of viability (Stanley and Linskens 1974). As the metabolic activity of the pollen continues, though at a very reduced rate, even after shedding (Wilson et al. 1979), endogenous substrates are expected to be used up gradually and result in loss of viability. However, there are no clear evidences to show that loss of respiratory substrates is the primary cause for loss of viability, particularly in short-lived pollen. Pollen grains of cereals, which are short-lived, contain abundant reserve metabolites even after losing viability (Hoekstra and Bruinsma 1980). Loss of respiratory substrates, however, may be a factor in the loss of viability during long-term storage. Similarly, there are no direct data to indicate that inactivation of enzymes or deficiency of growth hormones are the causes for the loss of viability. Although a few investigators have reported changes in amino acid composition of stored pollen, there has been no clear correlation between the loss of viability and changes in amino acid content (Stanley 1971, Linskens and Pfahler 1973, Dashek and Harwood 1974).

Loss of membrane integrity

Many studies carried out during the last 20 years have highlighted the role of plasma membrane in maintaining pollen viability (Heslop-Harrison 1979a, Shivanna and Heslop-Harrison 1981, Heslop-Harrison et al. 1984, Jain and Shivanna 1987a,b, 1989). Irreversible loss of membrane integrity seems to be the primary cause for the loss of pollen viability. A close correlation has been established between membrane integrity and germinability of pollen of a number of species (Fig. 3.6).

Pollen grains of many species subjected to desiccation for different periods showed (Shivanna and Heslop-Harrison 1981) a marked variation in their ability to withstand desiccation. Pollen of several species lost membrane integrity following 1–4 h desiccation (Fig.3.7A); such pollen grains invariably failed to germinate in vitro. However, pollen of some species such as *Cytisus* retained membrane integrity as well as germinability even after 24 h of desiccation (Fig. 3.7B). Pollen of some of the desiccation-sensitive species such as *Iris, Eleocharis* and *Lonicera*, recovered membrane integrity as well as in-vitro germinability when desiccated pollen was subjected to controlled hydration before testing (Fig. 3.8A, B). Thus, controlled hydration provides suitable conditions for restoration of membrane integrity. This accords with many studies reporting favourable effects of controlled hydration on in-vitro germination (see Chapter 4). However, desiccated pollen grains of *Secale* failed to restore membrane integrity even after controlled hydration (Fig. 3.8C, D).

These observations confirm that pollen grains of cereals in general are highly susceptible to

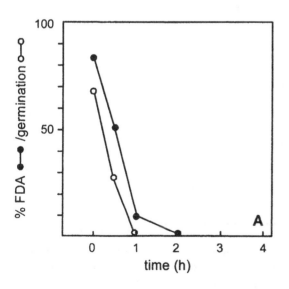
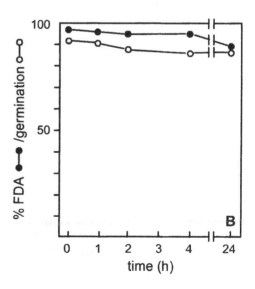

Fig. 3.7 Effect of desiccation on pollen viability of *Secale* (A) and *Cytisus* (B). Pollen grains of *Secale* were sensitive to desiccation whereas those of *Cytisus* were tolerant (after Shivanna and Heslop-Harrison 1981).

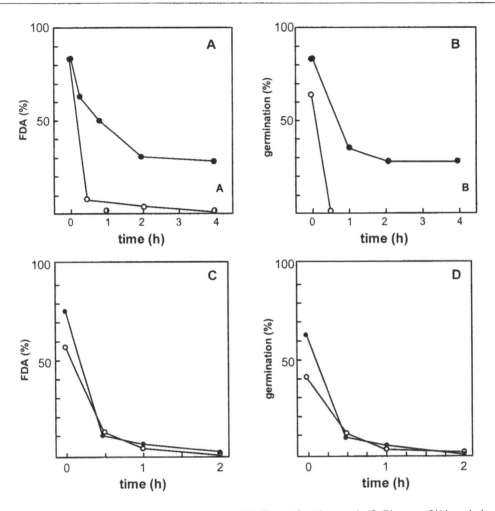

Fig. 3.8 Per cent viability of pollen grains of *Iris pseudacorus* (A, B) and *Secale cereale* (C, D) over a 2/4 h period assessed through FDA and in-vitro germination responses. Responses under low humidity (< 10%) (-o-), low humidity for a period shown followed by 1 h of controlled hydration over high humidity before the test (-•-). In both species, desiccation drastically reduced both membrane integrity and germinability; membrane integrity and germinability partially recovered following controlled hydration in *Iris* pollen, but not in *Secale* pollen (after Shivanna and Heslop-Harrison 1981).

uncontrolled desiccation. Rapid loss of viability in pollen of cereals is thus associated with irreversible loss of membrane integrity rather than loss of respiratory substrates. However, by using a 'pollen drier' which ensures gentle and uniform drying, it has been possible to reduce the moisture level of pollen of *Zea* and many other cereals to ca 20% without loss of viability (Barnabas 1985, 1994, Barnabas and Rajki 1976, 1981, Barnabas and Kovacs 1997). Pol-

len grains of *Pennisetum* are an exception to pollen of Poaceae; they can withstand desiccation similar to non-Poaceae pollen (Chaudhury and Shivanna 1986, 1987).

Differences among pollen of different species in their ability to retain functional membrane after desiccation seems to depend on the presence of sufficient amount of sugars which provide stability to the membranes under desiccated conditions (Hoekstra et al. 1989). In many other 'anhydrobiotic' organisms/systems (that survive extreme desiccation), trehalose has been shown to play a major role in maintaining membrane integrity (Crowe et al. 1984). Although the presence of trehalose has not been reported in any pollen species, many other soluble carbohydrates are generally present (Stanley and Linskens 1974) in pollen grains, which may impart desiccation tolerance.

Hoekstra and colleagues (1989) have shown that the difference between pollen of *Pennisetum* and other grasses is related to their sucrose level. They compared desiccation-induced responses in the pollen of *Zea* and *Pennisetum*; the pollen of both species contained sucrose as the sole soluble carbohydrate; in fresh pollen it was 4% in *Zea* and 14% in *Pennisetum*. Pollen of the two species showed no significant variations in the composition and content of membrane phospholipids before or after drying. They concluded that desiccation tolerance of *Pennisetum* pollen is correlated with the presence of a higher level of sucrose.

Quantitative changes in phospholipids

Many studies have reported successful storage of pollen grains in organic solvents (Jain and Shivanna 1987 a, b, 1989). Organic solvents with low dielectric constants (non-polar) such as n-hexane and cyclohexane are suitable for pollen storage. Pollen stored in such solvents showed very little leaching of sugars, free amino acids and phospholipids into organic solvents (Figs. 3.9, 3.10). Organic solvents with high dielectric constants (polar) such as isopropanol and methanol are not suitable for pollen storage; pollen stored in these solvents showed extensive leaching of sugars, amino acids and phospholipids (Jain and Shivanna 1987a, b). Phospholipids are associated with the membrane system. Thus, those organic solvents which do not remove membrane phospholipids do not affect membrane integrity and thus maintain viability; those which remove membrane phospholipids destroy membrane integrity leading to leakage of metabolites and loss of viability.

Quantitative analyses of phospholipids in pollen grains stored under different conditions have also shown the role of plasma membrane in maintaining pollen viability (Jain and Shivanna 1989). Pollen grains of *Crotalaria* stored under different conditions were monitored for their in vitro germinability, membrane integrity and changes in total and individual phospholipids. Irrespective of the storage conditions, there was a positive and significant correlation between loss of germinability and loss of membrane integrity as well as reduction in total (Fig. 3.11) and individual phospholipids, particularly phosphatidyl choline. These results also clearly show that phospholipid degradation and consequent loss of membrane integrity are associated with loss of viability. Similar results have been reported in aged seeds of many species (Roberts 1972). The causative factor of degradation of membrane phospholipids is not clear. Free radical-induced degradation of phospholipids in desiccated systems has been suggested as a causative factor (Senaratna et al. 1991).

Phase transition and membrane integrity

It is well known that when anhydrobiotic systems such as seeds and pollen (Simon 1974, Hoekstra 1984) are rapidly rehydrated, they leak soluble metabolites into the surrounding medium. This imbibitional leakage has been a major cause of damage to these systems. Many investigators have considered that the physicochemical changes of the plasma membrane are associated with imbibitional leakage. Simon (1974) postulated that desiccation of seeds (reducing the moisture level to <20%) brings about membrane changes from the bilamellar phase to a hexagonal phase (see also Heslop-Harrison 1979a, Shivanna and Johri 1985); rehydration restores the bilamellar phase and thus membrane integrity. The leakage observed in the early stages of imbibition occurs during the initial phases of this transition. It has been suggested that such changes in membrane structure occur in desiccated pollen also (Heslop-Harrison 1979a, Shivanna and Heslop-Harrison 1981).

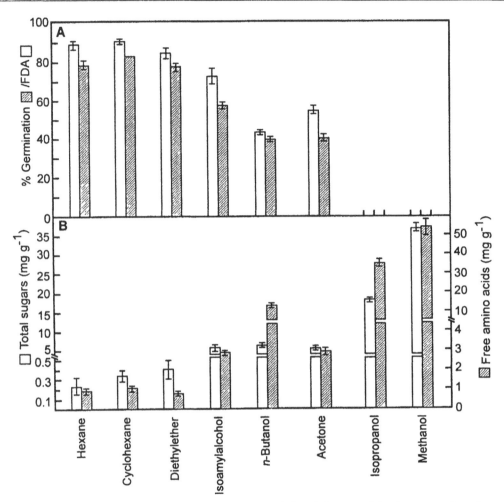

Fig. 3.9 In-vitro germination and FDA responses of pollen grains of *Crotalaria retusa* following storage in various organic solvents (for 6 months at –20°C) and extent of leaching of sugars and amino acids from stored pollen grains into the respective organic solvents during storage. A negative correlation was found between the extent of germination and the amounts of sugars and amino acids leached into the organic solvents (after Jain 1989).

Controlled hydration is more conducive for membrane restoration and thus prevents imbibitional leakage. Subsequent studies have not supported such a change in the membrane structure (bilamellar to hexagonal) in desiccated seeds/pollen (Thomson and Platt-Aloia 1982, Hoekstra 1984, Crowe et al. 1989).

Based on their studies of pollen of *Typha latifolia*, Crowe and co-authors (1989) put forward an alternative hypothesis for imbibitional leakage observed in desiccated pollen (Fig.

3.12). Phospholipids are present in the liquid crystalline phase in hydrated pollen. When pollen is dehydrated, phospholipids of the membrane enter the gel phase from the liquid crystalline phase. When such pollen grains are exposed to rapid hydration by placing them in liquid medium, they undergo leakage during transition of the membrane from gel phase to liquid crystalline phase (Fig. 3.12). However, when dehydrated pollen are exposed to controlled hydration (by exposing them to water vapour) before

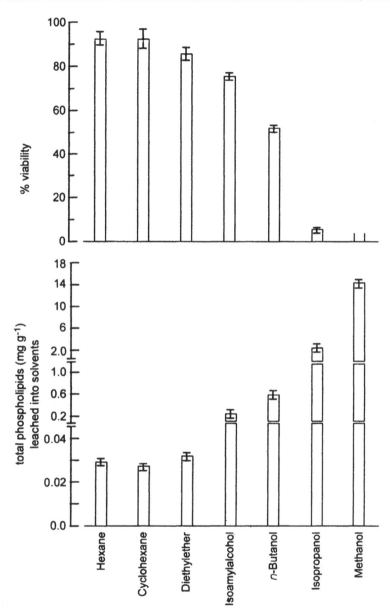

Fig. 3.10 Viability (% FDA) of *Crotalaria* pollen stored in various organic solvents for 6 months and the extent of leaching of total phospholipids into the organic solvents. A negative correlation was established between pollen viability and the amounts of phospholipids leached into the solvents (based on Jain 1989).

placement in the medium, phospholipids are transformed to the liquid crystalline phase during controlled hydration (in the absence of bulk water); such pollen show no imbibitional leak-

age when placed in the liquid medium (Fig. 3.12). These membrane changes are related to changes in transitional temperature of membrane phospholipids (for details see Crowe et al. 1989).

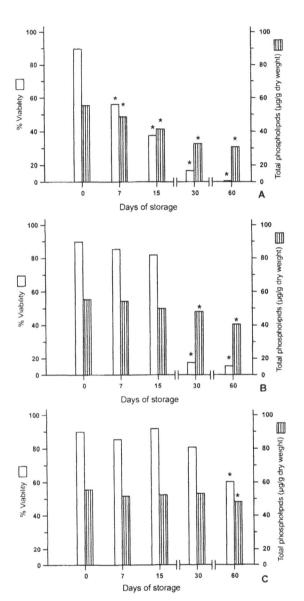

Fig. 3.11 In-vitro germination of fresh pollen grains and those stored for up to 60 days under uncontrolled laboratory conditions (A), under desiccation at laboratory temperature (B), and at −20°C (C), and their correlation with the amounts of total phospholipids extracted from such pollen. Loss of viability is associated with a significant reduction (∗) in the amount of phospholipids extracted from pollen (based on data presented in Jain and Shivanna 1989).

POLLEN VIGOUR

Pollen vigour refers to the speed of germination and the rate of pollen tube growth. It differs from viability since viable pollen samples may show differences in vigour. Until recently, pollen vigour was not taken into consideration in assessing the quality of pollen, particularly of aged and stored pollen, which is surprising given the extensive data available on seed vigour (Khan 1982, Priestley 1986). Seed vigour, assessed on the basis of the speed of germination, is considered a more reliable index of seed quality than their ability to germinate. Loss of vigour generally becomes evident before the loss of germinability. Seeds with reduced vigour have been shown to produce inferior plants with lower yields and are more susceptible to environmental stresses. Seed deterioration is thus assessed as the reciprocal of vigour rather than germination (Association of Official Seed Analysts, Seed Vigour Test Committee 1983). As seeds and pollen grains are very similar in many physiological manifestations, it was suggested that pollen grains also exhibit reduction in vigour before loss of viability (Shivanna and Cresti 1989, Shivanna et al. 1991a, b). Many investigations during the last 10 years have shown that ageing, and many environmental stresses, in particular desiccation, temperature and humidity affect pollen vigour before affecting viability (Shivanna et al. 1991a, b).

Tests for Vigour

In-vitro germination

Apart from its use in assessing pollen viability, in-vitro germination can also be used to assess pollen vigour (Shivanna and Cresti 1989). In viability tests using in vitro germination, the capacity of pollen to germinate is assessed without consideration of the time factor; the cultures are generally scored after maintaining them for a much longer period than that required for germination. To assess the vigour, however, germination is scored at intervals over a period and

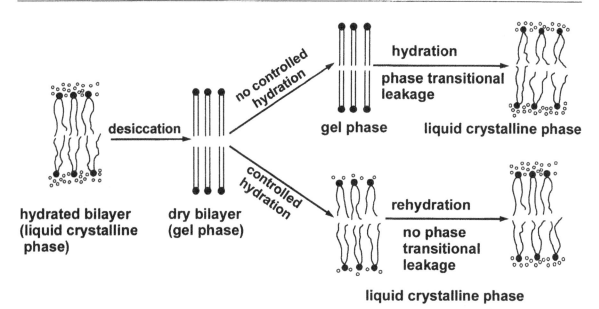

Fig. 3.12 Diagram of phase transition changes in membrane phospholipids following desiccation and hydration. During desiccation membrane phospholipids are transformed from liquid crystalline to gel phase. Such pollen samples when exposed to rapid hydration (by culturing them directly in liquid medium without exposing to controlled hydration) result in imbibitional leakage as a result of phase transition. When desiccated pollen are exposed to controlled hydration before culture, phase transition is completed during controlled hydration and thus imbibitional leakage is prevented when they are subsequently cultured in liquid medium. Water molecules are represented in hydrated bilayer by small open circles (after Crowe et al. 1989).

compared with the values obtained for control pollen (fresh pollen) (Fig. 3.13). Pollen grains with reduced vigour took a longer time to attain maximum germinability than did fresh pollen.

Semivivo technique

In the semivivo technique, the pollen sample to be tested is used to carry out controlled pollination (using unpollinated, compatible pistil). The pistil used for pollination can either be maintained on the plant (by taking precautions to prevent contamination) or be excised and maintained in the laboratory (Shivanna and Rangaswamy 1992). Pollinated pistils are maintained (for 3–6 h) until pollen grains germinate on the stigma and pollen tubes grow down for some length in the style. After suitable time of incubation, the style is cut ahead of the growing pollen tubes and the cut end of the style is implanted in the agar medium containing the components of the pollen germination medium (Fig.

3.14). The pollen tubes continue their growth and enter the agar medium through the cut end of the style (Shivanna et al. 1991a). The number of pollen tubes that emerge into the medium are counted and their length measured either in situ or after pulling out the implant and observing under a microscope. The semivivo technique requires some preliminary studies on pollen germination and pollen tube growth in vivo. Pollen vigour is assessed on the basis of the time taken for pollen tube emergence into the medium and the number of emerged pollen tubes.

In-vivo pollen germination and pollen tube growth

Pollen vigour can also be assessed by studying pollen germination and pollen tube growth at regular intervals in pistils pollinated with pollen samples. Comparison of the extent of pollen germination and pollen tube growth in pistils pollinated with the test pollen sample and those

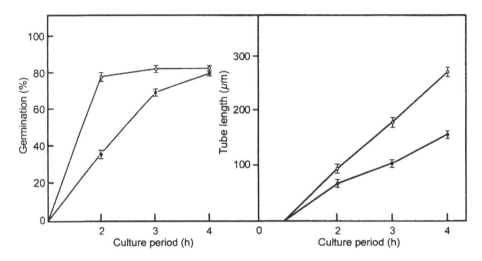

Fig. 3.13 In-vitro germination test to assess pollen vigour in *Nicotiana tabacum*. Maximum germination is achieved in about 2 h in fresh pollen (–o–), while stressed pollen (high humidity at 38°C) (–•–) requires 4 h. Differences in pollen tube lengths are apparent throughout the 4 h period (after Shivanna and Cresti 1989).

with fresh pollen indicate the vigour of the pollen sample. Another simple method for assessing pollen vigour is to excise the stigma and a part of the style at different intervals after pollination. Pollen tubes, which grow through the excision zone before excision, continue to grow and effect fertilization (see Jauh and Lord 1995). Thus, more vigorous tubes would grow through the excision zone earlier and result in seed-set while less vigorous pollen tubes may not grow or only a few may grow through the excision zone and this would result in no seed-set or reduced seed-set.

EFFECTS OF ENVIRONMENTAL STRESSES ON POLLEN QUALITY

Pollen grains can withstand higher temperatures because of their desiccated condition. Pollen of lily, apple and rose showed 50% germinability when exposed to 70°C for 4–8 h (Marcucci et al. 1982). A small proportion of *Eucalyptus rhodantha* pollen retained germinability even after exposure to 70°C for 24 h (Heslop-Harrison and Heslop-Harrison 1985).

High temperature stress up to 60°C to pollen of *Brassica*, *Petunia* and *Nicotiana* (Rao et al. 1992, 1995) did not affect pollen viability and vigour. A temperature of 75°C affected both viability and vigour to different degrees depending on duration of the treatment. In *Brassica* 75°C-treated pollen failed to set fruits and seeds, while in *Nicotiana* and *Petunia* they set fruits and seeds even after 12 h and 24 h of treatment respectively. These studies clearly show that dry pollen grains can withstand fairly high levels of heat stress. This is related to their ability to withstand desiccation. Pollen grains of many cereals which are shed in hydrated condition lose their ability to effect fertilization even under moderately high temperature stress of 40°C for 4 h (Dupuis 1992).

High temperature combined with high humidity affects both pollen viability and vigour. Exposure of pollen to either high humidity (>90% RH) or high temperature (38°/45°C) did not affect pollen viability or vigour (Shivanna and Cresti 1989, Shivanna et al. 1991a, b). Nor did the two stresses together (high RH + 38°/45°C) affect viability (FDA test) but significantly reduced vigour; pollen grains took a longer time to germinate in vitro and to emerge from the cut end of semivivo implanted styles (Table 3.1). Pollen grains subjected to high RH at 45°C failed to germinate in vitro. However, they germinated on the stigma, though pollen tubes reached the ovary ca 40 h later than those from fresh pollen (Shivanna et al. 1991, Bajaj et al. 1991). Such pollen samples were able to induce fruit- and seed-set.

Table 3.1: Details of viability (FDA test), in-vitro germination and vigour (semivivo method) of pollen samples subjected to high humidity and temperature stresses (based on data presented in Shivanna et al. 1991a, Bajaj et al. 1991)

Treatment	Viability (%) FDA	In-vitro germination (%)		Vigour (Number of pollen tubes/ pistil emerged from 10 mm semivivo implanted style after):	
		2 h	6 h	16 h	24 h
Fresh pollen	83.3	79.5	81.2	> 100	> 100
Pollen exposed to high humidity at:					
Lab. temperature	93.5	78.1	82.0	> 100	> 100
38°C	90.7	8.2	61.5	10.4	> 100
45°C	79.9	0	0	0	0

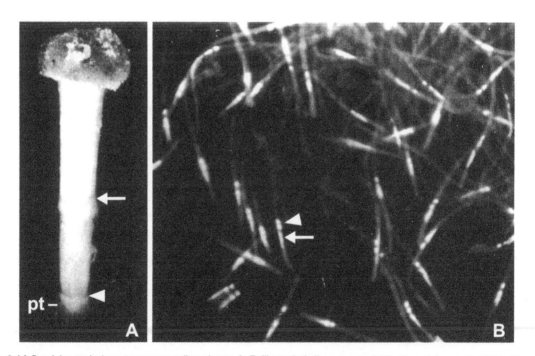

Fig. 3.14 Semivivo technique to assess pollen vigour. A. Pollinated pistil segment of *Nicotiana tabacum* implanted in agar medium (level of the medium marked by arrow) as seen 18 h after implant. Pollen tubes (pt) have emerged from the cut end of the style (arrowhead). B. Fluorescent micrograph of a group of pollen tubes emerged from the cut end of the style similar to the one shown in A, stained with DAPI, a DNA fluorochrome. Vegetative nucleus (arrow) and sperm cells (arrowhead) are obvious in many pollen tubes (after A. Shivanna and Rangaswamy 1992).

The major ultrastructural effects of high RH and temperature stresses were on the RER. Stacks of RER characteristically present in fresh pollen were dissociated in stressed pollen in which vigour was affected (Ciampolini et al. 1991). Similarly, exposure of pollen to alternating cycles of 1 h of high RH and 1 h of drying at 38°C also affected pollen vigour without affecting viability (Shivanna and Cresti 1989).

Storage also affects pollen vigour much earlier than viability. In *Malus domestica* (Bellani et al. 1984), pollen grains stored for one year showed no loss of germinability. However, their ability to incorporate nucleic acid precursors (^{3}H thymidine and ^{3}H uridine) was markedly reduced.

In *Crotalaria retusa* (Jain and Shivanna 1989) pollen grains stored in some of the organic solvents for 30 days showed no reduction in germinability but pollen tube length was notably reduced. In *Nicotiana* and *Agave* also (Shivanna et al. 1991 b), storage stress affected the vigour of the pollen much before viability.

Thus, as in seeds, in pollen also loss of vigour is a general response to environmental stresses. Consideration of pollen vigour is important in the light of recent investigations on pollen competition and selection (Ottaviano and Mulcahy 1989, Hormaza and Herrero 1992, Sari-Gorla and Frova 1997) as well as in establishment of pollen banks (see Chapter 13).

4

In-vitro Pollen Germination and Pollen Tube Growth

Germination of pollen is the first morphogenetic event in fulfilling its function of transport and discharge of sperm cells into the embryo sac. In-vivo studies on pollen germination are difficult to perform due to involvement of the pistillate tissue. Since pollen grains of a large number of species readily germinate in-vitro on a simple medium, in-vitro germination has been extensively used in studies on structural and physiological details of germination and tube growth. Two-celled pollen grains in general are more amenable to in-vitro germination compared to 3-celled pollen. A number of methods have been used for in-vitro germination and have been comprehensively described by Shivanna and Rangaswamy (1992). Details of the processes involved in pollen germination and tube growth are discussed in many reviews (Heslop-Harrison 1988, Steer and Steer 1989, Mascarenhas 1993, Pierson et al. 1994, Feijo et al. 1995, Derksen 1996, Taylor and Hepler 1997, Raghavan 1999).

GERMINATION REQUIREMENTS

Though there is species-to-species variability in the requirements for optimal germination, pollen grains of most species require full hydration, a carbohydrate source, boron and calcium for satisfactory germination and tube growth.

Hydration

As pollen grains are shed under desiccated conditions, hydration is a basic prerequisite for germination. In liquid medium hydration is rapid and completed within a few minutes. In semi-solid medium, it is slower and the rate of hydration depends on the moisture level of the medium. Many studies have shown that the rate of hydration is critical for optimal germination, particularly for desiccated pollen.

Controlled hydration/vapour phase hydration, achieved by exposing pollen grains to high humidity for about 30 min, is more favourable or even essential for germination in several species; this is particularly true for 3-celled and stored pollen (Hoekstra and Bruinsma 1975a, Bar-Shalom and Mattsson 1977, Shivanna and Heslop-Harrison 1981, Heslop-Harrison et al. 1984). For example, desiccated pollen of *Carex* (see Fig. 3.6) and *Eleocharis* germinated only after controlled hydration (Shivanna and Heslop-Harrison 1981). In *Chrysanthemum*, culture of pollen directly in the medium gave low germination and resulted in leakage of metabolites from the pollen; controlled hydration before culture gave better and consistent germination (Hoekstra and Bruinsma 1975a). Controlled hydration seems to provide better conditions for restoration of membrane integrity (see Chapter 3) and thus prevents leakage of metabolites when transferred to the culture medium (Shivanna and Heslop-Harrison 1981). In *Petunia*, controlled hydration affects rigidity of the pollen wall (Gilissen 1977); pollen grains exposed to controlled hydration before culture showed 3-fold increase in volume after culture compared to direct culture which showed only a 2-fold increase.

Carbohydrate Source

A suitable carbohydrate source in the medium is required for adequate pollen germination and tube growth. A carbohydrate source serves two functions: (i) maintains the required osmotic potential of the medium and (ii) serves as a substrate for pollen metabolism (Nygaard 1977). The former is perhaps more important than the latter in short-term cultures; pollen grains contain sufficient endogenous sugars for germination and early tube growth. In long-term cultures, exogenous carbohydrates are needed for continued tube growth. Sucrose has been the most commonly used carbohydrate source; in some species other sugars such as glucose, fructose and raffinose also support pollen germination and tube growth (Hrabetova and Tupy 1964, Chiang 1974). In general, 2-celled pollen require a lower sucrose level (10–15%), while 3-celled pollen require higher levels (>20%). Sucrose seems to be taken up in unhydrolyzed form along with the inward movement of protons (Baker 1978, Deshusses et al. 1981).

Pollen grains are desiccated units at the time of shedding and generally show higher osmotic potential compared to other cells. This is obviously an adaptation to take up water from the stigma following pollination. After pollen germination, the resultant pollen tubes are fully hydrated and their osmotic potential is expected to be lower than that of pollen grains; it would be comparable to the cells of the transmitting tissue. Thus, higher sucrose concentration in the medium, although required for germination to prevent bursting of pollen, may not be condu-

cive for pollen tube growth. Most investigators, however, have used the same high sucrose medium for germination as well as tube growth. In *Brassica*, reduction of sucrose concentration in the medium from 20% to 5% reduced germination but improved tube growth markedly (Fig. 4.1A) (Shivanna and Sawhney 1995). It may be worthwhile to reduce the osmoticum of the medium after pollen germination to achieve longer and faster tube growth.

Non-metabolizing osmotic agents such as mannitol and penterythritol also permit pollen germination in a few species. In recent years polyethylene glycol (PEG) of different molecular weights (4000–10,000) in combination with lower concentration of sucrose has been shown to improve in-vitro pollen germination and, more dramatically, pollen tube growth in several species (Fig. 4.1B) (Zhang and Croes 1982, Subbaiah 1984, Leduc et al. 1990, Read et al. 1993, Shivanna and Sawhney 1995, Shivanna et al. 1998, Tandon et al. 1999). Pollen tube growth in the PEG medium continued for much longer periods compared to control and pollen tubes were straight and smooth without abnormality (Read et al. 1993, Shivanna and Sawhney 1995, Tandon et al. 1999). In *Nicotiana tabacum* incorporation of PEG 6000 (12.5%) + 30 μM of $CuSO_4$ in a germination medium containing amino acids resulted in sustained growth of pollen tubes for over 48 h at 250 μm h^{-1} (Read et al. 1993).

The mechanism of action of PEG is not clear. Some of its positive effects as a component of the medium are that PEG is highly soluble in water, non-toxic, biologically stable and does not denature biological macromolecules (Powell 1980). PEG has been well recognized as a non-penetrating osmotic agent and it decreases the water potential of the culture medium. PEG has also been suggested to regulate the permeability of the plasma membrane of the growing tip (Read et al. 1993). Sucrose, on the other hand, enters the cell and increases the osmotic potential of the cell which may not be conducive for tube growth. Inclusion of PEG as osmotic

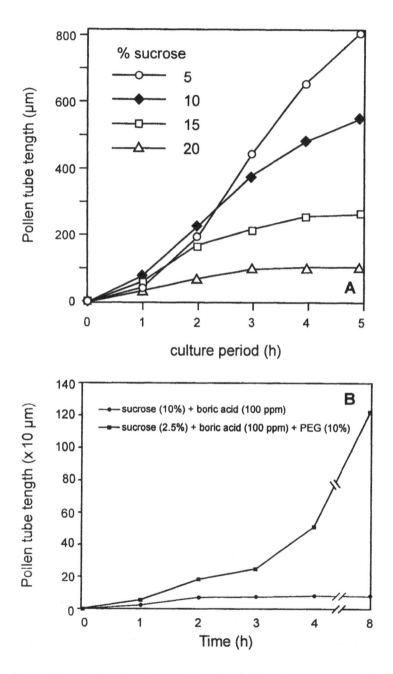

Fig. 4.1 A. Kinetics of pollen tube growth in *Brassica napus* in media of different sucrose concentrations. The medium contained boric acid 1.62 mM + calcium nitrate 1.69 mM + potassium nitrate 0.99 mM + magnesium sulphate 0.84 mM + TAPS 20 mM, pH 8.0. Pollen tube growth is inversely proportional to the sucrose concentration. B. Effect of polyethylene glycol on pollen tube growth in oil palm (A—after Shivanna and Sawhney 1995, B—Tandon et al. 1999).

agent in the medium permits marked reduction in sucrose level required for germination.

Boron

The stimulatory effect of boron on pollen germination and pollen tube growth has been reported in a large number of species since the first report as early as 1933 (Schmucker 1933). Boron is a regular component of all pollen germination media. Boric acid is generally used as the boron source. In the absence of boron, pollen grains generally show poor germination and a high degree of bursting. Pollen grains are believed to be deficient in boron which is compensated by high levels of boron in the stigma. In *Yucca* (Portnoi and Horovitz 1977), boron permitted optimal pollen germination over a wide range of sucrose concentrations; in the absence of boron, optimal germination was confined to a very narrow range of sucrose (Fig. 4.2). Also fructose, which alone did not support pollen germination, supported good germination in boron medium. Thus, the type and concentrations of sugars required for pollen germination are less specific in the presence of boron.

The role of boron in higher plants in general and pollen in particular is not clearly understood.

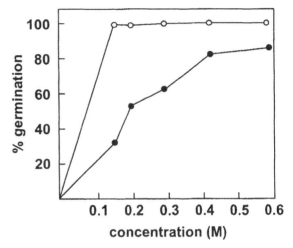

Fig. 4.2 Effect of boron on pollen germination in *Yucca* sp. Pollen germination in the presence (-o-) and absence (-•-) of boron in a medium containing various concentrations of sucrose. Sucrose concentration is not critical in the presence of boron (after Portnoi and Horovitz 1977).

Boron has been implicated in many biochemical and membrane functions of the pollen. Boron appears to play an important role in carbohydrate metabolism of the pollen (Dugger 1973). Boron seems to affect various pathways of carbohydrate metabolism (Fig. 4.3). A major metabolic activity of germinated pollen is the

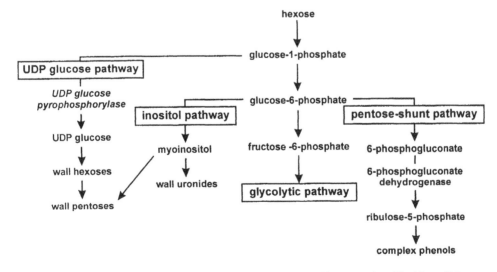

Fig. 4.3 Major pathways of carbohydrate metabolism reported to be affected by boron (modified from Shivanna and Johri 1985).

synthesis of pollen wall components to cope with the requirement for long pollen tubes needed to grow through the length of the style to reach the embryo sac. The inositol pathway in which myoinositol acts as an intermediate in converting hexoses to wall pentoses and uronides is an important route for the synthesis of wall polysaccharides (Chen et al. 1977, Dickinson et al. 1973, Maiti and Loewus 1978, Rosenfield et al. 1978). Myoinositol is taken up and used for tube wall biosynthesis; methylene-myoinositol, an antagonist of myoinositol, inhibits pollen germination and pollen tube elongation (Chen and Loewus 1977). It is suggested that boron stimulates conversion of myoinositol to wall polysaccharides, probably by complexing with a specific enzyme, thus enabling the enzyme to bind to inositol or inositol derivatives (Maiti and Loewus 1978).

Boron also seems to affect the availability of substrates for other pathways of carbohydrate metabolism (Fig. 4.3) (Birnbaum et al. 1977). Boron deficiency reduces UDP glucose pyrophosphorylase activity leading to reduction in synthesis of UDP glucose, a precursor of cellulose and UDP sugars of the cell wall. This results in the accumulation of glucose-1-phosphate (the precursor of UDP glucose), making it available for synthesis of phenols through the pentose-shunt pathway.

Boron has also been reported to complex with 6-phosphogluconate, the initial substrate in the pentose-shunt pathway, and inhibits the action of 6-phosphogluconate dehydrogenase (Lee and Aronoff 1967). In the absence of boron, the enzyme operates more efficiently, producing additional ribulose-5-phosphate (Fig. 4.3) which is the substrate for synthesis of complex phenols.

Boron affects β-glucosidase activity (Maevskaya et al. 1976) involved in maintaining an optimal phenolic aglycone-glycoside ratio. Glycosidation of phenolic compounds removes toxic effects of free phenols. Deficiency of boron would lead to accumulation of phenolic compounds, leading to growth inhibition (see Linskens and Kroh 1970).

In tobacco, boron greatly reduces the release of proteins from pollen tubes into the culture medium (Capkova-Balatkova et al. 1980) and increases the proportion of proteins that bind to the insoluble pollen tube fraction. The newly synthesized pollen tube wall contains up to 30% dry mass of proteins. Boron thus appears to prevent loss of proteins from pollen tube wall and increases their binding to the wall. Boron is also believed to stabilize newly synthesized cell wall at the tip and prevent pollen tubes from bursting.

Efficient operation of the membrane transport system seems to require boron (Pollard et al. 1977). Boron deficiency reduces the activity of several transport systems including membrane ATPase, which is known to be associated with transport of monovalent cations. Some studies have also shown stimulation of plasma membrane H^+-ATPase by boron (Obermeyer et al. 1996). This may play a role in pollen germination since an increase in H^+ transport has been suggested to initiate pollen germination and pollen tube growth (Feijo et al. 1995). According to the model put forward by Obermeyer and colleagues (1996) stimulation of plasma membrane H^+-ATPase by boron energizes the plasma membrane, resulting in the uptake of ions (mainly K^+) and consequent influx of water into pollen grain/pollen tube. The turgor pressure thus created drives pollen tube elongation.

Calcium

Calcium is another important inorganic requirement for pollen germination and pollen tube growth. The importance of calcium was discovered as a consequence of a well-known observation of Brink (1924) that pollen grains in larger populations show better germination and tube growth than those in smaller populations. This crowding effect/population effect was reported to be due to the release of a heat-stable, water-soluble substance termed pollen growth factor (PGF) from pollen grains into the medium; the effective concentration of PGF is attained only in larger populations (Brewbaker and Majumdar 1961). Subsequently PGF was identified to be calcium ion (Brewbaker and Kwack 1963). The amount of calcium present in the pollen is generally far less than that present in the vegetative parts and seeds (McLellan 1977). In some species calcium is not required for pollen germination and this is explained on the basis that pollen grains of such species contain higher levels of endogenous calcium.

Calcium is one of the most important cations affecting cell metabolism in a variety of ways. Many of the effects of calcium are mediated through Ca^{2+}-binding protein, calmodulin. In *Prunus dulcis* (Polito 1983), incorporation of exogenous calmodulin promotes in-vitro pollen germination; trifluoperazine or chloropromazine, inhibitors of calmodulin-calcium complex, inhibits pollen tube growth.

Ca^{2+} can be localized by using fluorescent probes such as Fure-2 and Quin 2 (for cytosolic-free Ca^{2+}), and chlortetracycline (for membrane-bound Ca^{2+}), and calmodulin by fluphenazine (Tirlapur and Willemse 1992). Both Ca^{2+} and calmodulin show localized distribution in the germ pore(s) (region of pollen tube emergence) of mature pollen (Tirlapur and Willemse 1992). Soon after formation of a short tube, Ca^{2+} distribution shows a tip-to-base gradient. The concentration of Ca^{2+} is around 3.10 μM at the very tip and decreases to < 0.2 μM within 20–30 μm from the tip (Pierson et al. 1994). This tip-focused gradient (of both free-calcium and membrane-bound calcium) is maintained throughout the growth of the pollen tube (see Steer and Steer 1989, Herth et al. 1990, Derksen et al. 1995).

Many functions have been attributed to calcium. It has been suggested that calcium gives rigidity to the pollen tube wall by binding pectic carboxyl groups (Kwack 1967). Calcium also seems to have a role in controlling permeability of the pollen membrane (Dickinson 1967). Absence of Ca^{2+} in the germination medium increases membrane permeability leading to loss of internal metabolites. The most important role of calcium has been shown to be in regulating tip growth of the pollen tube; this aspect is discussed below under the section pollen tube growth.

EFFECTS OF OTHER PHYSICAL AND CHEMICAL FACTORS

Only a few studies have been carried out on the effects of pH and light on pollen germination. In general, the pH of the medium does not appear to be critical for pollen germination and pollen tube growth. In *Malus* (Speranza and Calzoni 1980) and *Crotalaria* (Sharma and Shivanna 1983b), pollen germination is satisfactory over a wide range of pH (4–9) (Fig. 4.4). This seems to be related to the ability of pollen diffusates to shift the pH of the medium to around 6 (optimal for pollen germination) when pollen grains are cultured in acidic or alkaline media. When this pH shift is prevented by using buffered medium, optimal germination and tube growth occur only in media with pH around 6. Pollen grains of *Brassica*, however, show optimal germination only at alkaline pH around 8 (Roberts et al. 1983, Hodgkin 1983, Shivanna and Sawhney 1995).

Pollen grains of most species seem to germinate equally well in both light and dark conditions. Although a few studies have reported the effects of light quality (see Shivanna and Johri 1985), there is no clear evidence for the involvement of light and the phytochrome system in pollen germination and tube growth.

Innumerable studies have been carried out on the effects of different growth hormones, inhibitors, antimetabolites, growth regulators, steroids, pesticides, insecticides, fungicides, surfactants, pollutants and other toxic agents. Generally, these chemicals are not regularly required for in-vitro germination. Some, such as pesticides, insecticides and pollutants, show inhibitory effects. The literature on these aspects is summarized by Shivanna and Johri (1985). Spraying of agrochemicals containing any such inhibitory components on crop plants during flowering may reduce yield by inhibiting pollen germination on the stigma.

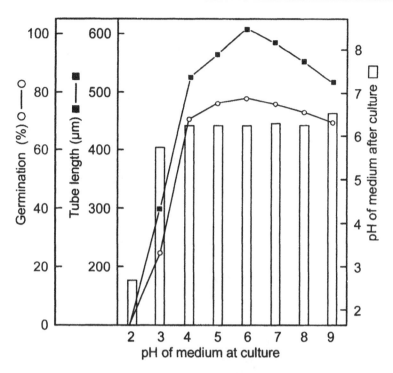

Fig. 4.4 *Crotolaria retusa*: Effects of pH on pollen germination and pollen tube growth in unbuffered medium. Vertical columns represent shift in pH of the medium after pollen culture (after Sharma and Shivanna 1983a).

Naturally occurring polyamines such as putrescine, spermidine and spermine which are known to play a role in a number of events associated with plant growth and development, seem to play a role in pollen germination and pollen tube growth also. Addition of polyamines such as spermidine or putrescine in the germination medium stimulated pollen tube growth (Bagni et al. 1981, Prakash et al. 1988, Rajam 1989). Inhibitors of polyamine synthesis/metabolism markedly inhibited pollen germination and pollen tube growth; this inhibition could be reversed by exogenous application of polyamines (Prakash et al. 1988, Rajam 1989).

PHASES OF GERMINATION AND TUBE GROWTH

Lag Phase

The time interval between pollen hydration and pollen tube emergence is termed the lag phase

and varies from species to species. In some species such as *Tradescantia* and *Impatiens*, it is limited to a few minutes while in others it extends from about 30 min to a few hours (Shivanna et al. 1991a, Mascarenhas 1993). A close correlation exists between duration of the lag phase and level of mitochondrial differentiation in the pollen at the time of dispersal (Hoekstra and Bruinsma 1978, Hoekstra 1979). Three-celled pollen grains generally contain highly differentiated mitochondria with well-developed cristae at the time of shedding and show a short lag phase (Fig. 4.5).

Two-celled pollen show variations in the extent of mitochondrial differentiation (Fig. 4.5) at the time of dispersal (Hoekstra 1979). In some 2-celled species such as *Tradescantia*, mitochondria are fully differentiated as in 3-celled pollen, and show a short lag phase. In other 2-celled species such as *Nicotiana* and *Typha*, mitochondria are not fully differentiated in

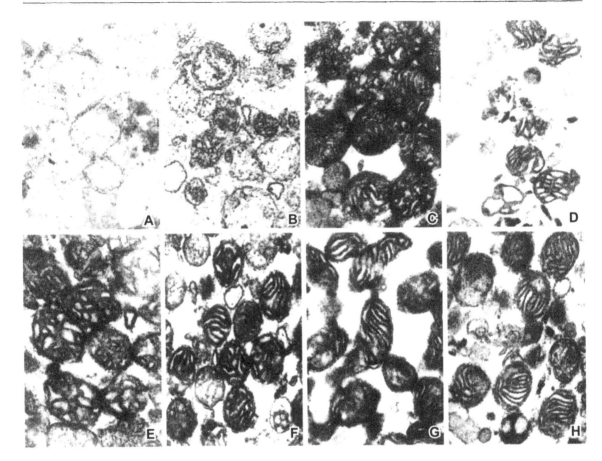

Fig. 4.5 Electron micrographs of mitochondrial fractions isolated from ungerminated (A, B, C, D) and germinated (E, F, G, H) pollen of 2- and 3-celled species. A, E—*Typha* (2-celled), B, F—*Nicotiana* (2-celled), C, G—*Tradescantia* (2-celled), D, H—*Aster* (3-celled) (after Hoekstra 1979).

mature pollen; they appear as spherical saccate structures without clearly discernible cristae. Pollen grains of such species have a prolonged lag phase since pollen tubes do not emerge until their mitochondria become fully differentiated (Fig. 4.5). This difference in lag phase also correlates with differences in their requirement for protein synthesis for germination. Pollen grains with short lag phase do not require protein synthesis for germination while those with long lag phase do.

Pollen Tube Emergence

Pollen tubes emerge from the germ pores. In species with more than one germ pore, gener-

ally a single pollen tube emerges through one of them. Occasionally more than one pollen tube arises from a pollen but only that one among them which receives the nuclei will continue growth; the others will abort. Germ pores are generally free from exine or covered with a very thin layer of exine and thus do not offer much resistance to the emerging tube. In operculate pollen a thick cap of exine is present in the centre of the germ pore and is connected to the margin through a thin sporopollenin membrane. In pollen grains of grasses and a few other species a lenticular thickening rich in pectic polysaccharides and termed 'Zwischenkorper', is present between the intine and the operculum

(Heslop-Harrison 1979c). During germination the Zwischenkorper gelatinizes and the gel pushes aside the operculum, facilitating emergence of the pollen tube.

The pollen tube emerges as an extension of the intine. The intine seems to become less rigid in the region of the germ pore, partly by the activity of wall-held hydrolases and partly by weakening of hydrogen bonds between microfibrils following hydration. Soon after emergence of the tube, new wall materials are inserted at the tip of the tube in an orderly fashion for continued tube growth. Pollen germination is characterized by conversion of quiescent Golgi bodies to active, vesicle-producing forms, and the formation of polysomes (Cresti et al. 1977). Incorporation of new wall material is invariably through the fusion of dictyosome-derived vesicles. (Dashek and Rosen 1966, van der Woude and Morre 1968, van der Woude et al. 1971, Picton and Steer 1981).

Following pollen tube emergence, a clear zonation is established at the growing tip (Rosen 1971, Cresti et al. 1977). The apical zone is made up of a large number of dictyosome-derived vesicles and is free of dictyosomes, mitochondria and RER. The subapical zone is rich in RER. The region behind the RER zone is rich in organelles, especially mitochondria and dictyosomes. In *Lilium*, the tip has been divided into five zones (Reiss and Hearth 1979): an apical vesicle-rich zone, a small zone containing ER and vesicles, a zone rich in mitochondria with vesicles and ER, a zone rich in dictyosomes, and a large zone containing all the organelle types and storage bodies. The pollen tube wall contains a considerable amount of proteins, including those rich in hydroxyproline. In *Lilium*, wall-bound proteins constitute 25% of the dry wall matter in pollen tubes grown in the styles but only about 3.2% in tubes grown in vitro (Li et al. 1983).

Pollen Tube Growth

Unlike many other cell types in which growth takes place all around the cell, the growth in pollen tubes is confined to 2–5 µm of the tip. Dictyosome vesicles are the source of wall and membrane components for the growing pollen tubes. Tip growth is maintained through transport of dictyosome vesicles to the tip of the pollen tube via active cytoskeleton, fusion of vesicles with the plasma membrane and discharge of their contents. The wall at the pollen tube tip appears homogeneous and its electron density is similar to that of the contents of the dictyosome vesicles. The tip is largely made up of pectins, which are esterified and hemicellulose (Li et al. 1994). As esterified pectins do not mutually interact, the wall remains flexible at the tip. The pectic polysaccharides of the tube behind the tip are acidic; they strongly interact in the presence of calcium to make the wall rigid and inflexible. A short distance from the tip the tube is surrounded by a cellulose wall. The wall gradually separates into an outer electron-translucent pectic layer and an inner stratified cellulose layer. In the non-growing region of the tube, a secondary wall, largely made up of callose (β, 1–3 glucan), is laid down. Callose plugs are also formed across the wall in the older part of the tube, which successively isolate older vacuolate parts of the pollen tube from the densely cytoplasmic tip region.

The growth of pollen tubes in vitro in species having a solid style, such as *Nicotiana* and *Petunia*, is not uniform; it shows ring-like annular pectic deposits separated by 5–7 µm distance (Li et al. 1992). The significance of these rings is not clear.

Pollen grains accumulate a considerable amount of proline (Stanley and Linskens 1974); in *Petunia*, proline forms 26% of the dry weight. The pollen tube wall contains hydroxyproline-rich proteins. Azetidine-2-carboxylic acid, a proline analogue, decreased ^{14}C-proline uptake and its incorporation into the cell wall (Dashek and Mills 1980). A part of the proline is used up as a substrate for respiration, and as a nitrogen source for a number of amino acids and other nitrogenous compounds (Britikov et al. 1970, Zhang and Croes 1983a,b). Part of the proline is utilized in the synthesis of hydroxyproline-rich wall protein (extensin) (Dashek and Rosen 1966, Dashek and Harwood 1974). However, only a small proportion of the proline pool is accounted for in these functions; in *Petunia* for example, only 5–6% is utilized during 3 h of culture (Zhang and Croes 1983a). The major pool of proline seems to be required for other function(s) yet not known. In *Lilium,* proline seems to provide resistance to temperature stresses (see Zhang and Croes 1983a). More critical studies are required for a better understanding of the role of proline in pollen biology.

Role of cytoskeleton and calcium in tip growth

The microtubules (MTs) and microfilaments (MFs) and their associated proteins make up the cytoskeleton elements (Table 4.1) of the pollen tube. The cytoskeleton elements not only provide rigidity to the cytoplasmic structure of the pollen tube, but also are involved in pollen function, especially in cytoplasmic streaming (Steer and Steer 1989, Derksen and Emons 1990, Pierson et al. 1986, Pierson and Cresti 1992, Astrom et al. 1995, Derksen et al. 1995, Taylor and Hepler 1997).

The MTs are composed of tubulin and the MFs of actin. Cytoskeletal elements, in particular MFs, are sensitive to chemical fixatives and are frequently lost during fixation. They are well preserved in cryopreserved and freeze-substituted preparations. Both MTs and MFs have been observed in properly fixed TEM prepara-

tions of pollen grains and pollen tubes by many investigators. The advent of fluorescent probes and antibodies for localizing cytoskeletal proteins (Table 4.1) has greatly facilitated studies on MTs and MFs in pollen grains and pollen tubes. MTs are readily identified through the use of anti-tubulin antibodies while MFs are identified by use of a fluorescent probe, phalloidin, as well as antiactin antibodies. The role of MTs and MFs in pollen tube growth has been studied through the use of their depolymerizing drugs—colchicine for MTs and cytochalasin B for MFs.

In pollen tubes, the MTs occur in bundles and are present in the form of strands parallel to the long axis in the cortical region (Fig. 4.6). MT filaments form cross-bridges with MFs as well as plasma membrane. At the pollen tube tip, however, MTs are less organized and randomly distributed (see Derksen et al. 1995). Pollen tube growth is relatively insensitive to colchicine and therefore it is presumed that MTs function mainly as skeletal elements and give stability to actin cytoskeleton; the evidence available so far does not indicate a direct role for them in pollen tube growth. Although MT-associated proteins (kinesin- and dynein-like proteins) (Tiezzi et al. 1992, Moscatelli et al. 1995) have been reported from pollen tubes, their functions in pollen tube growth have not yet been elucidated (Cai et al. 2001).

MFs are also distributed in the cortical region (Franke et al. 1972, Condeelis 1974, Tiwari and Polito 1988, Pierson and Cresti 1992) and form bridges with MTs. Apart from cortical MFs where they occur as single filaments, cytoplasmic MFs are present in bundles (Fig. 4.6) and their distribution reflects the pattern of cytoplasmic streaming. Myosin-like peptides (MF-associated proteins) (Tang et al. 1989) were localized in pollen tubes through immunological techniques (Kohno and Shimmen 1988, Kohno et al. 1992). Subsequently myosin was isolated from pollen tubes of *Lilium* (Yokota and Shimmen 1994, Yokota et al. 1999). Much evidence indicates that myosin-like peptides are located on the surface of secretory vesicles (see Derksen

Table 4.1 Details of cytoskeleton elements in pollen tubes

Features	Microtubules	Microfilaments
Chemical constituent	Tubulin	Actin
Probes for localization	Antitubulin	Antiactin, phalloidin
Depolymerizing agents	Colchicine, oryzalin	Cytochalasin B and D
Associated proteins	Kinesin, dyenin	Myosin, profilin

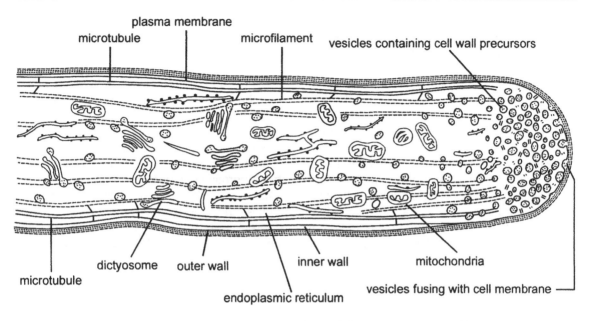

plasma membrane
microtubule
microfilament
vesicles containing cell wall precursors

microtubule
dictyosome
outer wall
inner wall
mitochondria
endoplasmic reticulum
vesicles fusing with cell membrane

Fig. 4.6 Diagram of pollen tube tip to indicate features relevant to tip growth. Dictyosome-derived vesicles move unidirectionally towards the tip of the tube through microfilament pathway, fuse with the plasma membrane and provide wall and membrane components to the growing pollen tube tip (modified from Mascarenhas 1993).

et al. 1995). These reports on myosin and sensitivity of the cytoplasmic streaming to cytochalasin B, a depolymerizing drug for MFs, indicate that actin-myosin interactions are involved in organelle movement, especially polar movement of the secretory vesicles essential for pollen tube growth. The generative cell is also encased in intertwining bundles of cytoplasmic MFs. Movement of generative cell/sperm cells and the vegetative nucleus are also considered to be mediated through actin-myosin force (Heslop-Harrison and Heslop-Harrison 1989, 1992).

Calcium is directly involved in the tip growth of pollen tubes (Steer and Steer 1989, Derksen et al. 1995). As pointed out earlier, Ca^{2+} shows a steep tip-to-base concentration gradient in growing pollen tubes (Herth et al. 1990). This Ca^{2+} gradient is essential for tip growth. The tip-focused gradient is maintained as a result of influx of exogenous calcium at the tip and its uptake behind it by a sink largely of ER (Jaffe et al. 1975). Calcium import occurs through calcium

channels. Blocking of Ca^{2+} channels with such blockers as gadolinium stops tip growth and precludes the flow of current from tip to base. Feijo and co-authors (1995) critically analyzed the role of ion currents, ion dynamics and ion transporters in pollen tube growth, and proposed a model to explain the role of ion channels and ion pumps in pollen germination and pollen tube growth. Cytoplasmic streaming ceases when the internal concentration of calcium reaches 10^{-5} M, which is much higher than that measured at the tip. Maintenance of calcium gradient is necessary for exocytosis (transport and fusion of dictyosome-derived vesicles at the tip releasing their contents), the principal mechanism of tip growth.

Tip-focused Ca^{2+} gradient is also necessary for structural organization of the cytoskeletal elements. Calcium ionophore (which releases calcium ions from intracellular sequestration sites and thus increases Ca^{2+} concentration in the cytoplasm) blocks cytoplasmic streaming and leads to fragmentation of MFs and depolymerization of MTs (Taylor and Hepler 1997). Lack of organized MTs and MFs prevents directed movement of vesicles and results in cessation of pollen tube growth. Thus Ca^{2+} is involved in the transport of vesicles and their organized fusion at the tip through its effect on the MF system.

The pollen tube tip generally shows diffuse fluorescence with probes for both MTs and MFs, indicating their presence in unpolymerized condition. Calmodulin has been shown to be present in pollen tubes of *Lilium* in association with myosin under optimal concentration of calcium (Yokota et al. 1999). At higher concentration of calcium ($< 10^{-6}$ M) calmodulin gets dissociated from myosin. Thus calcium seems to regulate the motility of myosin through calmodulin. In the extreme tip region of the pollen tube (vesicle zone), the concentration of calcium is more than 3 μM, which is not conducive for organization of MFs; this in turn results in lack of active cytoplasmic streaming. Calcium concentration gradually decreases to a level of 0.2 μM within 20 μm. When the Ca^{2+} concentration is experimentally reduced at the tip by microinjection of Ca^{2+} buffer or by treatment with caffeine (Pierson et al. 1994), cytoplasmic streaming is initiated even at the tip and tube elongation is blocked. It is suggested that high concentration of Ca^{2+} at the tip is responsible for absence of cytoplasmic streaming at the tip which allows accumulation and fusion of vesicles.

In-vitro vs in-vivo Tube Growth

In general, the rate and extent of pollen tube growth achieved in vitro is much less than in vivo. Pollen tubes in vivo grow straight with regularly spaced callose plugs; their tips are smooth and narrow. Two-celled pollen generally shows two phases of tube growth: first, the slow autotrophic phase in which endogenous nutrients are used and second, the faster heterotrophic phase in which the pollen tube utilizes the exogenous nutrients (Mulcahy and Mulcahy 1982, 1983). The division of the generative cell seems to separate these two phases. Three-celled pollen

grains do not show the first phase and exhibit a faster heterotrophic phase immediately.

Many differences have been reported between pollen tubes growing in vitro and those in vivo. Pollen tube growth in vitro is much slower than that in vivo and they often show twisted or curled tubes with swollen tips. The growth in vitro is confined to a few hours and the tube length is generally limited to a few mm. The callose plugs in pollen tubes are fewer or even absent and the generative cell generally does not undergo division (see Vasil 1987). Pollen tubes in vivo, on the other hand, continue to grow for a considerable period, producing long tubes, in some cases up to 10–20 cm. This has been explained to be the result of lack of suitable physiochemical conditions in vitro compared to those in vivo.

Many attempts have been made to improve the rate and extent of tube growth in vitro through manipulation of the medium. In *Nicotiana* (Tupy et al. 1983), incorporation of some of the amino acids such as glutamic acid and aspartic acid (which are the precursors to other amino acids) as well as casein hydrolysate (CH) stimulated pollen tube growth. Unlike the standard medium in which pollen tubes cease growth in about 10 h, in the CH-medium tube growth continued for about 24 h. Interestingly, the stimulatory effect of CH was observed only after about 10 h of culture. This seems to be related to the dependence of pollen tubes on an exogenous nitrogen source only during the second phase of growth.

Inclusion of reduced nitrogen source (amino acids or ammonium salts) in the medium enabled pollen tubes of tobacco to complete division of the generative cell (Read et al. 1993). Addition of 10–15% polyethylene glycol 6000 (PEG 6000) and $CuSO_4$ (30 μM), prevented bursting of tubes and resulted in the growth of straight, smooth, thin-walled tubes (Read et al. 1993). In the refined medium, pollen tube tips were comparable to those growing in the style and the tubes showed regular callose plugs. Growth was sustained for more than 48 h at a rate of 250 μm h⁻¹, ca 30–40% of the rate observed in vivo. The rate of tube growth was more or less constant from the beginning and thus did not follow the biphasic pattern expected for 2-celled pollen.

Pollen grains of 3-celled species such as *Brassica* are generally difficult to germinate and the medium used for many other species do not give consistent response. A revised medium formulated for in-vitro germination of *Brassica* (Hodgkin 1986, Hodgkin and Lyon 1983, Roberts et al. 1983) permits consistently good in vitro germination, but does not permit satisfactory tube growth. Incorporation of 10–15% PEG 4000 in the medium, containing 5 or 10% sucrose not only permitted good pollen germination but markedly increased both the rate and extent of pollen tube growth (Shivanna and Sawhney 1995). In the standard medium pollen tube growth was slow (ca 30 μm h^{-1}) and did not continue beyond 2–3 h. In the PEG medium, however, pollen tube growth was ca 300 μm h^{-1} and growth continued for over 10 h.

RELEASE OF METABOLITES

Pollen grains release a wide range of metabolites into the culture medium. These include sugars (Dickinson 1967), amino acids (Linskens and Schrauwen 1969, Zhang et al. 1982), proteins (Stanley and Linskens 1965), phenolics (Namboodiri and Bhaskar 1983), RNA (Tupy et al. 1974) and polyamines (Speranza and Calzoni 1980). Some of these are from the pollen grain wall, both exine and intine (Howlett et al. 1975, Heslop-Harrison et al. 1973, 1974, 1975, Knox et al. 1975), while others are from the pollen cytoplasm probably leached before restoration of membrane integrity (see Chapter 3). Proteins are also released from the pollen tube tip during tube growth. The proteins released into the medium are heterogeneous (Howlett et al. 1975, Ashford and Knox 1980) and include many enzymes such as esterases, phosphatases, nucleases (Matousek and Tupy 1983), and cutinases (Shayk et al. 1977).

Studies by Tupy and co-authors (1974) provide evidence for the release of RNA and even ribosomes from the cultured pollen of *Nicotiana*. The amount of RNA increased in the medium while the cytoplasmic RNA decreased steadily over the culture period; cytoplasmic RNA was practically absent by 16 h of culture. The loss of cytoplasmic RNA and ribosomes may be one of the causes for cessation of tube growth in vitro (Tupy et al. 1974). These studies also suggest the possibility of extra-cellular synthesis of proteins in the culture medium, which would then have important implications in pollen tube nutrition and their metabolism (Tupy et al. 1974, Speranza and Calzoni 1980).

The role of various metabolites leached into the culture medium during pollen germination and tube growth is not clearly understood. In *Lilium longiflorum* (Fett et al. 1976) and *Crotalaria retusa* (Shivanna 1978, Sharma and Shivanna 1983b), elution of pollen before culture did not affect pollen germination and pollen tube growth. In *Pyrus communis*, however, repeated elutions (15 s each) of pollen before culture reduced pollen germination (Stanley 1971). In *Petunia* also (Kirby and Vasil 1979), elution of pollen markedly reduced pollen germination. Addition of eluted substances back into the medium partially restored germination capacity of eluted pollen. A heat labile protein fraction of 50,000–100,000 daltons was shown to be responsible for this restoration. Studies on *Malus* (Speranza and Calzoni 1980), *Crotalaria* (Sharma and Shivanna 1983a) and *Lilium* (Southworth 1983) have shown that one of the functions of the metabolites released into the medium is in regulating the pH of the medium within the range suitable for germination (see page 67).

RNA AND PROTEIN SYNTHESIS

Both transcription and translation occur during later stages of pollen development, before their desiccation. Mature pollen stores a considerable amount of proteins and RNA. The pool of proteins and RNA of mature pollen is available during pollen germination and pollen tube growth (Mascarenhas and Bell 1970, Willing et al. 1988). Using in-situ hybridization combined with confocal microscopy, mRNAs have been localized in mature pollen grains (Reijnen et al.1991).

During pollen germination and pollen tube growth also both RNA and protein synthesis occur (Fig. 4.7) (Tupy et al. 1977, Suss and Tupy 1978, 1979, Ghosh and Shivanna 1983, Shivanna and Johri 1985, Mascarenhas 1993). Actinomycin D, an inhibitor of RNA synthesis, does not inhibit pollen germination and early pollen tube growth. However, cycloheximide, an inhibitor of translation, inhibits both these processes (Fig. 4.8) (Shivanna et al. 1974a, b, Ghosh and Shivanna 1983, Shivanna and Johri 1985). Thus, RNAs present in mature pollen are used to synthesize proteins during early stages of germination and tube growth. However, fresh transcription is required for continued growth of pollen tubes and for generative cell division in 2-celled pollen (LaFleur and Mascarenhas 1978). There are many species, especially 3-celled pollen, in which cycloheximide does not inhibit pollen germination and pollen tube growth (Hoekstra and Bruinsma 1979, Shivanna et al. 1974a). In *Impatiens* (Shivanna et al. 1974a), CHI does not affect germination and tube growth but inhibits the division of the generative cell (Fig. 4.9A). Generative cell division is not affected if CHI is added 90 min after culture (Fig. 4.9B), indicating that the proteins required for generative cell division are synthesized during the first 90 min of culture.

In pollen systems which have a short lag phase (3-celled systems and some 2-celled systems such as *Tradescantia*, *Impatiens*), proteins required for germination and early tube growth are already synthesized before pollen shedding. Such a highly developed metabolic state of pollen that enables early germination and rapid tube growth represents an advanced phylogenetic state (Hoekstra and Bruinsma 1978, 1979). In pollen systems which have a long lag phase (*Typha*, *Lilium*, *Nicotiana*, *Trigonella*) proteins required for germination and tube growth are synthesized only after hydration and thus are sensitive to CHI.

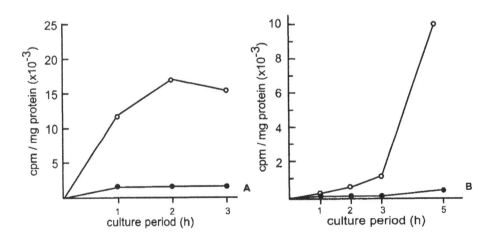

Fig. 4.7. Protein and RNA synthesis in cultured pollen grains of *Zephyranthes citrina*. A. Incorporation of ^{3}H leucine into TCA precipitable protein fraction in the presence (-•-) and absence (-o-) of cycloheximide (5 μg ml^{-1}). B. Incorporation of ^{3}H uracil into acid precipitable RNA fraction in the presence (-•-) and absence (-o-) of actinomycin D (100 μg ml^{-1}) (after Ghosh and Shivanna 1983).

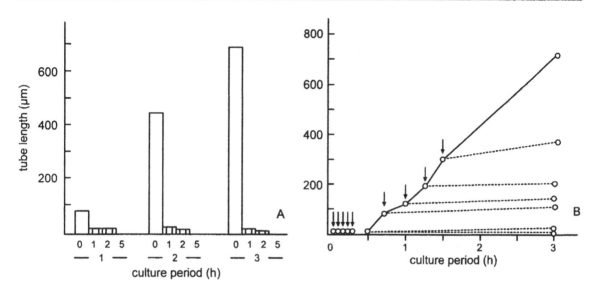

Fig. 4.8 *Zephyranthes citrina*: A. Effect of cycloheximide (CHI—0, 1, 2, 5 μg ml⁻¹) on pollen tube growth over a 3-h period. B. Effect of delayed addition of 10 mg ml⁻¹ of CHI on further growth of pollen tubes. Arrows indicate time of addition of CHI and dotted lines subsequent growth of pollen tubes. Pollen tube growth was arrested soon after addition of CHI (after Ghosh and Shivanna 1983).

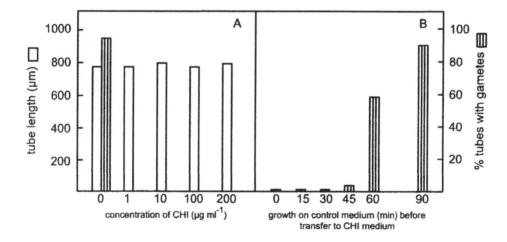

Fig. 4.9 *Impatiens balsamina*: A. Effect of cycloheximide (CHI) on pollen tube growth and gamete formation. CHI did not affect pollen tube growth but did inhibit gamete formation. B. Per cent pollen tubes showing gamete formation when pollen tubes were grown first in control medium for different periods before transfer to CHI-medium. Gametes formed in cultures maintained on CHI-free medium for 60 – 90 min (after Shivanna and Johri 1985).

Analysis of proteins in *Tradescantia* showed no differences in protein profiles in the absence or presence of actinomycin D (Mascarenhas et al. 1974; Mascarenhas and Mermelstein 1981), suggesting that the genes active during pollen maturation continued to be expressed during germination. However, recent studies indicate that novel proteins are produced in germinating pollen (Capkova et al. 1988, 1994, Taylor and Hepler 1997). In *Nicotiana tabacum*, two new-wall bound glycoproteins of 69 and 66 kD are synthesized during pollen germination and tube growth; these are absent in ungerminated pollen (Capkova et al. 1994). The 69 kD protein is glycosylated with glucose and/or mannose and becomes the most abundant protein in pollen tubes. This is not an extensin-like protein as it shows relative abundance of serine, glutamine and glycine but not of hydroxyproline.

IN-VITRO POLLEN GERMINATION· ASSAY TO STUDY THE EFFECTS OF TOXIC CHEMICALS

Pollen grains are simple haploid organisms. They can be collected in sufficiently large quantities and germinated in vitro in a simple nutrient medium. Since pollen germination experiments can be completed in a few hours, aseptic conditions are not generally required for conducting germination experiments. Also, pollen grains of many species can be conveniently stored in viable condition for a considerable length of time and used for experimental studies throughout the year. Because of these advantages, pollen grains provide a convenient system to study a range of problems such as intracellular differentiation, cytoskeleton elements and polarity in cell biology (Heslop-Harrison and Heslop-Harrison 1989, Mascarenhas 1989, Derksen 1996, Taylor and Hepler 1997). Pollen grains are highly sensitive to many toxic chemicals and pollutants (see Shivanna and Johri 1985). They can be used as convenient systems for monitoring cytotoxic effects of bioactive chemicals such as herbicides, pesticides and pollutants (Kappler and Kristen 1987, Kristen and Kappler 1990, Pfahler 1992, see also Wolters and Martens 1987).

A simple photometric method to study the extent of pollen tube growth has been developed (Kappler and Kristen 1988, Kristen and Kappler 1990). Since pollen tube growth involves a steady incorporation of wall material into the growing pollen tubes, determination of the amount of tube wall material in pollen suspension would give the extent of tube growth. Alcian blue specifically binds to water-insoluble carbohydrates (WIC) of the pollen tubes. The quantity of WIC in pollen tubes is directly proportional to the extent of tube growth. The photometric method is based on photometric estimation of the amount of alcian blue bound to the pollen wall over a period of time (see Shivanna and Rangaswamy 1992). The inhibitory effect of a toxic chemical on pollen tube elongation is expressed in the numerical value: ED 50 (the concentration of the test substance that reduces pollen tube growth by 50% of the control).

5

Pollen Sterility

Inability of a plant to produce functional pollen grains is termed pollen sterility. Development of functional pollen involves a series of sequential events starting from the initiation of anthers and culminating in anther dehiscence (see Chapter 1). In pollen sterile plants, functional pollen grains do not form due to a deviation in any one of the sequential processes involved in pollen development. Table 5.1 gives a broad classification of pollen sterility. Non-selective sterility operates during both pollen and ovule development

Table 5.1 Classification of gamete sterility (based on Shivanna and Johri 1985)

I . Non-selective (affects the development of both male and female gametes):
   A. Genetic
      1. Meiotic abnormalities as a result of polyploidy or hybridization; univalent and multivalent formation is common
      2. Chromosomal deficiencies and rearrangements resulting in genomically unbalanced gametes
      3. Presence of lethal genes; usually recessive and inducing sterility under a homozygous condition
   B. Environmental: any environmental factor not conducive for the development of gametes
II. Selective (affects development of pollen grains; development of female gamete is normal)
   A. Genic: usually under the control of a recessive gene; induces male sterility under homozygous condition
   B. Cytoplasmic: determined by cytoplasmic factors in association with nuclear gene(s)
   C. Environmental: any environmental factor not conducive for pollen development

---

whereas selective sterility operates only during pollen development. Because of the importance of pollen sterility in hybrid seed production (see Chapter 12), extensive data have accumulated on various aspects of pollen sterility.

GENIC MALE STERILITY

Genic male sterility (GMS) is due to mutation of nuclear gene(s) (ms genes). GMS is reported in almost all major crop species (Kaul 1988). Mutation of any gene required for pollen development results in pollen sterility. A number of non-allelic loci (for e.g. ca 21 in maize, 16 in soybean and 12 in tomato) have been shown to induce pollen sterility (Sawhney 1997). In most species the ms gene is recessive and induces pollen sterility in a homozygous condition (ms ms). The genetics of GMS and its maintenance are presented in Figure 5.1. In a few species GMS is controlled by a dominant MS gene (Rao et al. 1990). GMS is generally spontaneous in origin; in many species it has also been induced through mutagenic agents (Kaul 1988) and recently through recombinant DNA technology (see p. 80)

Phenotypic Effects of GMS

The phenotypic effects of ms genes are highly variable. Some ms genes impair development of stamens. For example, in the stamenless-1

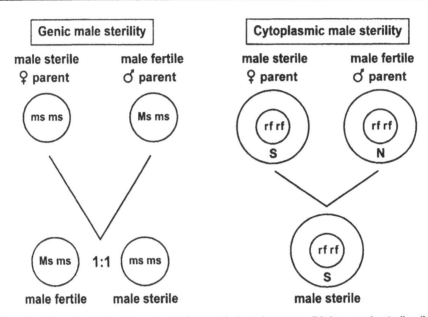

Fig. 5.1 Genetics of genic and cytoplasmic male sterility and their maintenance (Ms/ms—male sterile alleles, rf—fertility restorer recessive allele, S and N—sterile and normal cytoplasms).

(*sl-1*) mutant of tomato and apetala-3 (*ap-3*) of *Arabidopsis*, the stamens are modified into carpels. Antherless (at) mutant of maize has normal filaments but lacks anthers. Many other mutations affect meiosis or post-meiotic development of microspores. In *ms-3, ms-15* and *ms-29* mutants of tomato, and *ms-8* and *ms-9* mutants of maize, meiosis is not initiated and degeneration of microspore mother cells results (Rick 1948, Albertsen and Phillips 1981). *Arabidopsis* mutants, *ms-3, ms-4, ms-5* and *ms-15*, show normal meiosis but the tetrads break down (Chaudhury et al. 1992). In *ms-1* mutant of *Arabidopsis*, microspores are released from the tetrads but become abnormal (Chaudhury 1993). Many post-meiotic mutants such as *ms-2, ms-7, ms-12* and *ms-13* have also been reported in maize (Albertsen and Phillips 1981). In some mutants (e.g. *ms-11* of *Arabidopsis*) pollen development is normal but pollen grains are not released due to failure of anthers to dehisce (Dawson et al. 1993). In a few mutants, apparently normal pollen grains are released but they are not functional on the stigma due to a lack of flavonols in the pollen coat substance (Coe et al. 1981) or insufficient tryphine (Preuss et al. 1993). Attempts are underway to clone many of the ms genes. Cloning and characterization of these genes would facilitate not only the induction of GMS, but also a better understanding of the mechanism of ms genes in inducing pollen sterility.

Structural and Biochemical Changes

A considerable number of investigations have been carried out on the structural details of GMS, both at the light and electron microscopic levels (see Kaul 1988). In most ms mutants, sterility is invariably associated with tapetal abnormality (Palmer et al. 1992). The tapetum does not undergo degeneration at the right time. In several taxa the tapetal cells enlarge prematurely, become highly vacuolate and show delay in degeneration (Fig. 5.2) (Shukla and Sawhney 1993). These studies highlight not only the role of tapetum in pollen development but also the possibility of ms genes inducing pollen sterility through their action on the tapetum.

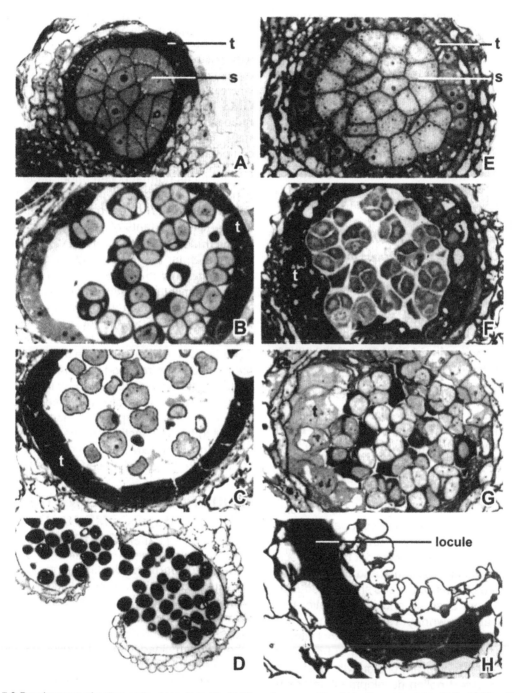

Fig. 5.2 Development of pollen grains in male fertile (A-D) and comparable stages in genic male sterile (E-H) anthers of *Brassica napus*. In fertile line, sporogenous cells (s) complete meiosis and give rise to microspore tetrads which develop into fertile pollen grains (D). In male sterile line, microspore tetrads degenerate and result in the collapse of anther locule. The tapetum (t) in male sterile line shows vacuolation and enlargement (after Shukla and Sawhney 1993).

Many studies have also demonstrated quantitative and qualitative differences in amino acids, proteins and many enzymes between sterile and fertile anthers. In a number of species the level of proline, leucine, isoleucine, phenylalanine and valine is reduced while that of asparagine, glycine, arginine and aspartic acid is increased in sterile anthers (see Kaul 1988). Similarly sterile anthers generally contain lower protein content and fewer polypeptide bands than fertile anthers (Bhadula and Sawhney 1987).

Mistiming of callase activation in anther locules has been demonstrated in a GMS line of *Petunia* (Izhar and Frankel 1971). The activity of many enzymes such as esterases, amylases, peroxidases, phosphatases and cytochrome oxidases has been shown to decrease in sterile anthers (see Kaul 1988, Bhadula and Sawhney 1987). Although these studies indicate the effects of ms genes in protein metabolism, it is not clear whether these changes are the cause for or the result of pollen sterility.

Hormonal Changes

Some studies have shown that pollen sterility is associated with changes in levels of one or more plant growth substances (see Sawhney and Shukla 1994). GMS was associated with a deficiency of gibberellins in maize (Nakajima et al. 1991), and a lower level of indoleacetic acid (Hamdi et al., 1987) and higher levels of cytokinins (Louis et al. 1990) in *Mercurialis*. In *sl-2* mutant of tomato, the concentration of gibberellins and cytokinins was lower but that of abscisic acid was higher (Sawhney and Shukla 1994); a GMS line of *Brassica napus* showed reduced levels of both abscisic and indoleacetic acid (Shukla and Sawhney 1994, Sawhney and Shukla 1994). These studies indicate that altered balance of endogenous growth substances rather than the level of any one growth substance is likely to affect pollen development. Interestingly, exogenous application of gibberellins restores pollen fertility in some mutants such as *sl-1, sl-2, ms-15* and *ms-33* mutants of tomato,

and GMS lines of barley and *Cosmos* (see Sawhney and Shukla 1994).

Effects of Temperature and Photoperiod

The expression of many of the male sterile mutants is strongly influenced by environmental factors, particularly temperature and photoperiod. Variable male sterile mutant of tomato (*vms*) is male sterile when grown at 30°C and above, but produces normal pollen under greenhouse conditions (Rick and Boynton 1967). Similarly, cooler temperatures restore pollen fertility in several GMS mutants of tomato (Schmidt and Schmidt 1981, Sawhney 1983). However, many other ms mutants are insensitive to temperature changes.

A GMS mutant of rice (subsp. *japonica* cv Nongken 58) is photoperiod-sensitive; it is sterile under long days (> 14 h) but fertile under short days (< 13 h) (Shi 1986). In tomato a male sterile mutant *7B-1* produced fertile pollen when the photoperiod was reduced to 8–10 h (Sawhney 1997). Restoration of pollen fertility in GMS lines through manipulation of environmental conditions or hormonal application has been suggested to be an effective approach for maintenance and use of GMS lines in commercial production of hybrid seeds (see Chapter 12).

INDUCTION OF POLLEN STERILITY THROUGH RECOMBINANT DNA TECHNOLOGY

Recent studies have identified and characterized many genes that are expressed exclusively or at elevated levels in anthers (see Chapter 1). Some of the anther-specific genes are expressed only in the diploid tapetum while many others are confined to haploid pollen (Goldberg et al. 1993, McCormick et al. 1993, Williams et al. 1997). These advances have led to use of recombinant DNA technology approaches for induction of pollen sterility. It has been possible to achieve targeted breakdown of developing pollen by tagging a cytotoxic gene under the control of a tapetum- or pollen-specific promoter.

One of the early attempts in this direction was to target the function of the tapetum by placing a cytotoxic gene under the control of a tapetum-specific promoter. This was elegantly done by Mariani and co-authors (1990) of Plant Genetic Systems (Belgium), by constructing a chimeric gene containing the 5′ regulatory (promoter) region of *TA29* gene (expressed specifically in the tapetum) and a gene encoding ribonuclease. One of these genes, *RNase-T1*, from the fungus *Aspergillus oryzae* was chemically synthesized and the other was barnase, a natural gene from the bacterium *Bacillus amyloliquefaciens*, which produces RNase. The chimeric genes (*TA29-RNase* and *TA29-barnase*) were introduced individually into tobacco. In many of the transformants RNase production in the tapetum led to precocious degeneration of tapetal cells and resulted in pollen sterility (Fig. 5.3) (Mariani et al. 1990). These transformants were normal in all other characters including female fertility. Subsequently, *TA29-barnase* gene-induced male sterility was reported in *Brassica napus* (Mariani et al. 1990), maize (Mariani et al. 1992) and several vegetable species (Reynaerts et al. 1993). *TA29-barnase* is a dominant nuclear encoded gene. Absence of a chimeric gene results in pollen fertility.

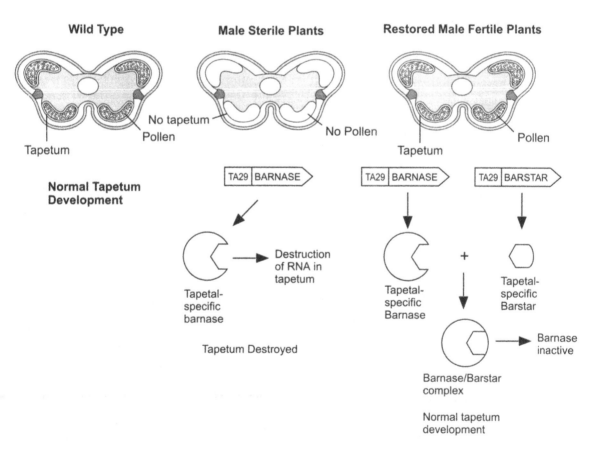

Fig. 5.3 Diagram of induction of pollen-sterile and restorer-lines using barnase and barstar approach (after Goldberg et al. 1993).

Following the success of Mariani and colleagues (1990), a number of investigators have used this approach to induce pollen sterility. Many chimaeric genes, containing a gene encoding a cytotoxic protein and a gene expressed specifically in the tapetum or the pollen, have been introduced to induce pollen sterility in several species such as tobacco (Koltunow et al. 1990, Worrall et al. 1992, Paul et al. 1992, Spena et al. 1994, Hernould et al. 1993, An et al. 1994) and *Arabidopsis* (Day et al. 1994). Some investigators have used antisense RNA technology (van der Meer et al. 1992, Xu et al. 1995a) or cosuppression of endogenous genes that are essential for pollen development or function (Taylor and Jorgensen 1992). Introduction of chimeric gene (containing tapetum specific promotor A3 or A9 and *β-1,3-glucanase* gene resulted in premature degeneration of callose wall around meiocytes (Worrall et al. 1992). These r-DNA approaches have opened up exciting possibilities for manipulating pollen sterility in crop species.

For effective use of pollen sterility developed through r-DNA technology in commercial production of hybrid seeds, suitable restorers are needed (see Chapter 12). Recombinant-DNA technology has been successfully used to construct restorers also (Mariani et al. 1992). Another gene, barstar from *B. amyloliquefaciens*, encoding a protein which specifically forms a stable complex with *barnase* was introduced into oilseed rape under the same tapetum-specific promoter *TA29*. The aim was to activate the *barstar* gene in the tapetum at the same time as the *barnase* gene in the progeny derived from barnase and barstar crosses. The accumulation of barstar molecules in the tapetal cells in at least equal amounts as barnase molecules would result in inactivation of barnase RNase. Crosses between TA 29-barstar and TA 29-barnase plants produced progeny in which both the genes were expressed as expected and were male fertile (Fig. 5.3). Restorers have also been developed in tobacco and maize using the same approach (Mariani et al. 1992). These successes have provided effective tools for development of pollination control systems for production of hybrid seeds through r-DNA technology (see Chapter 12).

CYTOPLASMIC MALE STERILITY

Cytoplasmic male sterility (CMS) is determined by cytoplasmic factors. As the cytoplasm in most flowering plants is inherited through the egg, CMS is also inherited through the female parent. In many CMS lines the expression of pollen sterility by cytoplasmic factors is controlled by one or a few nuclear genes termed Rf (restorer of fertility) genes. Male sterility results when sterile cytoplasm (S) is associated with a homozygous recessive restorer gene (rf rf) (Fig. 5.1). The presence of a homozygous dominant (Rf Rf) or heterozygous (Rf rf) restorer gene restores pollen fertility in plants carrying (S) cytoplasm. Such lines which restore pollen fertility in a sterile line are called restorer lines. The Rf gene has no effect on plants with normal (N) cytoplasm. Thus, male sterility in a CMS line can be restored by crossing it with a suitable restorer line.

CMS has been reported in over 150 species (Laser and Lersten 1972, Kaul 1988). It arises spontaneously or as a result of hybridization followed by a series of backcrosses, resulting in the combination of the nuclear genome of one species and the cytoplasm of another species (alloplasmics) (Edwardson 1970, Hanson 1991, Rao et al. 1993, Prakash and Chopra 1990, Chopra et al.1996, Malik et al. 1999). Production of alloplasmics through conventional or somatic hybridization has been one of the effective methods for the development of CMS lines in *Brassica* (Figs. 5.4, 5.5) and many other crop species (see McVetty 1997, Prakash et al. 1998).

Different CMS lines have been characterized on the basis of their restorer genes. For example in maize, pollen fertility in CMS-T line is restored by two restorer genes, *Rf 1* and *Rf 2*, the CMS-S line is restored by *Rf 3* and CMS-C by *Rf 4*. In the CMS *axillaris* line of *Petunia*, there are three

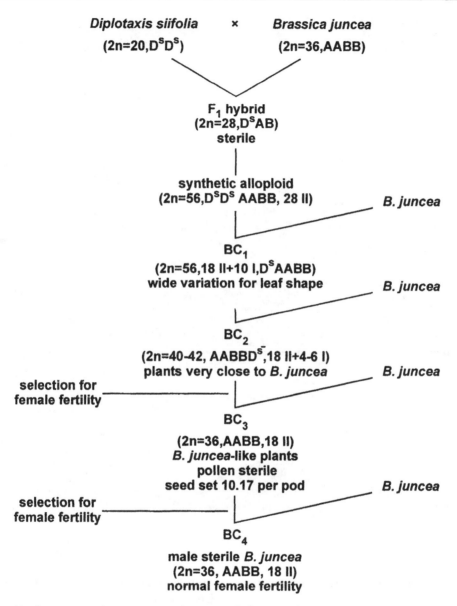

Fig. 5.4 Protocol for development of a new cytoplasmic male-sterile line through wide hybridization. Repeated backcrosses of the alloploid to the cultivar resulted in alloplasmic (cytoplasm of the wild species and nuclear genome of the cultivar) CMS line. Chromosome number, nuclear genome (A/B/D^s), and number of univalents (I) and bivalents (II) observed during meiosis in each BC generation are indicated in parentheses (based on Rao 1995).

(*Rf 1*, *Rf 2* and *Rf 3*) restorer genes. Restorer genes have not yet been isolated for several CMS lines, and many investigators restrict the term CMS to such systems; those with restorer lines are grouped under cytoplasmic-genic male sterility. However, no such distinction is made in the present discussion as the two types show no basic differences and also restorers are likely to be isolated for other CMS lines in the coming years.

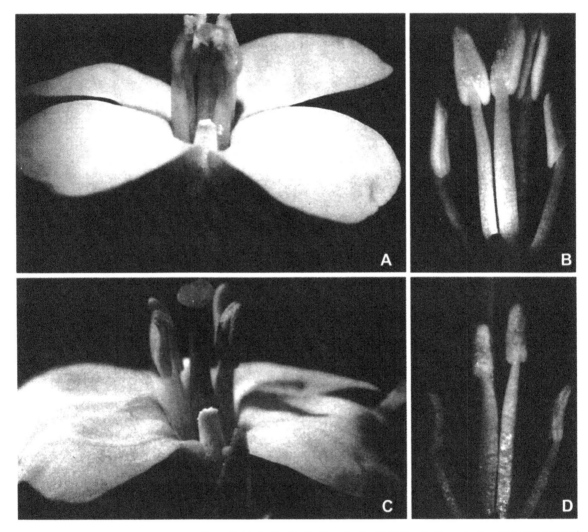

Fig. 5.5 Flowers of *Brassica juncea* cv Pusa Bold. A, B—Fertile line. C, D—CMS line in the cytoplasmic background of *Diplotaxis erucoides*. In B and D sepals and petals have been removed to show stamens (after Malik 1998).

Alterations in Mitochondrial Genome

Analyses of organelle DNA through molecular biology techniques in a number of CMS systems have shown that CMS is the result of mutation(s) in the mitochondrial genome (Table 5.2). These mutations are largely the result of genome rearrangement. Studies on chloroplast DNA (cpDNA) have revealed no significant and consistent differences between fertile and CMS lines (McVetty 1997). Analyses of cpDNA in CMS lines derived

from somatic hybrids have also shown that cp DNA has no role in CMS. Unlike the mitochondrial DNA (mtDNA), cp DNA derived from a sterile parent does not induce CMS in the progeny (Hanson 1991). Most of the CMS lines of different crop species can be characterized by analysis of mtDNA or their translation products (Forde et al. 1978). This method is much faster and more convenient compared to the traditional method of identifying CMS lines through crossing with the restorer lines. CMS lines of corn, in particular

Table 5.2 Characterization of mitochondrial genome in some of the CMS systems

Zea mays	
CMS-T	A unique mitochondrial gene *T-urf 13* originated through recombinational events involving other mitochondrial genes is associated with CMS. *T-urf 13* encodes a non-essential 13-kD protein (URF 13) which is a component of the inner mitochondrial membrane (Forde et al. 1978, Dewey et al. 1986, 1987). Restorer gene *Rf 1* decreases the abundance of URF 13 proteins by 80% (Dewey et al. 1987, Kennell and Pring 1989). Reduction in URF-13 protein is essential for restoration of pollen fertility. The role of *Rf 2* (also needed for fertility restoration) is not clear; it does not alter the expression of *T-urf 13*.
CMS-S	Two plasmid-like linear strands, S1 and S2 of approx. 6.2 and 5.2 kb long, respectively, are associated with CMS (Pring et al. 1977). CMS mitochondria produce a unique polypeptide but its production is not affected by a restorer gene (*Rf 3*).
CMS-C	Two circular DNA molecules of approx. 1.57 and 1.42 kb respectively, are associated with CMS. CMS mitochondria produce a unique 17.5 kD polypeptide which does not respond to a restorer gene (*Rf 4*) (Levings and Pring 1979, Kemble et al. 1980)
Petunia hybrida	
axillaris CMS	A unique mitochondrial locus, *pcf*, is associated with *axillaris* CMS. A 25 kD protein (PCF protein) associated with *pcf* is present in CMS but not in fertile lines (Hanson 1991). PCF protein is much reduced in restored lines (Nivison and Hanson 1989).
Sorghum vulgare	
Milo CMS	A 65 kD variant polypeptide not found in normal cytoplasm is present. The polypeptide is greatly reduced in restored cytoplasm (Dixon and Leaver 1982).
Oryza sativa WA CMS	A 2100 bp linear plasmid is present in CMS mt DNA but not in normal mt DNA (Mignouna et al. 1987).
Helianthus annus	Two major rearrangements, a 12 kb inversion pet-CMS and a 5 kb insertion/deletion are found in CMS mt genome (Crouzillat et al. 1987, Sicuella and Palmer 1988). A unique 16 kD protein is present in sterile as well as restored lines but not in fertile line (Horn et al. 1991).
Brassica napus	
Ogu-CMS	A unique 2.5 kb mt fragment not present in normal lines is present in CMS lines (Erickson et al. 1986, Bonhomme et al. 1991).
Pol-CMS	A novel unidentified reading frame upstream of the *atp 6* gene is present (Erickson et al. 1986, Singh and Brown 1991, Handa and Nakajima 1992, Handa 1993).
Vicia faba	CMS correlated with the presence of well-defined cytoplasmic spheroidal bodies (CSBs) in the cytoplasm not found in fertile lines, isogenic maintainer lines, revertants and restored lines (Edwardson et al. 1976, Scalla et al. 1981). CSBs contain double-stranded RNA and RNA-dependent RNA polymerase (replicase) (Lefebvre et al. 1990). Minor differences are present between the restriction patterns of mt DNA of CMS and maintainer line (Bounty and Briquet 1982) but are not related to the CMS trait (Turpen et al. 1988).
Phaseolus vulgaris	CMS is associated with the presence of a unique transcriptionally active 3 kb mitochondrial sequence, pvs. Fertility restoration results in the loss of pvs (Johns et al. 1992).

CMS-T, and the CMS *axillaris* line of *Petunia* have been thoroughly investigated.

Structural and Biochemical Changes

Pollen abortion in CMS anthers generally occurs after meiosis (Laser and Lersten 1972). In some CMS lines, abortion is delayed until the 2-celled pollen stage. Detailed studies on pollen development in a large number of species have invariably shown the association of CMS with tapetal abnormality (see for details Shivanna and Johri 1985, Kaul 1988, McVetty 1997).

In general, the tapetum in sterile anthers remains intact longer or breaks down earlier than in fertile plants (Laser and Lersten 1972, Frankel and Galun 1977, Kaul 1988). When the tapetal cells persist for a longer time, they tend to show vacuolation and enlargement and often fill most of the anther locule. In *Petunia*, the tapetal cells in CMS lines synthesize far less DNA and show lower levels of RER than in fertile lines. These differences are evident at a very early stage of anther development when all other cell components including organelles are normal (Liu et al. 1987). In CMS-T of maize, mitochondrial breakdown in the tapetum is the first visible abnormality (Warmke and Lee 1977).

In both CMS-T and fertile anthers of maize, there is a significant increase in the number and decrease in the size of mitochondria during meiosis in meiocytes and tapetal cells (Lee and Warmke 1979, Warmke and Lee 1978) (Table 5.3). This increase in mitochondrial number precedes the degeneration of meiocytes in CMS-T anthers. Mitochondria in the megasporocyte and nucellar cells remain relatively constant in size and number during meiosis in both fertile and CMS plants. The tapetum in CMS-S of maize, however, does not show mitochondrial breakdown but shows other abnormalities such as vacuolation and enlargement of tapetal cells (Lee et al. 1979).

In several CMS lines, the development of a callose wall around the microspore mother cells is abnormal; in some species the callose wall is not deposited around microspore mother cells, which results in the formation of naked protoplasts of meiocytes that soon degenerate (Overman and Warmke 1972). In *Petunia* CMS lines show mistiming of callase activity (Fig. 5.6); the

enzyme is activated earlier or later compared to fertile lines (Izhar and Frankel 1971, Bino 1985a, b). Activation of callase in both sterile and fertile anthers is associated with a drop in pH in the anther locule from over 7 to around 6 that is optimal for callase activity. It is suggested that the time of activation of callase is controlled by regulation of pH in the anther locule (Izhar and Frankel 1971). These deviations also reflect malfunctioning of the tapetum as the latter is involved in callase activation (see Chapter 1). CMS anthers of barley show thick deposits of sporopollenin on the tapetal cell surface and on the exine of degenerated microspores (Ahokas 1980).

As in GMS, in CMS lines also many studies have documented a number of variations in several amino acids, proteins and enzymes between sterile and fertile anthers (see Shivanna and Johri 1985). A decrease in proline (Khoo and Stinson 1957) and an increase in glycine (Brooks 1962) and asparagine (Izhar and Frankel 1973) in CMS anthers have been reported. Comparisons of electrophoretic patterns of proteins between fertile and sterile anthers have also shown differences in the banding patterns; sterile anthers show fewer bands (Ahokas 1980).

Alterations in the activities of cytochrome oxidase and malate dehydrogenase have been reported in CMS anthers of corn (Ohmasa et al. 1976). In *Petunia*, consistent differences in the partitioning of electron transport through the cytochrome oxidase and alternative pathway have been detected between fertile and sterile anthers (Connett and Hanson 1990). CMS lines showed consistent reduction in an alternative

Table 5.3 Mean values of mitochondrial numbers per cell for sporogenous tissue and tapetum (based on data presented in Lee and Warmke 1979)

Tissue	Developmental stage				
	Pre-callose S1	Tetrad S4	Intermediate microspore S6	Ratio	
				S4/S1	S6/S1
Sporogenous tissue	6,677	125,543	179,053	18.8	26.8
Tapetum	216	8,779	9,619	40.6	44.5

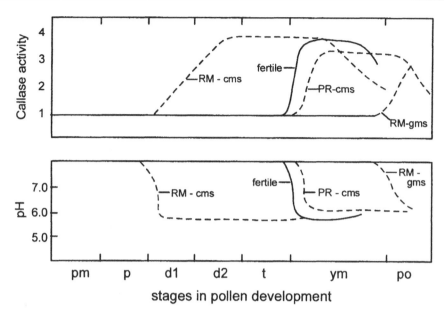

Fig. 5.6 *Petunia*: Drop in pH and increase in callase activity in anthers of one pollen fertile and three pollen sterile (RM - cms, RM - gms, PR - cms) lines during microsporogenesis. Activation of callase is associated with the drop in pH. Meiotic stages: pm—premeiotic, p—prophase, d1—diakinesis 1, d2—diakinesis 2, t—tetrads, ym—young microspores, po—pollen grains (after Izhar and Frankel 1971).

oxidative pathway; in restored lines the activity of this pathway was restored to the level of fertile lines. However, in *Nicotiana* no differences in electron transport were detected in leaves and suspension cultures of CMS and fertile lines (Hakansson et al. 1990).

Mechanism of Cytoplasmic Male Sterility

Structural and physiological evidences available so far have clearly shown that the CMS is mediated through a non-functioning tapetum. Comparison of mt DNA and its translation products between CMS, normal and restored lines particularly in CMS-T of corn, indicates that mitochondrial dysfunction is the primary cause for sterility. Studies carried out soon after an epidemic of Southern corn leaf blight in the USA (1969–70) caused by *Bipolaris maydis,* clearly showed that CMS-T cytoplasm is highly susceptible to the pathotoxin of *B. maydis*. Mitochondria from CMS-T plants showed irreversible swelling of the membrane and changes in res-

piratory rate and oxidative phosphorylation in the presence of the pathotoxin (Comstock et al. 1973). Mitochondria from other CMS and normal cytoplasms were not seriously affected by the pathotoxin. These studies clearly indicated that susceptibility of CMS-T to *B. maydis* race T is caused by mitochondrial sensitivity of CMS-T to a host-specific pathotoxin (BmT toxin) produced by the pathogen, while the mitochondria from disease-resistant lines were insensitive to the pathotoxin.

Subsequent studies (Dewey et al. 1986, 1987) showed that a mitochondrial gene designated T-*urf 13* is responsible for the CMS and disease susceptibility in CMS-T maize. T-*urf 13* encodes a 13 kD polypeptide (URF 13) which is uniquely associated with CMS-T type cytoplasm. One of the two restorers (Rf 1) needed to restore fertility deceases URF protein by about 80% (Forde et al. 1978, Dewey et al. 1986, 1987, Kennell and Pring 1989). Cell cultures of CMS-T maize grown in the presence of BmT toxin yielded toxin-resistant calli from which a considerable number of plants were regenerated. All these regenerants turned out to be disease resistant and male fertile (Gengenbach et al. 1973); analyses of their mt DNA showed deletion or mutation of *T-urf 13* gene. These studies on CMS-T revertants recovered in cell cultures provide strong evidence for the involvement of *T-urf 13* in CMS and disease susceptibility.

Expression of *T-urf 13* in *E. coli* (Braun et al. 1990) resulted in the permeabilization of periplasmic membrane and loss of respiratory activity in the presence of fungal toxin from *B. maydis*, similar to the responses of maize CMS-T mitochondria to the toxin (Dewey et al. 1986). Similar results have been reported for the insecticide methomyl to which maize CMS-T plants are sensitive (Koeppe et al. 1978). Expression of *T-urf 13* in yeast also induced sensitivity to the fungal toxin and the insecticide (Huang et al. 1990).

The fungal toxin binds to the URF 13 protein in maize mitochondria (Braun et al. 1990); this increases the permeability of mitochondrial membranes (Holden and Sze 1984) and affects respiration (Flavell 1975, Holden and Sze 1984). Although these studies clearly show that disruption of mitochondrial membrane is the primary effect of the URF 13 protein in the presence of fungal toxin, it does not explain the effects of URF 13 protein on mitochondria of uninfected CMS plants (which do not contain the toxin). Also, it does not explain normal functioning of mitochondria in the egg and somatic tissues of the CMS plants. A few hypotheses have been posited to explain the role of URF 13 in CMS anthers.

According to the hypothesis of Flavell (1974), put forward even before the *T-urf 13* gene was identified, the anthers produce a substance that mimics the effects of fungal toxin and methomyl. This hypothetical substance is produced in both normal and CMS anthers during pollen development. However, it does not affect the functioning of normal mitochondria but only the altered mitochondria (containing URF 13 protein). As the hypothetical substance is confined to anthers, the mitochondria of the egg and somatic tissues, though containing URF 13, are not affected. Restorer genes could make the mitochondria functional either by altering the binding sites of the substance on the mitochondria or by affecting production and transfer of the substance. However, no direct evidence is available thus far for this hypothesis and attempts to isolate such an anther-specific substance have not been successful (Levings 1993).

The demand on the mitochondria in anthers is greatest during pollen development. A 40-fold increase in number of mitochondria in the tapetal cells and a 20-fold increase in meiocytes (see p. 86) is likely to reflect this extra demand for energy. Rapid replication of mitochondria results in stress on the mitochondrial DNA or its protein synthesis machinery. In the mitochondria of CMS anthers, URF 13 protein is likely to impair the functioning of mitochondria (probably of electron transport or phosphorylation process) and results in the breakdown of tapetal cells; contrarily, mitochondria from normal anthers are capable of functioning under this stress (Levings 1993).

URF 13 may be overexpressed in tapetal cells, reaching toxic levels, and results in mitochondrial dysfunctions. Increased levels of mitochondrial biogenesis may be the cause for overexpression of URF 13 (Levings 1993).

In *Petunia*, PCF protein seems to be the cause for sterility. PCF protein is found in membrane and soluble fractions (Nivison and Hanson 1989). PCF protein is not present in fertile lines and is significantly reduced in restored lines. Differences between CMS and fertile anthers in the partitioning of electron transport through the cytochrome oxidase and alternate pathway have been reported. One of the important deviations in the functioning of the tapetum in CMS *Petunia* is the mistiming of callase activity (see p. 87). The relationship between PCF protein, alteration in electron transport partitioning and mistiming of callase activity is yet to be understood.

In other CMS systems, only limited information relevant to the mechanism of CMS is available. In a few species such as *Sorghum vulgare* and *Brassica napus*, although variations in mt DNA and the presence of unique proteins associated with CMS have been reported, there is no evidence to establish their role in CMS. Failure of the restorer genes to reduce unique proteins in some of the CMS systems indicates that they may not have a direct role in the expression of CMS in such systems. Similarly the presence of plasmid-like DNA fragments in CMS-S of corn and of rice is not directly related to CMS as the restored lines also show these plasmids.

ENVIRONMENTALLY INDUCED POLLEN STERILITY

Environmental factors, in particular temperature, significantly affect the expression of sterility in many CMS systems. For example, in a CMS line of *Petunia* sterile plants become fertile at lower temperatures (Izhar 1975). However, at higher temperatures even the restored plants become sterile (see also Duvick 1965, Edwardson and Warmke 1967, Banga 1992).

Even normal fertile lines of many species become pollen sterile under high and/or low temperatures. In cotton (Meyer 1969), high temperature (38°C) produced more and more sterile anthers in fertile lines. In spring wheat higher temperatures reduce pollen fertility in several cultivars (Welsh and Klatt 1971). In rice (Ito 1978) and *Sorghum* (Brooking 1979), exposure of plants to low temperatures during meiosis resulted in pollen sterility.

Water deficit during meiosis also induced pollen sterility (Morgan 1980, Saini and Aspinall 1981). This may have been the result of accumulation of abscisic acid in the spikelets under stress conditions (Morgan 1980).

Deficiency of micronutrients such as boron and copper induces pollen sterility. Under copper deficiency, the anthers of wheat showed a wide range of abnormalities in tapetal cells and pollen grains lacked starch grains (Jewell et al. 1988). Wheat plants grown under copper deficiency showed 100% male sterility without affecting ovule fertility (Graham 1975, Azouaou and Souvre 1993). Temporary copper deficiency significantly reduced pollen viability and the number of proline-rich pollen grains (Azouaou and Souvre 1993).

CHEMICALY INDUCED POLLEN STERILITY

Because of the potential application of pollen sterility in hybrid seed production, a large number of chemicals have been tested for their ability to induce pollen sterility. Such chemicals have been designated as 'gametocides' (Mohan Ram and Rustagi 1966, Chauhan and Kinoshita 1982) or 'chemical hybridizing agents' (CHAs) (McRae 1985, Cross and Ladyman 1991). A large number of reports appeared during the 1960s and 1970s on the gametocidal effects of a number of chemicals. Of these, promising results were reported with Mendok/FW-450 (sodium salt of 2,3-dichloroisobutyric acid), dalapon (2,2-dichloropropionic acid), gibberellic acid, and ethephon (for details see Mohan Ram and Rustagi 1966, Meyer 1969, Rustagi and Mohan Ram 1971, Fairey and Stoskopf 1975, Hughes and Bodden 1978, Chauhan and Kinoshita 1982, Colhoun and Steer 1983, Shivanna and Johri 1985, Graham 1986).

Most of these studies were confined largely to studying the extent of pollen sterility induced by these chemicals and hardly any data were collected on the mode of their action. Also most of these chemicals were reported to induce a considerable degree of female sterility and/or other undesirable morphological abnormalities at the concentration effective to induce dependable pollen sterility. Therefore, none of them was found ideal for commercial application. In recent years, however, many multinational companies have initiated research to develop CHAs particularly for wheat and a few CHAs have shown the potential for commercial application (Cross et al. 1989, 1992, Schulz et al. 1993, Cross and Schulz 1997). Details of these recent studies are presented in Chapter 12.

6

Pistil

The pistil is the female partner in sexual repro-duction of flowering plants; it consists of the stigma, style and ovary. The process of fertiliza-tion in flowering plants is more complex com-pared to other groups of plants, primarily be-cause of the involvement of the pistil. Although fertilization takes place in the ovules located in-side the ovary, the stigma and style play a cru-cial role in fertilization processes (see Chapter 8). The structural features of the stigma and style are adapted to receive pollen grains and to fa-cilitate their germination and subsequent growth of pollen tubes. A thorough knowledge of the structure of the pistil is essential for understand-ing pollen-pistil interaction and fertilization. This chapter gives a brief account of the structural details of the pistil in relation to its role in pollen germination and pollen tube growth. A brief de-scription of the ovule and the embryo sac is also given.

STIGMA

The stigma is the recipient of pollen. It shows wide variations in morphology (Heslop-Harrison and Shivanna 1977, Heslop-Harrison 1981, Cresti et al. 1992). Usually, the stigma is classi-fied into dry and wet types based on the pres-ence or absence of secretory fluid, the stigmatic exudate, at the time of pollination (Fig. 6.1A-D). Each of these types is further divided on the

basis of distribution of the receptive surface and the presence or absence of papillae (Table 6.1). Considerable diversity is evident in the morphol-ogy of stigmatic papillae. The papillae may be unicellular and multicellular; the latter may be uniseriate or multiseriate.

Irrespective of the morphological variations of the stigma, the receptive surface invariably

Table 6.1 Types of stigma (some examples of each group are given in parentheses (based on Heslop-Harrison and Shivanna 1977, Shivanna et al. 1997)

Dry stigma (without apparent fluid secretion)
Group I: Plumose—receptive surface dispersed on multiseriate branches (cereals)
Group II: Receptive surface confined to the stigma
    A: Surface non-papillate (Acanthaceae)
    B: Surface distinctly papillate
        1. Papillae unicellular (Brassicaceae, Asteraceae)
        2. Papillae multicellular
           a. Papillae uniseriate (Amaranthaceae)
           b. Papillae multiseriate (Bromeliaceae, Oxalidaceae)

Wet stigma (fluid secretion present on the stigma surface during receptive period)
Group III: Receptive surface papillate (some Rosaceae, some Liliaceae)
Group IV: Receptive surface non-papillate (Apiaceae)
Group V: Receptive surface covered with copious exudate in which detached secretory cells of the stigma are suspended (some Orchidaceae)

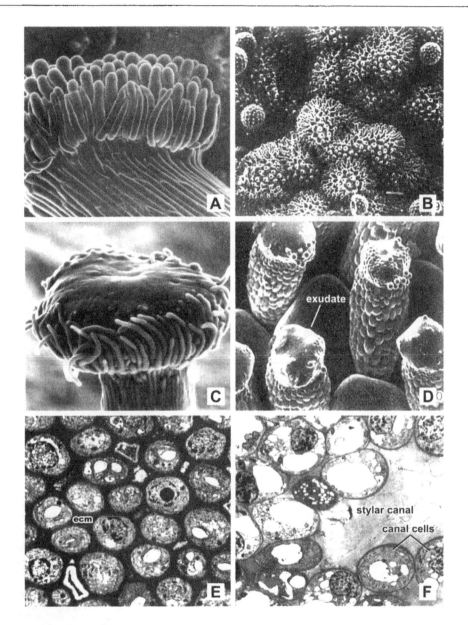

Fig. 6.1 A-D. Scanning electron micrographs of dry and wet stigmas. *Genista genefolia* (A) and *Ipomoea purpuria* (B)—dry papillate stigma types. *Tephrosia* sp. (C) and *Fragaria vesca* (D)—wet papillate types. E, F. Transmission electron micrographs of transections of style. E—*Lycopersicon peruvianum* (solid style); only a part of the transmitting tissue is shown; ecm – extracellular matrix. F—a part of the stylar canal together with canal cells of *Sternbergia lutea* (hollow style) (A, B—after Shivanna and Owens 1989, C-F—after Shivanna et al. 1997)

contains extracellular components (Mattsson et al. 1974, Heslop-Harrison and Shivanna 1977, Heslop-Harrison and Heslop-Harrison 1985a, Shivanna and Johri 1985, Knox et al. 1986, Raghavan 1999). These components are highly heterogeneous and include lipids, proteins and glycoproteins, a variety of carbohydrates, amino acids and phenols. A number of enzymes, predominantly non-specific esterases and phosphatases, have also been localized on the stigma surface (Heslop-Harrison and Shivanna 1977, Ghosh and Shivanna 1984a). Cytochemical demonstration of non-specific esterases (Shivanna and Rangaswamy 1992) has become a standard method of localization of the receptive surface of the stigma.

In the dry stigma, extracellular components are present in the form of a thin extracuticular hydrated layer, termed the pellicle (Fig. 6.2A, B) (Mattsson et al. 1974). The pellicle components originate from the epidermal cells of the stigma and/or stigmatic papillae and are extruded onto the surface of the stigma/stigmatic papillae through discontinuities present in the cuticle (Fig. 6.3) Heslop-Harrison and Heslop-Harrison 1982a). In some species the pellicle binds to lectins (Clarke et al. 1979). Arabinogalactans, a group of carbohydrates with adhesive properties, are also present in the pellicle (Clarke et al.

1979). The pellicle is also present in marine angiosperms in which the stigma remains submerged in water (Pettitt 1976, 1980).

In the wet stigma, the amount of exudate that accumulates on the stigma surface is highly variable; it may be confined to just the interstices of the papillae or may flood the entire surface of the stigma. In some legumes (Owens 1989, Kenrick and Knox 1989) and orchids (Calder and Slater 1985, Slater and Calder 1990), the stigma is cup- or funnel-shaped. The inner receptive surface is non-papillate and is lined with many layers of secretory cells covered with fluid secretion. In orchids, the cells of the secretory zone in the cup-shaped stigma are loosely arranged and often detach and float freely in the stigma secretion.

The exudate may be lipoidal as in *Petunia* and *Oenothera* or aqueous as in *Lilium*. The exudate generally contains lipids, amino acids, proteins, carbohydrates and phenolic compounds, and responds to assays for esterases (Fig. 6.2C), and phosphatases. Sucrose or free

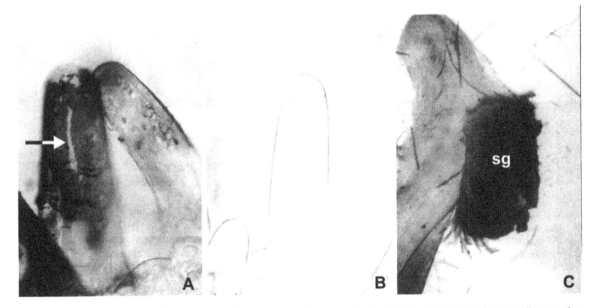

Fig. 6.2 Histochemical localization of non-specific esterases using α-naphthyl acetate as a substrate in a coupling reaction with fast blue B. A, B. Stigmatic papillae of *Acidanthera bicolor* (dry stigma) with substrate (A) and without (B) substrate (control). The pellicle is seen in A as a dark sheath but is torn at places (arrow) to reveal the underlying cuticle. C. *Vigna unguiculata* (wet stigma, sg) with substrate. Esterase activity is clearly seen in the exudate (after Shivanna 1979).

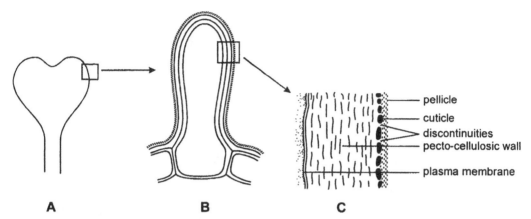

Fig. 6.3 Diagram of the pellicle in a dry type of stigma. A. Stigma. B. One of the papillae magnified. C. Part of the cell wall magnified to show its details with pellicle and cuticle (after Shivanna 1979).

monosaccharides occur in the exudate of only a few species (Konar and Linskens 1966a, b); sugars are often present as phenolic glycosides (Martin 1969) or as a part of glycoproteins (Labarca et al. 1970). In *Lilium*, the carbohydrate components are made up of arabinose (26%), rhamnose (6%), galactose (55%), and glucuronic acid (11%) (Aspinall and Rosell 1978). Proteinase inhibitors have been reported in the stigma of tobacco (Atkinson et al. 1993).

Studies on the developing wet type of stigma have shown that the young stigma is free from the exudate but like the dry stigma contains the pellicle (Fig. 6.4A) (Konar and Linskens 1966a, Shivanna and Sastri 1981). As the flower bud matures, the exudate originates from the epidermal and subjacent cell layers (Konar and Linskens 1966a, Dumas et al. 1978) and accumulates in the intercellular spaces of the stigmatic tissue and below the cuticle-pellicle layer of the epidermal cells (Fig. 6.4B, C). Continual secretion of the exudate exerts pressure on the cuticle-pellicle layer which eventually gets disrupted, thus releasing the exudate onto the stigma surface (Fig. 6.4D).

In many species, the stigmatic tissue has become necrotic by the time the flower opens (Dickinson and Lewis 1975, Herrero and Dickinson 1979, Kristen et al. 1979, Cresti et al. 1980). Electron microscopic studies indicate that the exudate is secreted by the ER through exocytosis (Kristen et al. 1979). Towards the later stages, the exudate is also secreted through holocrine secretion following degeneration of the protoplasts of the secretory cells (Kristen et al. 1979).

Different components of the stigma exudate, apart from their role in pollen-pistil interaction, have been implicated in other important functions. Lipids are considered to prevent excessive evaporation and wetting by acting as a liquid cuticle. Phenolics and proteinase inhibitors have been suggested to give protection against microbial pathogens and insects. The exudate has also been reported to serve as a nutrient source for pollinating insects.

STYLE

The style is basically of two types—solid (closed) and hollow (open) (Shivanna and Johri 1985). In the solid style, found in the majority of dicotyledons, a core of transmitting tissue traverses the whole length of the style (Fig. 6.1E, 6.5A-C). In the hollow style, the style is traversed by a canal, the stylar canal, bordered by one or a few layers of glandular cells, the canal cells (Fig. 6.1F, 6.5D-F).

The transmitting tissue (TT) of the solid style is made up of elongated cells connected end to

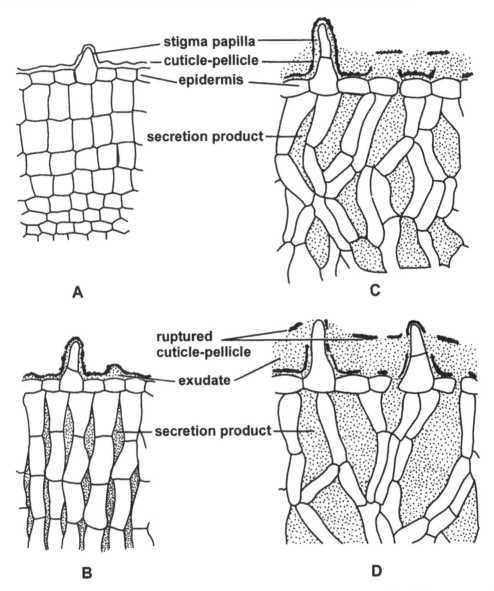

Fig. 6.4 Diagram of the origin of stigmatic exudate in taxa characterized by wet stigma and solid style. In very young stage (A), only a thin cuticle-pellicle layer is present. Secretion products accumulate between cuticle-pellicle layer and epidermis as well as in the intercellular spaces (B). Continued accumulation of secretion products results in enlargement of intercellular spaces and rupturing of cuticle-pellicle layer (C). Mature stigma shown in D (after Sastri 1979) .

end through plasmodesmata. In transection the cells of the TT are circular with massive inter-cellular spaces filled with extracellular matrix (ECM) (Fig. 6.1E, 6.5B, C). The cells of the TT exhibit normal electron microscopic profiles with numerous mitochondria, active dictyosomes, RER, plastids and ribosomes (see Shivanna and Johri 1985). The ECM, secreted by the cells of the TT, is predominantly made up of pectin but also contains proteins, glycoproteins and often lipids; it responds to assays for many enzymes such as esterases, acid phosphatases and peroxidases (Dickinson et al. 1982, Ghosh and Shivanna 1984a). Both ER and dictyosomes have been implicated in the secretion of the ECM (Heslop-Harrison and Heslop-Harrison 1982b). Recently a number of proline-rich proteins, specific to the TT, termed transmitting tissue specific (TTS) proteins have been localized in the ECM (Gasser and Robinson-Beers 1993, Cheung et al. 1993, Wang et al. 1993); much evidence indicates that they play an important role in pollen-pistil interaction (see Chapter 8).

In hollow styles the stylar canal originating from the stigma surface traverses the whole length of the style and joins the ovarian cavity.

The canal cells in young flower buds are lined with a layer of cuticle on the inner tangential wall. The secretion products from the canal cells accumulate below the cuticle. In some species such as *Lilium*, the cuticle is disrupted during development of the pistil and the stylar canal filled with the secretion product (Dickinson et al. 1982, Miki-Hirosige et al. 1987). In others such as *Gladiolus*, *Crocus* and *Sternbergia* the cuticle remains intact but is pushed towards the centre of the canal by the secretion (Fig. 6.5E, F) (Heslop-Harrison 1977, Clarke et al. 1977, Ciampolini et al. 1990).

In a few hollow-styled species such as *Lilium* (Rosen and Thomas 1970, Dickinson et al. 1982) and *Boswellia* (Sunnichan 1998), the inner

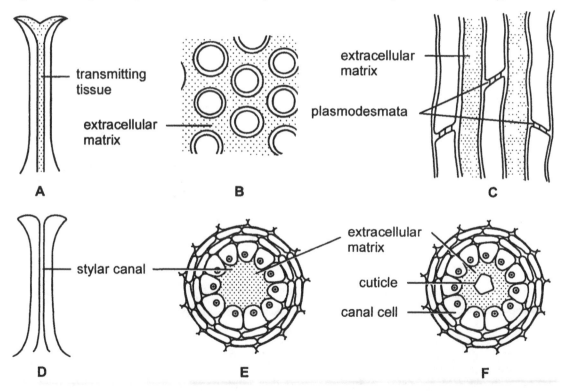

Fig. 6.5 Diagram of structure of the style. A-C. Solid style. A—longitudinal section. B and C—enlarged views in transverse and longitudinal sections of part of the transmitting tissue. Note plasmodesmatal connections in the transverse walls of cells but not along the longitudinal walls in C. D-F. Hollow style. D—Longitudinal section of style to show continuous stylar canal from stigma to ovary. E and F—are transverse sections in taxa in which the cuticle bordering the canal cells is completely disrupted and the stylar canal is filled with extracellular matrix (E) and in taxa in which the cuticle remains intact and extracellular matrix accumulates between the canal cells and the cuticle (F) (modified from Shivanna and Johri 1985).

tangential wall of the canal cells shows wall ingrowths characteristic of transfer cells (Fig. 6.6A-C). As in solid-styled systems, the secretion in the stylar canal is rich in carbohydrates and proteins and shows esterase and acid phosphatase activity (Tilton and Horner 1980). Lipids have also been reported in stylar secretions of some species.

In papilionoid legumes, although the style is hollow, it is not in continuity with the stigmatic surface (see Shivanna and Owens 1989). The stigma is solid, comparable to solid-styled systems; the upper part of the hollow style is derived secondarily by dissolution of the cells of the TT (Fig. 6.7), while the lower part represents extension of the ovarian cavity (Ghosh and Shivanna 1982a, Lord and Heslop-Harrison 1984). In some legumes, the lowermost part of the stylar canal is loosely filled by papillae formed by elongation of the canal cells (Fig. 6.7 E) (Malti and Shivanna 1984, Shivanna and Owens 1989). Elongation of the canal cells into the stylar canal at the base of the style is observed in many other hollow-styled species such as oil palm (Tandon et al. 2001). The TT (in solid-styled species) and canal cells (in hollow-styled species) continue into the ovary along the placenta, thus providing continuity of the path for pollen tube growth.

The most important structural feature of the stigma and style relevant to pollen-pistil interaction is the presence of extracellular components, particularly of proteins and glycoproteins, on the surface of the stigma and in the path of pollen tube growth in the pistil. These components come into direct contact with the pollen grains

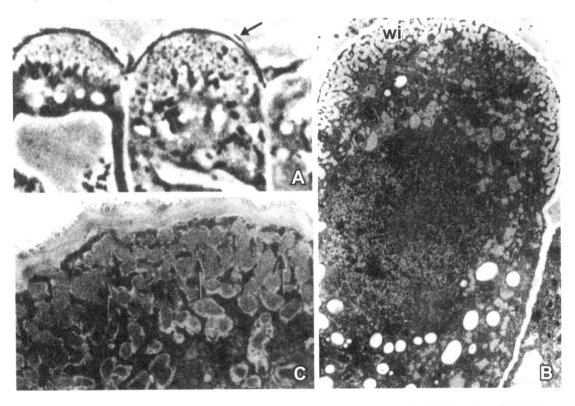

Fig. 6.6 Details of canal cells of *Lilium longiflorum* in transections. A. Near mature canal cells. Disruption of cuticle (arrow) seen. B. Electron micrograph of a mature canal cell to show wall ingrowths (wi) on the inner tangential wall. C. Part of the inner wall magnified to show details (arrows point to wall ingrowths) (after Dickinson et al. 1982).

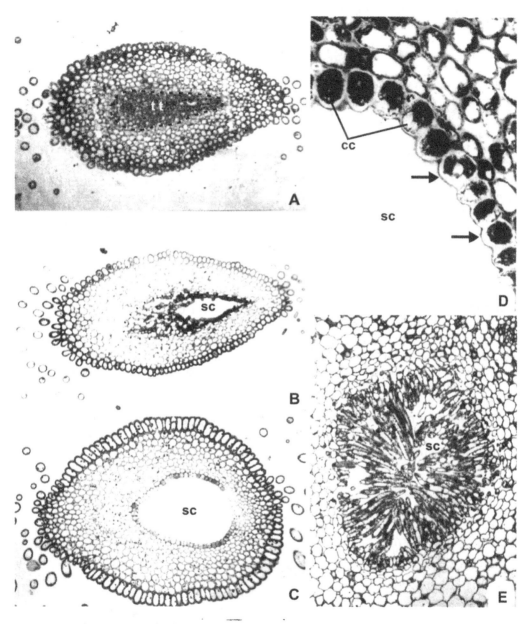

Fig. 6.7 Transections of the style of *Crotalaria retusa* at different levels. A. At base of stigma. A strand of transmitting tissue fills the central part and the stylar canal has not yet developed. B. Just below the stigma. A stylar canal (sc) has appeared on one side of the transmitting tissue by cell degeneration. In A and B the boundary of the transmitting tissue (tt) is shown in broken lines. C. Middle part of style. The stylar canal has enlarged and the transmitting tissue has become reduced to a single layer bordering the canal. D. Part of the section comparable to that shown in C stained with Coomassie blue magnified to show richly proteinaceous canal cells (cc) and a thin proteinaceous layer (arrows) on the wall facing the canal. E. Base of style; stylar canal is loosely filled with long unicellular papillae formed by elongation of the canal cells (after Malti and Shivanna 1984).

and pollen tubes and play an important role in pollen-pistil interaction (see Chapter 8).

OVULE AND EMBRYO SAC

The ovules develop on the placenta. Extensive studies have been carried out on the structural details of the ovule and the embryo sac (Maheshwari 1950, 1963, Tilton 1981a, b, Tilton and Lersten 1981a, b, Johri 1984, Gasser and Robinson-Beers 1993, Russell 1993). Only a brief overview of the ovule and the embryo sac is presented here.

The ovule (Fig. 6.8, 6.10A) essentially consists of one or two outer coverings (the integuments), the nucellus and the embryo sac. The integuments do not cover the nucellus completely but leave a narrow passage, the micropyle, at the tip. The ovule is attached to the placenta through a short stalk, the funiculus. In members of a few families such as Polygonaceae, Urticaceae and Piperaceae, the ovule is straight, and the funiculus and the micropyle lie in the same axis; such ovules are referred to as orthotropous or atropous (Fig. 6.8). More commonly, the ovule shows curvature to different degrees as a result of unequal growth of the funiculus. The most common is the anatropous condition in which the ovule is completely inverted and thus the micropyle is positioned close to the placenta (Fig. 6.8). When the curvature of the ovule is partial so that the micropyle is at a right angle to the funiculus, the ovule is referred to as hemianatropous. In a campylotropous ovule the curvature is less pronounced than in an anatropous ovule. When the curvature of the ovule affects the nucellus also, it resembles a horseshoe and the ovule is called amphitropous. In a circinotropous ovule, found in members of Cactaceae, the curvature continues beyond that in an anatropous condition and the micropyle points upwards again.

The nucellar tissue may be massive (crassinucellate) or confined to a few layers (tenuinucellate). In the tenuinucellate type, the nucellar cells surrounding the embryo sac degenerate during ovule development and the mature embryo sac comes into contact with the inner layer of the integument which differentiates into a specialized layer, the endothelium or integumentary tapetum. The cells of the endothelium are radially elongated and densely cytoplasmic with prominent nucleus.

There is great diversity in the development and structure of the mature embryo sac. Embryo sacs have been classified into monosporic, bisporic and tetrasporic types based on the number of megaspore nuclei taking part in embryo sac development; they are diagrammatically depicted in Figure 6.9. In the monosporic type, meiosis in the megaspore mother cell results in a linear tetrad of which only the chalazal or the micropylar megaspore functions. In the bisporic type, the first division of the megaspore mother cell results in a dyad; either the chalazal or the micropylar cell of the dyad degenerates and the other gives rise to the embryo sac. In the tetrasporic type, meiosis results in the formation of four megaspore nuclei without accompanying wall formation and all four megaspore nuclei participate in formation of the embryo sac (Fig. 6.9). Thus all the nuclei in monosporic types are homozygous but heterozygous in bi- and tetrasporic types. The embryo sac types are further divided on the basis of number of nuclear

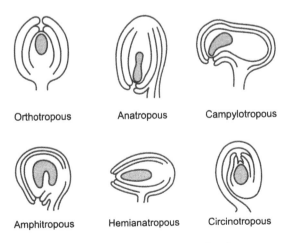

Orthotropous   Anatropous   Campylotropous

Amphitropous   Hemianatropous   Circinotropous

Fig. 6.8 Diagram of various types of ovules.

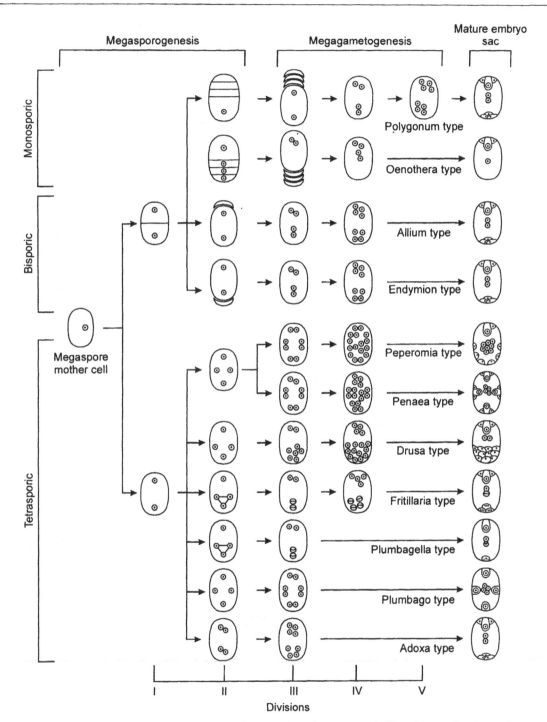

Fig. 6.9 Diagram of the development of mono-, bi- and tetrasporic embryo sacs and different types of mature embryo sacs with their cellular organization.

divisions involved and organization of the mature embryo sac (Fig. 6.9). A brief description of only the most common type of embryo sac, the polygonum type, is given below.

A megaspore mother cell differentiates in the hypodermal region at the micropylar end of the nucellus. The megaspore mother cell is distinguishable from other cells by its large size, dense cytoplasm and prominent nucleus. It undergoes meiotic division and gives rise to a linear row of four megaspores. Callose is deposited along the wall of the megaspore mother cell and along the transverse walls of the resultant megaspores, similar to the callose wall that develops around microspores. Three of the megaspores situated in the micropylar end degenerate and the chalazal megaspore enlarges and gives rise to the embryo sac. The callose wall disappears around the functional chalazal megaspore while it may

persist for a longer time around the degenerating megaspores. The chalazal megaspore eventually gives rise to the mature 8-nucleate, 7-celled embryo sac by undergoing three mitotic divisions of the nucleus and subsequent cellularization.

The mature embryo sac (Fig. 6.10 A) consists of the central cell, a 3-celled egg apparatus at the micropylar end (Fig. 6.10 B-D) and three antipodal cells at the chalazal end. The central cell is the largest cell of the embryo sac and contains two nuclei, the polar nuclei, towards the micropylar end. The central cell is highly vacuolated, and the cytoplasm is confined to a thin layer along the wall and around the polar nuclei. The polar nuclei eventually fuse to form the secondary nucleus. The central cell lacks the cell wall at the micropylar end where it borders the egg cell and the synergids (Fig. 6.10 B). In a few

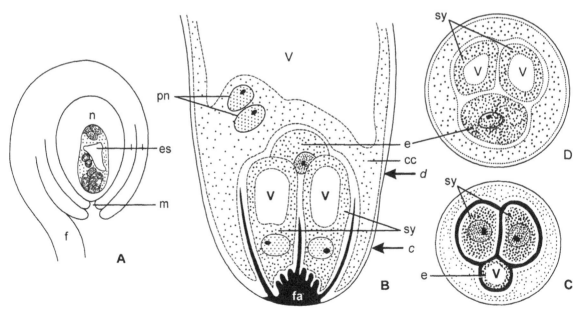

Fig. 6.10 Diagram of ovule and Polygonum type of embryo sac. Cellulosic wall is shown in black and plasma membrane in dotted lines. A. Longisection of an anatropous ovule with 8-nucleate, 7-celled embryo sac. B. Details of the micropylar part of the embryo sac in longisection. C and D. Transections of the embryo sac at levels marked *c* and *d* in Fig. B, to show the arrangement of cells of the egg apparatus. Cellulosic wall present around egg and synergids in Fig. C whereas in Fig. D the egg, synergids and central cell are separated by their respective membranes. (cc—central cell, e—egg, es—embryo sac, f—funiculus, fa—filiform apparatus, m—micropyle, n—nucellus, pn—polar nuclei, sy—synergids, v—vacuole) (after Shivanna et al. 1997).

species, wall ingrowths, characteristic of transfer cells, develop on the lateral wall of the central cell at the micropylar and/or chalazal region (Kapil and Bhatnagar 1975). The central cell does not show high metabolic activity, as indicated by ultrastructural and cytochemical studies. The antipodal cells in most species are ephemeral and have degenerated by the time fertilization is effected.

The egg apparatus consists of three pear-shaped cells (two synergids and one egg cell), attached to the embryo sac wall at the micropylar tip (Fig. 6.10B-D). The egg cell is distinctly polar, containing a large vacuole at the micropylar end, and most of the cytoplasm and the nucleus at the chalazal end. The cell wall of the egg attenuates towards the chalazal pole and is absent or incomplete at the chalazal end. The egg is metabolically less active compared to other cells of the embryo sac.

The synergids are also polarized cells; they are densely cytoplasmic at the micropylar end and highly vacuolated at the chalazal end. The synergids develop extensive wall ingrowths into the cytoplasm in the micropylar region. This thickened wall region is referred to as the filiform apparatus and enormously increases the surface area of the plasma membrane. The synergids are metabolically the most active cells in the embryo sac. They contain a large number of mitochondria and dictyosomes, abundant ribosomes and extensive endoplasmic reticulum. The filiform apparatus plays an important role in the entry of pollen tubes (see Chapter 8). The cell wall of the synergids also attenuates towards the chalazal pole, and is incomplete or absent at the chalazal end. Thus, the chalazal end of the egg apparatus lacks a cell wall; in this region the egg cell, synergids and central cell are separated by their respective plasma membranes. The egg, the two synergids and the central cell together have been referred to as the 'female germ unit' (Huang and Russell 1992) since all of them together play a direct role in pollen tube entry, discharge of sperm cells and double fertilization. Frequently, one of the synergids degenerates prior to arrival of the pollen tube in the embryo sac.

7

Pollination

Pollination is the transfer of pollen grains from the anther to the stigma. In seed plants pollination is a prerequisite for fruit- and seed-development and is the basis of genetic exchange between plants and recombination within plants. Flowering plants have evolved an amazing array of adaptations to achieve pollination. The evolution and diversity of flowering plants and advanced orders of insects (Lepidoptera, Diptera and Hymenoptera) are believed to be the result of over 140 million years of co-evolution (Endress 1994).

Pollination is a critical factor for sustainable agriculture and for commercial production of hybrid seeds. Pollination biology has been an area of considerable interest for more than a century. A large number of books and monographs have been published over the years on various aspects of pollination biology (Knuth 1906–1909, McGregor 1976, Faegri and van der Pijl 1979, Real 1983, Richards 1986, Bawa et al. 1993, Free 1993, Roubik 1995). The techniques of pollination biology have been covered comprehensively by Jones and Little (1983), Dafni (1992), Kearns and Inouye (1993), Dafni and Firmage (2000).

Table 7.1 describes the types of pollination and Figure 7.1 depicts them. The major advantage of self-pollination is its certainty. However, continued self-pollination over many generations results in inbreeding depression. For effective self-pollination (autogamy), the flowers have to be bisexual, anther dehiscence and stigma receptivity have to be synchronous, and the position of the stigma and anthers should be close to one another so that pollen grains readily come into contact with the stigma soon after anther dehiscence. In most crop species the flowers open and the sex organs in the flower (i.e., the stamen and pistil) become visible; such flowers are called chasmogamous. In plants such as *Commelina benghalensis* and species of *Viola*, a proportion of flowers remain closed; such flowers are termed cleistogamous (Lord 1981).

Many species produce both chasmogamous and cleistogamous flowers depending on environmental conditions, especially temperature and light duration. For example, *Ruellia* (Acanthaceae) produces chasmogamous flowers during summer and cleistogamous flowers during winter under Delhi (India) conditions. In

Table 7.1 Pollination types

- *Autogamy*: Transfer of pollen grains from the anther to the stigma of the *same* flower
- *Geitonogamy*: Transfer of pollen grains from the anther to the stigma of *another* flower of the *same plant* or *another plant* of the same clone (genet)
- *Xenogamy*: Transfer of pollen grains from the anther to the stigma of a *different plant* (not of clonal material)
- *Allogamy*: Transfer of pollen grains from the anther to the stigma of *another* flower of the *same* or *another* plant (includes both geitonogamy and xenogamy)

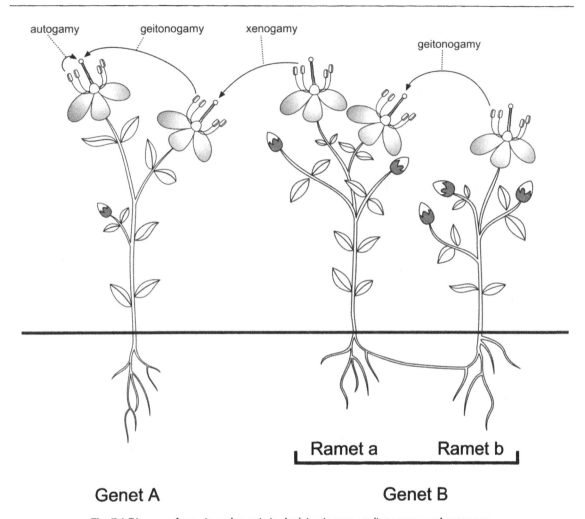

autogamy geitonogamy xenogamy geitonogamy

Ramet a Ramet b

Genet A   Genet B

Fig. 7.1 Diagram of genets and ramets to depict autogamy, geitonogamy and xenogamy.

cleistogamous flowers, self-pollination is the norm, whereas chasmogamous flowers may be predominantly self-pollinated or cross-pollinated. Some species bearing chasmogamous flowers such as *Utricularia* (Khosla et al. 1998) and *Polypleurum* (Khosla et al. 2000) also show obligate self-pollination. Pollen grains germinate in situ inside the anthers before dehiscence. Because of the physical contact between the stigma and the anthers, dehiscence of anthers results in deposition of in-situ germinated pollen mass on the stigma before opening of the flower. This is often referred to as pre-anthesis cleistogamy (Lord 1981). A unique method of self-fertilization, termed 'internal geitonogamy' has been reported in some members of Malpighiaceae (Anderson 1980) and Callitrichaceae (Philbrick 1984, Philbrick and Anderson 1992); (details are discussed in Chapter 8).

Cross-pollination results in genetic heterogeneity and thus cross-pollinated species show wider adaptations. Cross-pollination involves considerable wastage of resources because of its uncertainty and thus cross-pollinated species have to produce much more pollen to compensate the wastage.

BREEDING SYSTEMS

A breeding system is defined as all aspects of sex expression that affect the relative genetic contributions to the next generation of individuals within a species (Wyatt 1983). The system includes pollination mechanisms and pollen movements. Flowering plants are highly variable in the way in which they express sexuality (sexual systems). Traditionally, breeding systems have been treated in relation to the mechanism(s) which promote(s) or reduce(s) outcrossing (Dafni 1992).

Outbreeding Devices

1. Dichogamy: In dichogamous species, anther dehiscence and stigma receptivity are temporally separated. The anthers and the stigma mature at different periods; the time gap between the two may vary from one day to many days. In some species (e.g. members of Asteraceae, Campanulaceae and Lamiaceae), the anthers dehisce before the stigma becomes receptive (protandry) and thus autogamy is prevented. In certain other species, such as *Magnolia* and *Mirabilis* the stigma becomes receptive before the anthers dehisce (protogyny) (for further details see Lloyd and Webb 1986, Bertin and Newman 1993).

2. Herkogamy: Herkogamous species show spatial separation of the anthers and the stigma. Their relative position is such that autogamy cannot occur (Webb and Lloyd 1986). The stigma often projects beyond the level of the anthers, and therefore the pollen of the same flower cannot land on the stigma. In many legumes, the stigma is surrounded by a tuft of trichomes which prevent the pollen of the same flower from reaching the stigma.

3. Self-incompatibility: In many species, self-pollinations do not result in fertilization. This is because pollen germination on the stigma or the growth of pollen tubes in the stigma or style is inhibited. For effective fertilization, pollen has to come from another plant. Self-incompatibility (SI) is genetically controlled and is widespread in flowering plants (de Nettancourt 1977). Details of self-incompatibility are discussed in Chapter 9.

4. Dicliny: In diclinous species, the flowers are unisexual. In some species such as *Cucurbita*, male and female flowers are borne on the same plant (monoecious), in others (e.g. papaya, mulberry, *Cannabis*) male and female flowers are borne on different plants (dioecious).

Most of the outbreeding devices explained above, although effective in preventing autogamy, cannot prevent geitonogamy (i.e., pollination involving pollen from other flowers of the same plant). Only self-incompatibility and the dioecious form of dicliny prevent both auto- and geitonogamy. Further, in most of the species which bear bisexual flowers (except those that are self-incompatible), generally a combination of self- and cross-pollinations occurs to different degrees; often a species can be predominantly an inbreeder or an outbreeder (Richards 1986, Koul et al. 1993). Some species exhibit more than one outbreeding device. *Anthocercis gracilis*, for example, shows protogyny as well as self-incompatibility (Stace 1995).

Secondary Pollen Presentation

Pollen is generally presented in situ on the dehisced anthers for transfer by pollinating agents. However, in several species, following anther dehiscence, pollen is deposited on a specific floral structure, termed the pollen presenter, from which pollen is distributed for cross-pollination (Howell et al. 1993, Ladd 1994). Secondary pollen presentation is defined as 'the location of pollen for biotic vectors on flower organs other than anthers'.

Secondary pollen presentation systems are diverse in structure and origin and have been reported in 5 monocot and 20 dicot families (Howell et al. 1993, Ladd 1994). In most species, pollen presenters develop from stylar tissue. Most of the taxa with a secondary pollen presentation mechanism avoid self-pollination either by protandry or by preventing self-pollen from coming into contact with the stigma because of the spatial separation of the stigma, and pollen presenters. In protandrous species the stigma becomes receptive one to several days after anthesis, by which time most of the pollen from pollen presenters has been moved by pollinators (see Howell et al. 1993, Ladd 1994).

Pollen:Ovule Ratio

The number of pollen grains produced for each ovule of a flower has been reported to reflect the breeding system of the species (Cruden 1977). Pollen:ovule ratio gives an estimate of the outcrossing level (Table 7.2).

Table 7.2 Pollen:ovule ratio in various breeding systems (based on Cruden 1977)

Breeding system	Pollen-ovule ratio (range)
Cleistogamy	2.7–5.4
Obligate autogamy	8.1–39
Facultative autogamy	31.9–396
Facultative xenogamy	244.7–2588
Obligate xenogamy	2108–195,525

MODES OF POLLINATION

Basically three pollinating agents (two abiotic, wind and water, and one biotic, i.e., many species of animals) are involved in effecting pollination (Frankel and Galun 1977, Faegri and van der Pijl 1979, Real 1983, Dafni 1992, Endress 1994). Insects are considered the original pollinators of early flowering plants. Those species which use non-insect pollinators, are believed to be derived (Endress 1994).

Anemophily

Wind is the major of the two abiotic pollinating agents (Whitehead 1983). Wind pollination, termed anemophily, is prevalent in dry environment—open grasslands, savannahs, semi-arid areas and zones above the timber-line. Members of Poaceae (cereals, millets and bamboos), Cyperaceae, Juncaceae, amaranths and spinach are predominantly anemophilous.

Anemophilous flowers are generally small, inconspicuous, emit no scent and do not produce nectar. The perianth is reduced because visual attractants are not required. Pollen-producing flowers are often clustered as catkins in anemophilous species. Anemophilous flowers generally have well-exposed stamens; the anthers are generally suspended from long filaments and hang freely from the flowers. In wind-pollinated flowers the stigmas are generally dry, have a large surface, and are often feathery, which helps them to effectively intercept windborne pollen. The number of ovules per flower is fewer and often solitary. Pollination in anemophilous plants is a chance factor and thus inefficient compared to that in zoophilous plants.

Anemophilous plants produce an enormous amount of pollen; this is achieved by an increase in size of anthers or in number of stamens per flower, and in species which bear unisexual flowers, the number of male flowers per plant is increased. Pollen grains are light and non-sticky, and have a high surface-to-volume ratio. The distance which the pollen has to travel to reach a stigma depends on pollen longevity and prevailing wind currents (Kevan 1997). Prevalence of insect pollination in fossil records of early angiosperms, lack of wind pollination in primitive species, and the occurrence of wind pollination in several more specialized groups of angiosperms support the view that wind pollination is secondarily derived (Endress 1994).

Hydrophily

In hydrophilous pollination, water acts as the vector for pollen transport. Hydrophily is limited to 31 genera of 11 families, largely monocotyledons (McConchie 1983). Although the majority of aquatic flowering plants bear flowers above the water level and are hence insect- or wind-pollinated, marine angiosperms are exclusively hydrophilous.

Hydrophilous species share some traits with anemophilous species because both wind and water are wasteful abiotic agents. Both anemophilous and hydrophilous species generally bear less conspicuous unisexual flowers with reduced perianth and produce abundant pollen and a solitary ovule per flower but as the stigma is large, rigid and often branched, this increases the surface of the receptive area for pollen.

Hydrophily is of two types—ephydrophily and hyphydrophily. In ephydrophilous plants, pollination takes place on the water surface and thus pollen movement is two dimensional. In hyphydrophily, pollination takes place below the water surface and hence the pollen moves in three dimensions. Hyphydrophily is often referred to as submarine pollination (Cox 1988, 1993; Cox and Knox 1988, Cox and Humphries 1993). Hyphydrophily is reported in only 18 submerged genera, 17 of which are monocots and 12 marine (Dafni 1992, Mohan Ram and Khosla 1998).

Many ephydrophilous species such as *Vallisneria,* release male flowers on the surface of the water which glide with the breeze. Two sterile anthers function as tiny sails and two fertile anthers hold the spherical pollen above the water surface. The floating female flower anchored by a long pedicel creates a slight depression on the water surface. As the male flower reaches the female flower it falls into this depression caused by the latter and thus pollen comes into contact with the stigma; pollen grains do not come into contact with water. Similarly, in *Hydrilla* and *Enhalus* also, pollen grains land on the stigma without coming into contact with water.

In several other ephydrophilous taxa the pollen grains are released on the water surface where they form extended rafts ('search vehicles', Cox 1993). The search vehicles are transported on the water surface and may be either filiform pollen (*Thalassodendron*) or pollen contained in mucilaginous tubes or mats (*Halophila, Thalassia, Elodia, Amphibolis* and *Ruppia*). Pollination is effected when these search vehicles come into contact with filamentous stigma floating on the water surface. Dispersal of pollen grains in groups seems to increase the possibility of their coming into contact with the target (stigma) in two-dimensional movements (Cox 1993).

In hyphydrophilous species such as *Phyllospadix* and *Zostera*, pollination takes place below the water surface and thus requires greater modification and adaptation of reproductive structures for pollen release, transport and capture in water (Cox 1988, 1993, Ackerman 1997a, b). Pollen grains in several hyphydrophilous species are long and filamentous; if spherical, they are generally embedded in elongated strands of mucilage. Longer pollen grains seem to be more efficient in 3-dimensional search of the stigma than the spherical pollen (Cox 1993).

Zoophily

The number of species pollinated by animals far exceeds that pollinated by wind and water and the adaptations that zoophilous plants show are also very diverse (McGregor 1976, Richards 1986, Roubik 1995). Zoophilous species develop effective floral devices to attract pollinating agents. Floral attractants, rewards and flower modifications largely determine the class of animal species able to visit the flowers, and the degree of species specificity (Faegri and van der Pijl 1979).

Among the animals, insects are the most important pollinators (entomophily); of these the most common insects are the bees (melittophily) (Fig. 7.2A), beetles (cantharophily), moths (phalaenophily) and butterflies (psychophily). In many instances pollinators transport pollen on particular parts of their body (Fig. 7.2B, C); they are basically of three types: nototribic (on their back), sternotribic (on the underside), and pleurotribic (on the flanks).

Bees are the prime pollinators; they are involved in pollination of most field and orchard crops. Social bees are especially versatile as they are able to exploit a broad range of different flower forms. Bees live on nectar; they

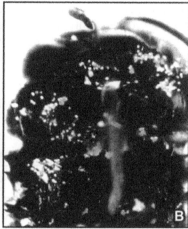
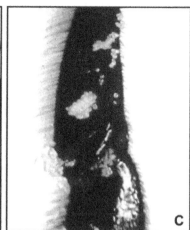

Fig. 7.2 A. Asian honey-bee (*Apis dorsata*) amid flowers of *Acacia senegal*. B, C.Ventral (abaxial) view of thorax (B) and tibia (C) of *A. dorsata* following its visit to flowers to show a large number of pollen grains on the body parts (A—after Tandon 1997, B, C—after Sunnichan 1998).

collect pollen also to feed their larvae. Bees show great variation in body size and length of proboscis. Bees are good at recognizing colours, scents and contours; unlike humans, bees can perceive ultraviolet light, but they cannot visualize shades of red, which appear black to them. Bee-pollinated flowers are showy, brightly coloured (mostly yellow) and often exhibit distinctive markings on the petals (honey guides); honey guides help the bees reach the source of nectar easily. Both nectar and pollen serve as rewards for the bees. The nectary in bee-pollinated flowers is usually situated at the base of the corolla tube and is therefore accessible only to those bee species with a tongue of the right length. Pollen adheres to the bristles of the bees or the bees pack it in special baskets on their legs.

Flowers pollinated by butterflies (*Oenothera, Amaryllis, Geranium*) are similar to bee-pollinated flowers. They attract insects by a combination of sight and scent as in bee-pollinated flowers. Butterflies are active during the day and have good vision but a weak sense of smell. Moths are generally nocturnal and have a good sense of smell. Moth-pollinated flowers are typically white or pale and mostly open and exude

odour at night. The nectar in butterfly- and moth-pollinated flowers is found at the base of a long slender corolla tube and is thus accessible to only butterflies and moths.

Flowers of many plant species emit unpleasant odours which serve as chemical attractants for insects within the order Coleoptera and Diptera. These odours are generally attributed to the presence of amine-containing compounds. Many beetles and flies are attracted to the flowers of those species emitting foul and foetid odours. Species of *Magnolia*, lily and wild rose are examples of beetle-pollinated species (Bernhardt 2000). In beetles, the olfactory sense is much better developed than the sense of sight. Thus, beetle-pollinated flowers are often white or pale and emit strong odours attractive to beetles but rather unpleasant to humans and distinctly different from the pleasant odours emitted by flowers pollinated by bees and butterflies. Beetles derive most of their nourishment from fruits and saps; often they feed on floral parts, especially the petals.

Although bees, beetles, moths and butterflies are the major pollinating insects, many other insects in particular thrips (Mathur and Mohan Ram 1986, Ananthakrishnan 1993), mosquitoes and some flies also effect pollination in some plant species. Pollen is the major food source for thrips, which also use flowers as breeding places. Nectar may not be present in thrips-pollinated flowers.

Birds and bats (Stiles 1976, Subramanya and Radhamani 1993) likewise bring about pollination in several floral species, particularly in the tropics. Many of the bird-pollinated species belong to Proteaceae, Myrtaceae, Leguminosae, Bignoniaceae, Verbenaceae, and Loranthaceae. Some well-known examples of bird-pollinated flowers are species of *Butea, Bombax, Spathodea, Fuchsia* and *Dendrophthoe*. Bird-pollinated flowers are generally bright coloured; reds are the predominant colours and this could well be a device to exclude insect visitors (bees and many other insects cannot see red) that would act as pollen robbers. Nectar is the main reward for the bird pollinator; the flowers produce copious nectar, generally of low viscosity and low amino acid content. Bird-pollinated flowers are scentless and generally have a long tubular corolla; birds with long beaks and tongues easily reach the source of nectar. The floral organs are generally robust. While feeding on the nectar, the anthers and stigma touch the nape of the bird and pollination is thereby achieved (Fig. 7.3). Birds are attracted largely through colour since their sense of smell is not strongly developed. Sunbirds and honey-eaters are

Fig. 7.3 Diagram of a male purple sunbird foraging the flower of *Butea monosperma* for nectar. The anthers and stigma touch the nape of the bird, which brings about pollination (after Tandon 1997).

the principal pollinators in the Old World, and humming-birds in the New World. Humming-birds usually hover during their floral visits and do not need a perch (landing platform). However, birds of the Old World need a perch (Endress 1994).

Bat pollination (chiropterophily) is an exclusive tropical phenomenon. About 60 families of flowering plants are reported to contain bat-pollinated species (Endress 1994). Bats are nocturnally active; in bat-pollinated species anthesis generally occurs during the night and the flowers emit a strong odour. Bats are attracted to flowers largely through the sense of smell. Bat-pollinated flowers are similar to bird-pollinated flowers in many features; they produce mucilaginous nectar and often copious pollen. Bat-pollinated plants are generally trees; some of the well-known examples are: *Kigelia, Oroxylum, Mucuna* and *Adansonia*. While the bat feeds on nectar, pollen grains adhere to its face and get transferred to the stigmas of other flowers visited.

Many species have recently been shown to be pollinated by non-flying mammals, in particular marsupials, rodents and primates (Kress 1993, Endress 1994). Such flowers bear inflorescences near the ground and floral organs are highly lignified and produce copious sucrose-rich nectar. Some examples of mammal-pollinated species are: *Symphonia* (Symphoniaceae), *Ravenalia* (Strelitziaceae), *Parkia* (Mimosaceae), and *Dryandra, Banksia* and *Protea* (Proteaceae). Kress (1993) has presented evidence for pollination of *Ravenalia* by lemurs (a primate) in Madagascar. In a recent study even giraffes have been implicated in pollinations of *Acacia* (du Toit 1992). In *Butea monosperma*, apart from bird, squirrel also brings about pollination (Dr Rajesh Tandon, unpublished)

An unusual pollination mechanism through a bird's tongue has been reported in a member of Asclepiadaceae, *Microloma sagittaria*, from South Africa (Pauw 1998). The flowers of this species exhibit many features such as odourless red flowers and copious quantities of dilute nectar, suggestive of bird pollination. The flowers do not open except for five tiny slit-like pores. Sunbirds visit these flowers regularly to feed on the nectar, probing the slits with the tip of their bill and extending their tongue to the bottom of the flower where the nectar is located. During this process pollinia get attached to the bird's tongue. They are carried inside the bird's mouth to the next flower where they are mechanically detached during feeding and pollination thereby effected.

EVOLUTIONARY SIGNIFICANCE OF INSECT POLLINATION

The evolutionary success of angiosperms has been attributed to the evolution of pollination by insects. It is thought that in the early stages, the insects probably feeding on leaves or other secretions of the plants would have come into contact with the protein-rich pollen grains. The insects would have started visiting more frequently for pollen and thus bringing about cross-pollination inadvertently. The more attractive the devices plants developed to lure insects, the more frequent would have been their visits, resulting in more effective pollination of their flowers, leading to the production of a larger number of vigorous seeds. Thus, any change in the structure of the flower that made insect visits more frequent or more efficient in bringing about pollination would offer an immediate selective advantage. A wide range of diversification and specialization is seen in the present-day flowering plants for attracting pollinators (through colour, scent, nectar, etc.) and in positioning the anthers and stigma for effective pollination.

Flowers of many species attract a wider variety of pollinators. A disadvantage for such species is that the pollen may not land on the right stigma when the pollinator subsequently visits flower of a different species. There are a number of instances in which the plant species and specific pollinators are totally dependent on each other as a result of co-evolution (the mutual evolutionary influence between the plant and its animal pollinator); each species involved serves as a source of natural selective pressure on the other. Co-evolution has resulted in the development of flower adaptation to such an extent that the flowers of a particular species can be visited by only one specialized visitor.

The number of species whose flowers permit visits by several effective pollinators (polyphilic) is much higher than those adapted to a very narrow group of pollinators (oligophilic). The number of species adapted to a single pollinator (monophilic) is very low (Faegri and van der Pijl 1979). The relationship between figs (*Ficus* spp.) and their species-specific pollinators (wasps) (see Patel et al. 1993) is one such example of co-evolution.

The flowers in figs are borne in urn-shaped inflorescences called syconia. Each syconium opens through a small opening at the tip, the ostiole. The syconium bears a mixture of male and female flowers. All the syconia on a tree are synchronized. When the female flowers on the syconium reach receptivity, they emit volatile chemicals which attract pollinators. The species-specific female wasp enters through the ostiole and pollinates female flowers; it lays eggs in a proportion of female flowers and dies inside the syconium. During the next few weeks the wasp larvae develop in the ovules of oviposited flowers, feed on the ovules and develop into mature wasps. The male wasps inseminate the females. Female flowers not used for oviposition develop seeds. By this time the male flowers of the syconium mature and release pollen. Female wasps whose lifespan extends 1–3 days get loaded with pollen and exit the syconium through a hole made by the males. Wingless male wasps die in the syconium. The female wasps loaded with pollen fly off in search of other receptive syconia to oviposit. Individual fig trees generally flower synchronously but show population level asynchrony in such a way that a few trees releasing the wasps and a few trees receptive to wasps are found all the year round so that species-specific wasp populations are maintained.

The loss of a certain percentage of pistils to ovipositing wasps is a price the plant must pay to ensure pollination.

The mechanism by which only a proportion of female flowers is used for oviposition, leaving the remainder for seed development is not yet clear. It was believed that dimorphism in style length of female flowers apportion the flowers for seed and wasp production. According to early investigators, monoecious figs bear two types of female flowers—short-styled used for oviposition and long-styled for seed production. Since the ovipositor of wasps does not reach the ovary of long-styled flowers, wasps can lay their eggs only in short-styled flowers and the long-styled flowers produce seeds (Galil and Eisikowitch 1968). However, recent studies of Ganeshaiah and his associates (Kathuria et al. 1995) have not supported this concept; they did not find such a bimodal distribution of female flowers for style length in any of the seven monoecious species they investigated.

FLORAL ATTRACTANTS AND REWARDS

Zoophilous plants have to fulfil the following requirements to attract pollinators:
a) advertise the presence of a reward
b) supply some reward, usually food
c) position the anthers and stigma so that they come into contact with the body of the pollinator to facilitate transfer of pollen.

The attractants/advertisements are largely visual and olfactory. Visual advertisements include flower colour, size and shape. Colour differences between flowers allow pollinators to discriminate between species and varieties. Flowers of many species show a colour pattern (nectar guides) in UV range which directs the pollinators to the floral reward in such a way that pollination is assured.

The colour of flowers in many species changes in response to pollination. For example, in *Lantana camara* yellow flowers will change to red after pollination (Mathur and Mohan Ram 1978). Such colour changes are used as cues by pollinators in deciding whether to visit or ignore the flower. In *Lantana*, thrips (Mathur and Mohan Ram 1986) and butterflies (Weiss 1991) forage on yellow flowers but not on red ones devoid of nectar. Flower size and form also contribute to the visibility of flowers. The larger the flower or group of flowers arranged in an inflorescence, the greater the distance from which it can be detected.

Fragrance is the other most important floral attractant. Flower scents are largely volatile derivatives of alcohols, esters, aldehydes and ketones. Some floral scents are distinctive and some are mimetic with the smell of dung or mammal musk or insect pheromones. In the orchid *Ophrys,* the shape and colour of the flower and the scent mimic the female insect and its pheromone respectively, and induce pseudocopulation of the male bee or wasp with the flower during which pollination occurs. For bees, colour is the main long-range attractant. Olfactory stimuli become increasingly important at closer range.

Many studies have clearly shown that pollen grains emit odours that differ from those of other flower parts and the pollen of other species (Dobson 1988, Dobson et al. 1996). Pollenkitt is the main source of volatile substances and plays a role in guiding pollen-foraging insects to flowers (Dobson 1988). Insects are able to discriminate between odours of different pollen (Schmidt 1982). Pollen odour is used by pollen-foraging insects not only to discriminate between plant species, but also to assess availability of pollen between individual flowers; this allows the pollinator to restrict its visit to the rewarding flowers (see Dobson et al. 1996). The adaptive advantage of pollen odour seems to be related to the conflicting pressures of protecting pollen from overexploitation by other insects while concomitantly advertising the availability of pollen as reward.

Floral rewards cater to the essential need of the pollinator to ensure repeated visitation. Pollen and nectar are the main nutritive rewards.

Pollen is highly nutritious with ca 25% carbohydrates, 25% proteins, 10% amino acids and 5% lipids. It is also rich in vitamins and minerals (Schmidt and Buchmann 1992). Pollen grains of zoophilous spp. are generally ornamented with an oily coating. In some species they are held in clumps by viscin threads (e.g. Onagraceae), in yet others they are released as polyads (Ericaceae, Onagraceae, Mimosoideae) or in complex pollinia (Asclepiadaceae, Orchidaceae).

Starch and oil are the major calorific reserves of pollen; smaller pollen tends to have oil and larger pollen starch (see Chapter 2). Pollen rich in starch tends to have a low lipid content and is of lesser value for insects as a food source (Baker and Baker 1979). Plants which offer pollen as the main or only source of energy tend to have oil-rich pollen. Bees gather pollen in special parts on their bodies, the pollen baskets. The pollen baskets of honey-bees and bumble-bees are on the hind legs, but on leaf-cutting bees under the abdomen. The insects carry the harvested pollen to their nests.

Nectar is largely a sugary solution and includes a minor proportion of amino acids, organic acids and minerals. Nectar is the main fuel for movement of anthophilous animals. Sucrose, glucose and fructose are the major sugars of nectar. The volume of nectar and concentration of sugars, sugars ratios and amino acid content are relevant for pollinators.

The reward is not confined only to nutrients. Non-nutritive rewards include nest materials, shelter and warm resting places, sexual attractants, mating and ovipositing sites (see Patel et al. 1993, Kathuria et al. 1995, Seymour and Schultze-Motel 1996). There are many insects, e.g. beetles, which chew petals and/or ovarian tissues of flowers. Such flowers tend to have fleshy floral parts.

Floral Thermogenicity

The flowers of many species belonging to class Arecidae (Araceae, Aracaceae, Cyclantheraceae) and Magnoliidae (Annonacae,

Magnoliaceae, Aristolachiaceae, Nelumbaceae and Nympheaceae) (Endress 1994, Dieringer et al. 1999, Seymour and Schultze-Motel 1996, 1997) produce heat in the inflorescence/flower at the time of flower opening. Thermogenic flowers are always protogynous and the peak of heat production coincides with the receptivity of female flowers. Often the difference between ambient temperature and the temperature around the flowers is over 15°C. This heat production is believed to be associated with a large increase in the cyanide-insensitive non-phosphorylating electron transport pathway that is unique to plant mitochondria (Meeuse 1975). In *Sauromatum guttatum,* one of the arum lilies, this heat production is regulated by salicilic acid (Raskin et al. 1987). Thermogenic members are pollinated either by flies or scarab beetles. The heat facilitates volatilization of amines and indoles (higher boiling point compounds) whose putrescent odour attracts insect pollinators (Smith and Meeuse 1966, Uemura et al. 1993).

POLLEN TRAVEL AND GENE FLOW

The movement of alleles physically through space is called gene flow. The distance for which pollen travels from its source before it lands on the stigma is frequently used as a measure of gene flow. Studies on pollen flow and related aspects are important not only in understanding fundamental aspects of pollen biology, but also in applied areas concerned with contamination of hybrid seed production plots with unwanted pollen and escape of engineered genes from transgenic plants.

Escape of engineered genes from transgenics into other cultivars and related species through pollen is one of the major risks in commercial cultivation of transgenics (Bhatia and Mitra 1998). One of the basic requirements for commercial release of transgenics is to provide all relevant data on possible escape of engineered genes through pollen to the regulatory authorities. The extent of pollen flow depends on a number of factors, some intrinsic to the pollen of a given species and others on the prevailing environmental conditions.

The following are some of the basic studies required to develop effective strategies to prevent gene escape:

- Spatial and temporal distribution of other cultivars and related species in the area.
- Duration of pollen viability of transgenic plants.
- Duration of stigma receptivity of other cultivars and related species growing in the area.
- Distance of pollen travel of transgenic plants through biotic and abiotic means.
- Compatibility relationships between the transgenic and other cultivars/species.
- Possibility of hybrid seed development, hybrid seedling establishment and formation of backcross seeds with other cultivars/species.

Many of the aforesaid aspects, such as pollen viability and stigma receptivity are straightforward and can be determined accurately using simple and dependable techniques (see Chapters 3 and 8). Accurate determination of pollen travel is more difficult. The extent of pollen flow depends on the kind of pollinators and on the densities of plants. In anemophilous plants pollen travels in the air. Extensive studies have been carried out on the biometeorological and biophysical aspects of pollen dispersal (Niklas 1985, Di-Giovanni and Kevan 1991). Pollen grains particularly of anemophilous species travel for long distances even up to the polar regions. Pollen grains of many species belonging to North-American flora around Great Lakes have been isolated among mosses from North-West Greenland (Linskens 1995). Pollen grains have likewise been recovered from moss samples collected from Antarctica, which is over 2000 km from the nearest landmass with angiosperm plants (Linskens et al. 1991). In zoophilous species, pollen movement depends on the movement of pollinators. Movements of a pollinator between flowers is influenced by the amount of reward present (Richards 1986).

Several techniques are available to estimate the distance of pollen travel (for details see Dafni 1992, Kearns and Inouye 1993). The following are some of the more important ones:

1. Vital staining of pollen: Pollen grains of a tagged plant are labelled with a vital stain such as neutral red, Bismarck blue, orange G and Evans blue by injecting 1–2 µl of the stain into an about-to-dehisce anther. Alternately pollen grains can be labelled by spraying the stain on dry pollen. Movement of labelled pollen grains is traced by examining the stigmas of the surrounding plants in the population under a microscope.

2. Fluorescent powdered dyes as pollen mimics: A fluorescent dye (such as Hellecone, Saturn yellow and Arc yellow, available commercially; see Kearns and Inouye 1993) is applied to dehiscing anthers of a labelled plant with an automizer. Dye movements are traced by examining the stigmas of the surrounding plants under a fluorescence microscope.

3. Radioactive labelling of pollen: Labelled $^{14}CO_2$ fed to leaves of a tagged plant and allowed to enter pollen grains through metabolic pathways. Pollen movement is followed by collecting stigmas of the surrounding plants and checking for the presence of the label using a scintillation counter.

4. Genetic markers: Use of phenotypic or molecular markers to identify the progeny sired by the pollen of a marked plant. This is the most authentic technique for studying gene flow. Of the various markers, the phenotypic are the simplest and most convenient.

Often the actual gene flow may depart from pollen flow because of the influence of incompatibility, gamete competition and pollen viability. Studies related to compatibility and possible gene flow through hybridization require more elaborate experimentation, extending for several generations.

POLLINATION POSTULATES

To confirm the role of a vector in pollination, the following four postulates need to be demonstrated (Cox and Knox 1988):

1. Pollen transfer from anther to vector
2. Pollen transport by vector
3. Pollen transfer from vector to stigma
4. Fertilization from vector-deposited pollen.

Dafni (1992) has added the following four additional postulates to pollination ecology:

5. Flower advertisements perceived and used by pollinators
6. Flower reward consumed/used by vector as an integral part of the pollination process
7. Relative contribution of pollen and ovules to the next generation as a result of pollination
8. Interrelationships between different vectors involved in pollination.

Effective techniques for demonstrating the above postulates are detailed in Dafni (1992) and Kearns and Inouye (1993).

POLLINATION EFFICIENCY

The term pollination efficiency has been used ambiguously by pollination biologists (Dafni 1992, Inouye et al. 1994, Davis 1997). Many studies evaluate pollination efficiency on the basis of quantitative aspects of pollination events, such as number of pollen grains deposited on the stigma after a single visit, number of pollen grains deposited per stigma as a fraction of the number of pollen grains removed (Snow and Roubik 1987), and proportion of stigmas touched per insect visit (Robinson 1979). Many other studies consider pollination efficiency on the basis of per cent fruit- and seed-set as a consequence of pollination (for details see Dafni 1992, Davis 1997).

8

Pollen-Pistil Interaction and Fertilization

Pollen-pistil interaction involves a series of dialogues between the male gametophyte and the sporophytic tissues of the stigma and style. These interactions result in generation of appropriate signals which elicit the required responses in the pollen and/or pistil. Figure 8.1 presents the major events involved during pollen-pistil interaction. Only a beginning has been made to understand some of the details of events and the components involved in these interactions.

Successful completion of pollen-pistil interaction is an essential requirement for fertilization and seed-set. Any deviation in these sequential events prevents fertilization and consequently fruit- and seed-set. Early investigations of fertilization and seed development took pollen-pistil interaction for granted; there were hardly any studies on these aspects. However, during the last three decades there has been an increasing realization of the need for understanding the details of pollen-pistil interaction for its effective manipulation in crop production and improvement. This realization has resulted in extensive as well as intensive studies on the structural and functional aspects of pollen-pistil interaction. The topic has been covered periodically by many reviews (Shivanna and Johri 1985, Knox et al. 1986, Cresti et al. 1992, Russell and Dumas 1992, Dumas and Mogensen 1993, Reiser and Fischer 1993, Russell 1993, Shivanna and

Sawhney 1997, Raghavan 1999, de Graaf et al. 2001).

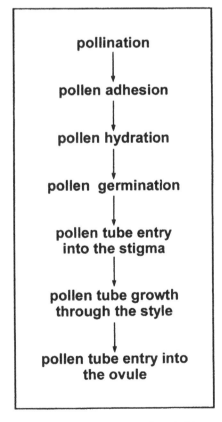

Fig. 8.1 Sequential events during pollen-pistil interaction.

SIGNIFICANCE OF POLLEN-PISTIL INTERACTION

One of the prerequisites for sexual reproduction in any organism is the ability of the gametes to establish recognition so as to facilitate fusion of only the right type of gametes. In lower groups of plants and animals, the male gametes, which are motile and released into the surrounding medium, come into direct contact with the egg cell; the gametes are directly involved in the recognition event (Fig. 8.2). In seed plants the gametes are not released into the surrounding medium but pollen grains act as vehicles for gamete transmission. In gymnosperms pollen grains are deposited in the pollen chamber and a short pollen tube is produced before the male gametes are released (Owens et al. 1998). In zooidogamous gymnosperms (cycads and *Ginkgo*) pollen tubes are not directly involved in the transfer of sperms to the egg; they function as haustorial organs to acquire nutrients from the surrounding sporophytic tissues of the ovule (Johri 1992). Members of Conifereae and Gnetales are siphonogamous; the pollen tubes undergo oriented growth, although for a short distance, towards the archegonia (the seat of the egg), and non-motile sperms are released (Friedman 1993). It is not yet known whether pollen recognition in gymnosperms is

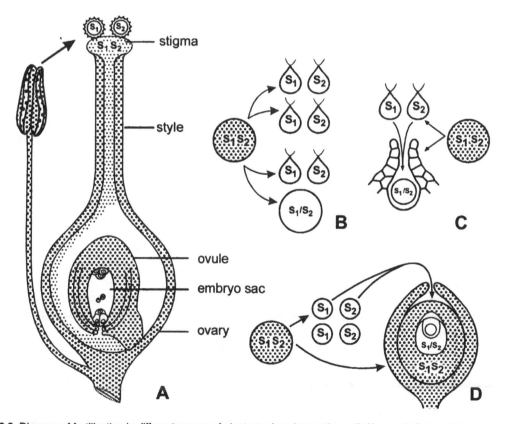

Fig. 8.2. Diagram of fertilization in different groups of plants to show interacting cells/tissues before and during fertilization, and efficacy of self-incompatibility (stippled areas represent sporophytic cell/tissue and open areas gametophytic cell/ tissue). Self-incompatible alleles, S_1 and S_2, are used in all groups of plants for uniformity and convenience. A. Flowering plants, B. Thallophytes, C. Bryophytes and Pteridophytes, D. Gymnosperms (redrawn from Shivanna and Johri 1985).

established at the level of gametes or during pollen tube interaction with a few-layered sporophytic tissue, the nucellus that lies between the pollen chamber and the archegonia.

In flowering plants, pollen grains are deposited on the stigma and pollen tubes carry the male gametes along the tissues of the stigma and style before releasing them near the egg. Thus in flowering plants, gamete recognition is far more elaborate and takes place not at the level of gametes, but during pollen-pistil interaction (Shivanna, 1979, 1982, Shivanna and Johri 1985). Only the right type of pollen is permitted to complete all post-pollination events while others are inhibited during germination or tube growth. This screening of pollen takes place for (i) compatibility and (ii) quality.

Screening for Compatibility

Unlike animals, which are mostly unisexual, a majority of plants are bisexual and there is ample scope for male and female gametes of the same plant to come together and fuse. To prevent such inbreeding almost all groups of plants have evolved outbreeding devices. Self-incompatibility (SI) is one such device (Whitehouse 1950). SI is controlled by one or a few genes each with two or multiple alleles (for details see Chapter 9). When the SI allele present in the male gamete matches with that of the female gamete, fusion is prevented (Fig. 8.2). In all lower groups of plants where SI recognition takes place at the level of gametes, SI cannot completely prevent the fusion of gametes produced by the same sporophyte. For example, S_1 and S_2 gametes produced by S_1S_2 sporophyte can fuse, although fusion among S_1 or S_2 gametes is prevented (Fig. 8.2) (Shivanna and Johri 1985). In flowering plants in which SI recognition takes place between the pollen and the pistil (sporophytic tissue) both S_1 and S_2 pollen are prevented as the pistil has S_1 as well as S_2 alleles (Fig. 8.2). In flowering plants, SI thus prevents inbreeding completely. Pollen grains of other species and genera (interspecific incompatibility) are also prevented in the pistil when the two parental species show prefertilization barriers.

Screening for Quality

Generally, the number of pollen grains deposited on the stigma is many times more than the number of ovules available for fertilization. Pollen grains are therefore subjected to competition and selection during pollen-pistil interaction (Mulcahy 1979, 1984). Those pollen grains which germinate early, and pollen tubes which grow faster in the style, enter the ovules and effect fertilization. Pollen grains which germinate later and those in which pollen tube growth is slow, are eliminated in this competition. This is often referred to as sexual selection (Tejaswini et al. 2001). As outcrossing takes place to a limited or greater extent in the majority of flowering plants, there is considerable genetic variability in a pollen population (in terms of speed of germination and rate of pollen tube growth) on which pollen competition and selection can operate. There is no scope for such competition among male gametes in lower plants. Any gamete which succeeds in establishing the first contact with the female gamete, is likely to achieve fertilization.

The basic requirement for operation of pollen competition and selection during pollen-pistil interaction is the expression of gametophytic genome of the pollen and control of pollen germination and pollen tube growth in the pistil by the genome of the pollen. Many studies have clearly demonstrated that (i) a large part of the gametophytic genome is expressed during pollen germination and pollen tube growth (Mascarenhas 1993) (see Chapter 1), and (ii) the genotype of the pollen determines, at least in part, the speed of germination and rate of pollen tube growth (Ottaviano and Mulcahy 1989, Hormaza and Herrero 1992, Sari-Gorla and Frova 1997).

The stigma in some taxa has developed devices to delay germination of pollen until a sufficient number of pollen grains land on the stigma.

Leucaena leucocephala (Ganeshaiah et al., 1986; Ganeshaiah and Shaanker, 1988) for example produces pods with more than 7–9 seeds; pods containing less than 7 seeds are not formed. This selective development of many-seeded pods is regulated by inhibition of pollen germination when the number of pollen grains on the stigma is less than a critical number (20–25). The inhibition is caused by a proteinaceous, pH-dependent inhibitor present in the stigma. Only when the pollen grain number exceeds the critical number, does the stigmatic pH increase and inactivate the inhibitor. At the critical threshold level (20–25 pollen grains) only 8–10 ovules are fertilized; thus 2 or 3 pollen grains compete to fertilize an ovule. This adaptation ensures competition among the pollen grains for fertilization. Adaptation of such number-dependent pollen germination may be prevalent in a wide range of multiovulate systems.

The adaptive significance of such pollen competition (sexual selection) in increasing the fitness of the progeny is well documented. Studies have been carried out by experimentally manipulating pollen competition (Tejaswini et al. 2001). For example in a few species the competition has been manipulated by altering the density of pollen on the stigma. The progeny derived from conditions of intense pollen competition (high-density pollination) are more vigorous and exhibit reduced genetic variability than those derived from conditions of little or no pollen competition (low-density pollination) (Davis et al. 1987, Schlichting et al. 1987).

Flowering plants, although last to evolve, became most successful and occupied a pre-eminent position among all groups of plants. Evolutionary success of angiosperms over other groups of plants has been an enigma. It has been suggested (Whitehouse 1950, Mulcahy 1979) that evolution of a most efficient outbreeding system (based on SI) and intense pollen competition during pollen-pistil interaction are responsible for the dramatic evolutionary success of flowering plants over other groups of plants.

POLLEN VIABILITY AND STIGMA RECEPTIVITY

Pollen viability and stigma receptivity are critical for effective initiation of pollen-pistil interaction. Details of pollen viability and the various methods used to assess pollen viability have been presented in Chapter 3. Stigma receptivity refers to the ability of the stigma to support germination of viable, compatible pollen. Details of stigma receptivity have been studied only in a limited number of species. In general, the stigma is receptive at the time of anthesis. However, in protogynous and protandrous species the stigma becomes receptive one or a few days before or after anthesis (see Chapter 7). The duration for which the stigma remains receptive is also variable; it may last just for a day or may remain receptive for many days. Generally, an unpollinated stigma remains receptive for a longer period. The stigma of unpollinated flowers of *Lilium* and *Petunia* remains receptive for over a week whereas that of pollinated flowers loses receptivity within 24 hours (Ascher and Peloquin 1966b, Shivanna and Rangaswamy, 1969). Similarly in many orchids also the stigma of unpollinated flowers remains receptive for many weeks. After pollination, the stigma loses turgidity following pollen germination and pollen tube growth and starts to senesce.

There are no satisfactory short-cut methods to determine stigma receptivity. As receptive stigmas invariably show the presence of several enzymes such as esterases, peroxidases and acid phosphatases, the presence of esterases/peroxidases on the stigma surface has often been considered to indicate receptivity. A few convenient methods based on demonstration of the presence of esterases/peroxidases have been described (Dafni 1992, Dafni and Maues 1998). Although receptive stigmas invariably show activity of many enzymes, this does not necessarily reflect receptivity since the surface enzymes in many species appear before the stigma supports pollen germination (Shivanna and Sastri 1981). Appearance of callose in ovules has been reported to indicate loss of viability in a few species (Dumas and Knox 1983). The only dependable method to study stigma receptivity is through controlled pollination and subsequent studies on pollen germination or fruit- and seed-set.

EVENTS ON STIGMA SURFACE

The stigma surface and especially the components of the exudate/pellicle play a vital role in post-pollination processes. Pollen adhesion, hydration, germination and pollen tube entry into the stigma are the major events that occur on the stigma surface.

Pollen Adhesion and Hydration

Following pollination, the first critical step in initiating post-pollination events is pollen adhesion and hydration. Electrostatic force seems to play a role in the initial capture of pollen (Corbet et al. 1982, Vaknin et al. 2001). Following pollen capture, retention of pollen depends on the extent of its adhesion to the stigma surface (Woittiez and Willemse 1979, Dumas and Gaude 1981).

Pollen adhesion essentially depends on the nature and extent of the components present on the surface of the stigma and of pollen. In species with a wet type stigma, pollen adhesion is not critical. Irrespective of the nature and extent of pollen coat substances, the exudate holds any pollen that lands on the stigma because of the stickiness and surface tension of the exudate. In dry-type of stigma, pollen adhesion is more critical and depends on the nature and extent of pollen coat substances and of the pellicle. Detailed analysis of the components of the stigma secretion of *Gladiolus* (Clarke et al. 1979) has shown the presence of ideal adhesive carbohydrate components, in particular the arabinogalactans on the stigma surface. Many of these components are also present on the stigma of other species, indicating the involvement of viscous polysaccharide slimes

and mucilages of the stigma surface to be a general feature in pollen adhesion (Knox et al. 1986).

In *Brassica*, two of the S-gene proteins, SLG and SLR 1 (see Chapter 9), have been implicated in pollen adhesion; these proteins have been shown to interact with pollen coat proteins in vitro (Doughty et al. 1993). In transgenic plants, down-regulation of SLR 1 by antisense suppression led to a reduction in adhesion of pollen (Luu et al. 1999). Similarly, treatment of wild type stigma of *Brassica* with either anti-SLG or anti-SLR antibodies resulted in a reduction in adhesive force of the stigma (Luu et al. 1999).

Recent studies have shown that pollen adhesion and hydration are more complex and both lipid and protein components present on pollen coat substances and/or stigma surface play a crucial role in these processes. In members of Brassicaceae, the presence of pollen coat substances seems to be essential for adhesion of pollen to the stigma surface. It has been shown that *Arabidopsis thaliana* stigma selectively binds its own pollen with higher affinity than pollen from related species (Zinkl et al. 1999), indicating some specificity in the adhesion process. Many mutants with defects in pollen coat substances have been isolated in this taxon (*lap* mutant – less adherent pollen) (Zinkl and Preuss 2000). In *Brassica*, pollen coat substances are released onto the stigma surface and form a meniscus at the interface (see Dickinson 1995). Apart from facilitating pollen adhesion, the meniscus enables interaction between pollen coat substances and the pellicle components; this interaction seems to provide appropriate signals that lead to the movement of water from the stigma into the pollen grain. Pollen grains from which pollen coat has been removed fail to hydrate (Doughty et al. 1993).

Recent studies have also indicated that long-chain lipids act as signals to stimulate pollen hydration. Mutants which have defects in pollen hydration (*cer* mutants and *pop 1* in *Arabidopsis*) have been shown to have defects in lipid biosynthesis (Preuss et al. 1993, Hulskamp et al.

1995). In *A. thaliana*, a mutation which resulted in loss of one of the most abundant pollen coat protein, oleosin-domain proteins GRP 17, delayed pollen hydration almost three-fold compared to wild-type pollen (Mayfield and Preuss 2000). Another outcome of recent investigations in *Brassica* has been the implication of aquaporins, water channel proteins, in pollen hydration (see Chapter 9) (Ikeda et al. 1997).

Pollen Germination and Pollen Tube Entry into Stigma

The stigma is traditionally considered to provide all the requirements, particularly of inorganic minerals, boron and calcium, needed for pollen grains to germinate. *Vitis vinifera* (Gartel 1974) plants grown in boron-sufficient soil contained 50–60 mg boron g^{-1} dry matter and their stigma secretion 2–5 mg g^{-1} dry matter; such stigmas support satisfactory pollen germination. When the plants were grown in boron-deficient soil, the concentration of boron in the stigma secretion was scarcely perceptible; such stigmas failed to support pollen germination. Considerable evidence has likewise shown that the stigma provides the calcium required by the pollen. Studies using chlorotetracycline fluorescence (Tirlapur and Shiggaon 1988) and energy dispersive X-ray analysis (Bednarska 1989) have shown high levels of calcium on the stigma surface and in papillary cells. Transport of calcium from the stigma to the pollen has been demonstrated through the use of radioactive labelled calcium (Bednarska 1991).

In dry-type of stigma, pellicle components seem to be involved in pollen germination. In *Gladiolus* (Knox et al. 1976) for example, washing of the stigma with a detergent, sodium deoxycholate, removed the ability of the stigma to support pollen germination. Treatment of stigma with a lectin, con A, which binds to the pellicle, however, did not inhibit pollen germination but prevented the entry of pollen tubes. In *Raphanus* (Shivanna et al. 1978b), enzymatic digestion of the pellicle reduced pollen germination and totally inhibited entry of the pollen tubes into the stigma. Thus, stigma surface receptors seem to be composed of many components, some of which are involved in germination and others in the entry of pollen tubes (Knox et al. 1976, Shivanna and Johri 1985).

Several studies using molecular and genetic approaches have yielded more specific information on the role of stigmatic and pollen components in various events that occur on the surface of the stigma (see Shivanna and Rangaswamy 1999). Flavonoids, which form a part of the pollenkitt, play an important role in pollen germination and pollen tube growth. Flavonoid-deficient mutants of maize (Coe et al. 1981) and *Petunia* (van der Meer et al. 1992) produce flavonoid-deficient pollen which cannot function on the stigma of the mutants. However, an exogenous supply of nanomolar concentration of flavonol aglycone, kaempferol, either to the pollen or to the stigma restored germination of flavonoid-deficient pollen on self-stigma and resulted in seed-set (Mo et al. 1992). Similarly flavonoid-deficient pollen could function on wild-type stigma (Taylor and Jorgensen 1992); apparently the required flavonoid is provided by the wild-type stigma (Mo et al. 1992).

Pollen tube entry into the stigma is another critical step in a series of sequential events leading to fertilization. In wet-type stigma, in which the cuticle of the stigma surface/papillae is disrupted during secretion of the exudate, there is no physical barrier to pollen tube entry into the intercellular spaces of the transmitting tissues of the stigma. In dry stigma, the cuticle provides a physical barrier to pollen tube entry. The pollen tube has to erode the cuticle at the region of contact by activation of cutinases. In many grasses (Sastri and Shivanna 1979, Heslop-Harrison 1979b, Heslop-Harrison and Heslop-Harrison 1987) the stigmatic surface, at the region of pollen tube entry, shows enhanced esterase activity, which probably represents cutinase activity. Long-chain lipids present on the pollen/stigma surface have also been shown to play an important role in pollen tube guidance into the stigma (see page 127 for details).

POLLINATION STIMULUS

Pollination initiates a number of changes in the pistil and other floral parts, especially petals. Many of these changes are initiated before pollen germination or much before the pollen tubes reach the ovary indicating transmission of a pollination-induced signal to the distal parts of the pistil and other floral parts.

One of the well-investigated pollination-induced changes is the increase in ethylene production in the stigma and later in the distal part of the style. Petal senescence is another important pollination-induced response (Gilissen and Hoekstra 1984, Whitehead et al. 1984) which occurs in association with increased ethylene production. Studies have been carried out on pollination-induced ethylene production and petal senescence in many species, e.g. *Petunia* (Hoekstra and Weges 1986), carnation (Larsen et al. 1995) and orchids (O'Neil et al. 1993). Petal senescence can be prevented by treatments that inhibit ethylene production, suggesting that ethylene production plays an important role in coordinating pollination-regulated petal senescence. Available evidence suggests that in orchids pollen-borne auxin is the pollination signal leading to ethylene production. In other species, e.g. *Petunia* and carnation, however, pollen-borne ACC alone or with another as yet unidentified pollen substance, may participate in the pollination signal (O'Neil 1997). Production of ethylene in the stigma and style may use pollen-borne ACC directly as a substrate or may depend on the rapid induction of ACC-synthase in the pistil. Amplification of the primary pollen signal requires secondary signals that emanate from the stigma. Interorgan transmission of the signals seems to be mediated through translocation of ACC or ethylene (O'Neil 1997)

In watermelon (Sedgley and Scholefield 1980), *Acacia* (Kenrick and Knox 1981, 1989) and oil-palm (Tandon et al. 2001), pollination stimulates another phase of stigmatic secretion. This is not specific to pollen of the same species; pollination with pollen from widely related species also brings about stimulation of post-pollination secretion in the stigma (Kenrick and Knox 1981). The significance of post-pollination secretion is not clear. It is suggested that it ensures sufficient medium for the hydration and germination of a large number of pollen grains. Post-pollination secretion of the stigma appears to be widespread as even dry stigma of many taxa become wet after pollination (Owens and Kimmins 1981).

Pollination results in accumulation of flavonols especially kaempferol in the stigma (Vogt et al. 1994); the accumulation reaches maximum within 24 h after pollination, by which time pollen grains germinate and pollen tubes reach the lower part of the style. Wounding of petals, sepals and stamens also results in the accumulation of kaempferol in the stigma in a manner identical to that of pollination, indicating that both pollination and wounding share elements of a common signal transduction pathway. Wounding of petals or stamens prior to pollination results in a significant increase in seed-set (Vogt et al. 1994).

Marked changes in the activity of several enzymes and in the pattern of RNA and protein synthesis (Linskens 1975) are some of the changes observed in the lower part of the pollinated pistil before pollen tubes reach this region. In pear (Herrero and Gascon 1987) pollination delays embryo sac degeneration, thus extending the period over which fertilization can be effected. Often, the stimulus is specific to self- and cross-pollination in self-incompatible species (van der Donk, 1974, Deurenberg 1976, Spanjers 1978). In many species pollination induces degeneration of one of the synergids, before the pollen tube reaches the ovule (Jensen 1973).

The observations described above clearly indicate that pollination stimulus is transferred to the ovary before pollen tubes reach it. In cotton, synergid degeneration occurs when ovules from unpollinated flowers are cultured on a GA_3-containing medium but not on IAA- or cytokinin-containing medium (Jensen and Ashton 1981).

On the basis of these studies it is suggested that pollen grains increase the GA_3 content in the pistil, which induces degeneration of one of the synergids. In many apomictic species in which the embryo develops without fertilization, pollination stimulates the egg to develop parthenogenetically (Koltunow 1993).

In most species, the ovules are fully differentiated by the time pollination occurs. Pollen tubes enter the ovules soon after reaching the ovary and effect fertilization. However, in many orchids the ovules are not yet ready to receive pollen tubes at the time of pollination or soon after pollen tubes reach the ovary (Wirth and Withner 1959, Withner et al. 1974, Arditti 1979, Goh and Arditti 1985, Zhang and O'Neil 1993). The ovary is immature and lacks ovules at the time of anthesis; ovule differentiation is initiated only after pollination. Thus the time lag between pollination and fertilization ranges from a few weeks to many months. In *Phalaenopsis* (Zhang and O'Neil 1993), morphological changes associated with ovule development are visible within 2 days after pollination while pollen germination is initiated only 5 days after pollination, indicating that the pollination signal reaches the ovary much before pollen germination. Although pollen tubes reach the ovary by 14 days after pollination, full differentiation of ovules and fertilization are completed only 80 days after pollination (Fig. 8.3) (Zhang and O'Neil, 1993). Evidence indicates that in orchids both auxin and ethylene contribute to regulation of ovule development following pollination (O'Neil et al. 1993).

POLLEN TUBE GROWTH THROUGH STYLE

Pollen germination on the stigma and pollen tube growth in the style has been extensively studied through aniline blue fluorescence of cleared whole-mount preparations of pollinated pistils (Fig. 8.4) In solid-styled pistils, pollen tubes enter the cuticle of the papillae, grow down the papillae between the cuticle and pectocellulosic wall (Fig. 8.5A, C) and enter the extracellular

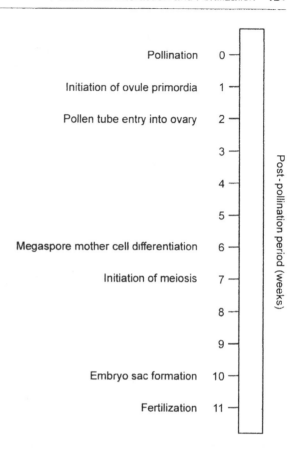

Fig. 8.3 Pollination regulated ovule development in an orchid (*Phalaenopsis*) (based on O'Neill 1997).

matrix (ECM) of the stigma. Further growth takes place through the ECM of the stigma and style (Fig. 8.5B). In hollow-styled systems such as *Lilium*, in which the stylar canal is filled with the secretion product, pollen tubes grow on the surface of the stigma, enter the stylar canal and grow down the surface of the canal cells. In heavily pollinated pistils, the entire stylar canal is often filled with a bundle of pollen tubes which can be separated from the stylar tissue (Malti and Shivanna 1984). In species such as *Gladiolus* (Clarke et al. 1977) and *Crocus* (Heslop-Harrison, 1977) in which a mucilaginous substance accumulates between the cuticle and cell wall, pollen tubes grow through the mucilage (Fig. 8.5D).

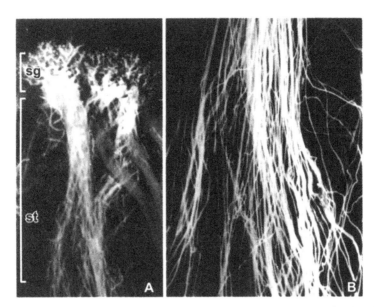

Fig. 8.4 Fluorescent micrographs of compatibly pollinated pistil of *Nicotiana tabacum* cleared and stained with decolourized aniline blue to show pollen germination on the stigma (sg) and pollen tube growth through the upper part of the style (st) (A). B. Pollen tube growth in lower part of the style.

The details of chemical interactions between the pollen and the stylar components are not clearly understood. The ECM through which the pollen tubes grow contains highly heterogeneous components secreted by the cells of the transmitting tissue (see Chapter 6). In many species, growth of the pollen tubes results in degeneration of adjoining cells of the stigma and transmitting tissue (Knox 1984b, Heslop-Harrison and Heslop-Harrison 1985a).

Studies by Sanders and Lord (1989) and Sanders et al. (1991) have indicated an active role for the ECM in pollen tube growth, similar to the ECM of animals. ECM in animal systems contains adhesion molecules, some of which, in particular fibronectin and vitronectin, have been implicated in cell migration (Sanders and Lord 1992). Using human vitronectin antibodies, vitronectin-like proteins have been detected in the stylar transmitting tract of *Vicia faba* (Sanders et al. 1991). Also, inert latex beads (6 μm in diameter) placed at the stigmatic end of the style in *Hemerocallis, Raphanus* and *Vicia* migrated at the same rate as pollen tube growth through

the transmitting tissue (Sanders and Lord 1989. It has been suggested that pollen tube growth is analogous to migrating cells in animal systems and that the stylar ECM actively facilitates pollen tube extension by way of the biochemical recognition-adhesion system (Sanders and Lord 1989, 1992, Lord 2001).

Recent studies have shown that transmission tissue-specific (TTS) (see Chapter 6) proteins encoded by two genes (*TTS 1* and *TTS 2*) play a vital role in pollen tube growth; they have been implicated in pollen tube adhesion in the style, nutrition and guidance. The levels of TTS proteins in *Nicotiana* (Wang et al. 1993) increased with flower development, reached maximum at anthesis and remained high for 2–3 days post-anthesis during which time the pollen tubes grew through the stigma and style. Pollination of buds containing low levels of TTS proteins increased their level by 2–3-fold. Thus, pollen tube growth through the style is associated with high levels of TTS proteins, suggesting that they may play an important role during pollen tube growth (Wang et al. 1993).

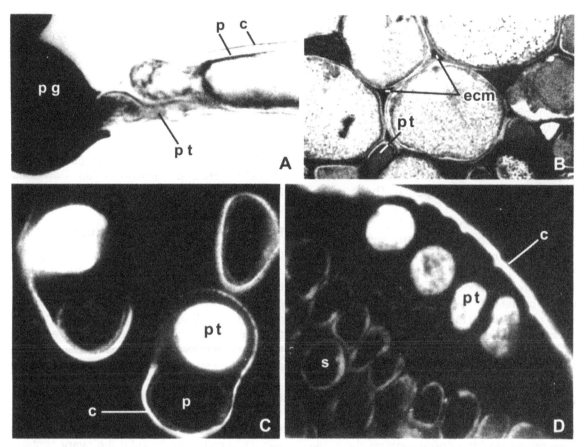

Fig. 8.5 Details of pollen tube entry into the stigma and growth of pollen tube in the style. A, B. *Primula vulgaris*. A. Pollen grain (pg) germinates and the resultant pollen tube (pt) penetrates the cuticle (c) at the tip of papilla (p) and grows down the papilla. B. Section of pollinated stigmatic head to show pollen tube (pt) growing in the extracellular matrix (ecm) of the transmitting tissue. C, D. *Crocus chrysanthus*. C. Fluorescence micrograph of auramine O stained stigma section to show passage of pollen tubes (pt) between the cuticle (c) and pecto-cellulosic wall of papilla (p). The cuticle extends considerably to accommodate the tubes. D. Section of lower non-papillate part of stigma to show several pollen tubes (pt) between the cuticle (c) and tissues of the stigma (s) (A—after Shivanna et al. 1981, B—after Heslop-Harrison et al. 1981, C, D—after Heslop-Harrison 1977).

Subsequently a TTS protein from *Nicotiana* stylar transmitting tissue was purified and characterized (Cheung et al. 1995). TTS protein has molecular weights ranging from 50–100 kD depending on the level of the style used for extraction; the lower the style level, the higher its molecular weight. Chemical deglycosylation results in a protein of uniformly 30 kD, indicating that differences in apparent molecular weights are due to differences in their levels of glycosylation. TTS proteins are proline-rich and belong to the

arabinogalactan protein family with galactose making up to 70% of the carbohydrate moiety.

A series of experiments were carried out to understand the role of TTS proteins in pollen tube growth (Cheung et al. 1995, Cheung 1996, Wu et al. 1995). TTS proteins markedly stimulated pollen tube growth in vitro. Chemically deglycosylated proteins were ineffective, however, in stimulating pollen tube growth. Transgenic plants with significantly reduced levels of TTS proteins (produced through antisense suppression or sense co-suppression) in the pistil showed reduced pollen tube growth rate. Such plants yielded small fruits with highly reduced seed number. Pollen from transgenic plants, however, yielded normal fruits and seeds when used for pollination of wild type plants. TTS proteins bind to the pollen tube surface and tube tip (Wu et al. 1995) and are incorporated into the pollen tube wall. TTS proteins are deglycosylated by the growing pollen tubes suggesting that these glycoproteins may provide nutrients for pollen tube growth. These results clearly indicate that TTS proteins are needed for normal pollen tube growth.

Although the presence of proteins in the wall of pollen grains and pollen tubes and their release during pollen germination and pollen tube growth are well documented, there is no information on the details of their interaction with the ECM of the pistil. Recently, a gene *Pex-1* that encodes a pollen specific extensin-like protein was identified in maize (Rubinstein et al. 1995). *Pex-1* gene is expressed most actively in mature pollen and Pex-1 protein accumulates in the pollen tube wall.

The amount of nutrients present in the pollen is limited and not sufficient to support pollen tube growth until the pollen reaches the ovule. In maize (Heslop-Harrison and Heslop-Harrison 1987), for example, calculations based on the volume of reserves present in pollen grains and the rates of incorporation of carbohydrates into the wall suggest that no more than 2 cm of growth can be sustained by the reserve present in the pollen whereas the tube may grow more than 10 cm. Similarly, endogenous reserves of *Sorghum* pollen (Heslop-Harrison et al. 1984) can account for only up to 500 μm growth. Pollen tubes therefore have to take up nutrients from the pistil during their growth. In *Lilium*, by incorporating labelled glucose and myoinositol into the pistils, Labarca and Loewus (1973) demonstrated that both these components are taken up by the growing pollen tubes from the pistil and are utilized for the synthesis of pollen wall material. Evidence for utilization of stylar carbohydrates by growing pollen tubes has also been obtained from labelled studies in *Nicotiana* (Tupy 1961), *Oenothera* (Kumar and Hecht 1970) and *Petunia* (Kroh and Helsper 1974).

The rate of pollen tube growth in the pistil is highly variable. In *Lilium* (van der Woude and Morre, 1968) pollen tube growth rate has been reported to be 2 mm h^{-1}, while in *Ananas* (Wee and Rao 1979) and many cereals (Shivanna et al. 1978a, Heslop-Harrison 1979a, Barnabas and Fridvalszky 1984) a growth rate of about 1 cm h^{-1} has been reported.

Regulation of Pollen Tube Number

Generally, the number of pollen grains germinating on the stigma and the number of pollen tubes entering the style is much more than the number of ovules available for fertilization. In many species there seems to be a mechanism to regulate the number of pollen tubes entering the ovary. In *Persea americana* (Sedgley 1976), a uniovulate species, on average over 66 pollen grains germinate on the stigma but most of the tubes cease growth after traversing various distances in the style and only one tube reaches the ovary. Some ovules contain twin embryo sacs; two pollen tubes enter the ovary containing ovules with twin embryo sacs. It is suggested that the embryo sac has control over growth of pollen tubes in the style.

It has been suggested that in members of Poaceae the architecture of the stigma is responsible for the limited number of pollen tubes reaching the ovary (Heslop-Harrison et al. 1985b). Individual trichomes (ultimate branches of the stigma) can support hydration of a limited number of pollen grains. The transmitting tissue of the stigmatic trichomes and of the main axis of the stigma, through which pollen tubes continue growth, is limited. These conditions create intense competition among pollen grains and pollen tubes. Only the tubes issued from those pollen grains which germinate early can enter the trichome and only the tubes which are fast growing can complete growth through the transmitting tract of the trichome and eventually the main axis.

Another significant device regulating pollen tube number in grasses is the development of an abscission zone at the base of the main axis (Heslop-Harrison et al. 1985a, Willingle and Mantle 1985). In corn, it lies just above the ovary. The abscission zone is not activated until pollination. Cortical cells of this zone lose turgidity within 6 h after pollination and in 24 h the cells becomes flaccid. The vascular tissue and the transmitting tract then disrupted. By this time a few pollen tubes will have grown through the abscission zone and subsequent entry of additional tubes into the ovary is effectively blocked. The abscission zone has also been suggested to provide a barrier to invading fungal pathogens entering through the stigma (Willingle and Mantle 1985).

In some members of Leguminosae (Caesalpinoideae), the transmitting tissue diminishes from the stigma to the base of the style (Owens 1989); this may impose restriction on the number of pollen tubes entering the ovary. In most of the taxa investigated, the micropyle of the ovule is in the form of a narrow canal. In sunflower, however, the micropyle is structurally similar to the transmitting tissue of the style (Yan et al. 1991); it is represented by a strand of transmitting tissue with intercellular spaces and narrows down towards the embryo sac. This has been suggested to be an adaptation to reduce the number of pollen tubes entering the embryo sac to one (Yan et al. 1991).

An interesting feature comparable to vaginal sealing observed in some animals has been reported in *Kleinhovia hospita* (Shaanker and Ganeshaiah 1989). In this species the number of seeds/ovary is much less than the number of ovules due to failure of fertilization in many ovules. This is caused by the production of IAA in early-formed zygotes and its diffusion into the style. This enhanced level of IAA inhibits growth of other tubes through the style and thus reduces the number of pollen tubes entering the ovary.

Internal Geitonogamy

A unique phenomenon in which pollen-pistil interaction is completely eliminated, termed internal geitonogamy, has been reported in a few species of Malpighiaceae (Anderson 1980) and Callitrichaceae (Philbrick 1984, Philbrick and Anderson 1992). These species produce both chasmogamous and cleistogamous flowers. Chasmogamous flowers show normal pollination and post-pollination events. However, cleistogamous flowers show remarkable deviation in achieving fertilization. In members of Malpighiaceae, the flowers are bisexual. Pollen grains germinate inside the indehiscent anthers; pollen tubes grow down the anther filament into the receptacle and then turn upwards and grow into the locule of the ovary and enter the micropyle (Anderson 1980).

In members of Callitrichaceae the flowers are monoecious. Staminate and pistillate flowers occur in the same leaf axil or in opposite axils of the same node. Several aquatic species of this family produce cleistogamous flowers which remain submerged (unlike chasmogamous flowers which open above the water level). In cleistogamous flowers, pollen grains germinate inside the anthers, pollen tubes grow through the filament, and after reaching the axil grow across the vegetative tissues and up into the ovary of the pistillate flower situated in the same axil (Fig. 8.6) (Philbrick 1984, Philbrick and Anderson 1992). The pollen tube may also grow across the stem into the pistillate flower of the opposite axil or along the internode to the pistillate flower located in the next internode (Fig. 8.6). Such reports are confined so far to studies on pollen tube growth using aniline blue fluorescence. No information is available on the anatomical details of the vegetative tissue through which pollen tubes grow. Also, fertilization through internal geitonogamy is yet to be demonstrated.

POLLEN TUBE GUIDANCE

Pollen tubes follow a predetermined path in the pistil. The distance between the stigma surface

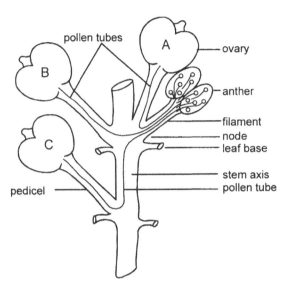

Fig. 8.6 Diagram of internal geitonogamy. Pollen grains germinate inside indehiscent anther and pollen tubes grow through the filament of the stamen and reach the ovule through the pedicel of the ovary through one of the three pathways: ovary located in the same axil as the stamen (A), the opposite leaf axil at the same node (B), or the adjacent node (C).

and the embryo sac varies from a few millimeters to more than 10 cm. Irrespective of this distance and whether the pistil is vertically upwards or drooping, pollen tube growth follows a precise pathway; pollen tubes enter the stigma, grow through the transmitting tissue or the canal of the style and enter the ovary. It has long been thought that pollen tubes are guided from the stigma to the ovary by a gradient of chemotropic substance present in the pistil (see Heslop-Harrison and Heslop-Harrison 1986). Many studies using in vitro assay have demonstrated a positive chemotropic effect of various parts of the pistil to pollen tubes. A few studies during the 1960s indicated that ionic calcium is the chemotropic agent and forms a gradient in the pistil (Mascarenhas and Machlis 1962,1964). Later studies, however, did not support this concept (see Rosen, 1971, Shivanna and Johri 1985, Heslop-Harrison and Heslop-Harrison 1986).

On the basis of structural details of the pistil through which the pollen tubes grow, a consensus developed during the 1970s and 1980s on pollen tube guidance in the pistil even in the absence of a chemotropic gradient (Heslop-Harrison and Heslop-Harrison 1986). The transmitting tract from the stigma to the ovary is generally made up of cylindrical elongated cells surrounded by the extracellular matrix (ECM). The cells are interconnected at the ends to form vertical files. The ECM provides a continuous path of least mechanical resistance along the transmitting tissue for pollen tube growth. Once the pollen tubes enter the transmitting tract, they follow this path unidirectionally from the stigma to the ovary (Jensen and Fischer 1968).

Many experimental studies have confirmed lack of chemotropic gradient in the style. In *Lilium*, pollen tubes grow without hindrance through morphologically inverted grafts of the style (Iwanami 1959). Further, when pollen grains are deposited in the stylar cavity (of *Lilium*) by making a window (Iwanami 1959) or in the stylar transmitting tissue (of tobacco) (Mulcahy and Mulcahy 1987), pollen tubes grow in both directions—towards the stigma as well as the ovary. Furthermore, when the ovarian end of an excised style of lily was pollinated, pollen tubes grew normally through the style towards the stigma (Ascher, 1977). These studies clearly demonstrate the absence of inherent polarization in the style.

In grasses, there is no organized transmitting tract from the individual trichomes to the transmitting strand of the main axis. Pollen tubes, after growing through the individual trichomes, grow across the cortex into the transmitting strand of the axis through the intercellular spaces of the cortical tissue. Heslop-Harrison and Heslop-Harrison (1986) suggested that chemotropic guidance might be involved in the growth of pollen tubes across the cortex towards the transmitting strand.

Recent studies of Cheung et al. (1995), however, have shown that TTS proteins of tobacco attract pollen tubes emerging from semivivo implanted pistils. Chemically deglycosylated proteins, on the other hand, were ineffective in attracting the pollen tubes, indicating that the sugar moiety of these proteins is responsible for this attraction. TTS proteins display a gradient of increasing glycosylation level in the transmitting tissue from the stigma to the ovary (Wu et al. 1995). It is suggested that this unique protein-based sugar gradient may play a role in pollen tube guidance. Given such data, the possible presence of a chemotropic gradient in the style has yet to be convincingly resolved.

Even the concept that the pollen tubes are guided by the architecture of the cells of the transmitting tissue envisages the presence of some factor(s) on the surface of the stigma to guide tube growth down into the stigma and inside the ovary, wherein the pollen tubes have to change their direction of growth by about 90°, from the placenta into the micropyle of the ovules. Until recently there was no information on these aspects. During the last few years integrated studies using a range of techniques have given some definitive information on at least a few systems.

On Stigma Surface

Studies carried out on tobacco, which produces copious lipidic exudate on the stigma, clearly suggest that the lipids present in the stigmatic exudate play an important role in pollen tube penetration into the stigmatic tissue. Transgenic tobacco plants, which do not produce stigmatic exudate due to the expression of a cytotoxic gene STIG 1-barnase, do not support pollen tube entry into the stigma (Goldman et al. 1995). However, this function of the stigma can be restored by application of the stigmatic exudate from wild tobacco plants as well as *Petunia*, which also produces lipidic stigmatic exudate (Wolters-Arts et al. 1998). However, the carbohydrate-rich aqueous exudate of the *Lilium* stigma does not restore pollen tube entry into the stigma in transgenic tobacco.

The lipidic exudate contains many saturated and unsaturated fatty acids largely as triacylglycerides. Application of purified oil such as olive oil (that has a fatty acid composition similar to that of stigma exudate) to transgenic tobacco restored the function of the stigma. Attempts were also made to study the effects of individual components present in the purified oil. Of the three triacylglycerides—triolein, trilinolein and trilinolenin—tested only trilinolein was effective in restoring the function of the stigma. These results show that of the triacylglycerides, trilinolein alone is sufficient and essential for normal functioning of the stigma.

In another experiment (Wolters-Arts et al. 1998) the oil containing unsaturated triacylglycerides was spread thinly on agarose gel (containing toluidine blue) prepared in pollen germination medium. Pollen grains were spread thinly on the oil phase. The distribution of toluidine blue in the oil phase showed the establishment of water gradient, thus providing directional supply of water to the pollen grains. Pollen grains nearest to the interface hydrated and germinated; the growth of their pollen tubes followed the directional supply of water and grew towards and into the agarose medium. When directional supply of water was disturbed by replacing the oil phase with an emulsion of unsaturated triacylglycerides and the germination medium, the directional growth of pollen tubes was also disturbed. Similarly, when the wet stigma of the wild-type tobacco (covered with the stigmatic exudate) was pollinated with pollen grains that had already been hydrated in the germination medium, the resultant pollen tubes failed to penetrate the stigma. Thus the directional growth of pollen tubes into the stigma occurs only when there is a directional supply of water to the pollen. The function of lipids in the stigmatic exudate seems to be establishment of such a water gradient as to ensure pollen tube growth directionally into the stigma.

Additional experiments carried out by Lush and colleagues (1998) further support the above concept. Dry pollen grains were dispersed in a drop of exudate/oil (0.2–1.0 μl) placed on a microslide. A smaller drop of an aqueous medium was injected into the centre of pollen cultures in the exudate/oil and the set-up topped with a cover glass. This established a two-phase pollen culture in which the central aqueous phase was surrounded by the exudate/oil. The period taken for pollen to hydrate and germinate was proportional to the distance between pollen and the interface; pollen grains closer to the interface hydrated faster than those lodged farther from the interface. Interestingly, 95% of the pollen tubes grew directly towards the interface and entered the aqueous medium. By deletion experiments involving different components of the germination medium, it was established that the directional supply of water to the pollen grains in the exudate/oil is the guidance factor for penetration of pollen tubes into the aqueous phase. In vivo also the lipidic exudate provides a water gradient from stigma surface cells and guides the pollen tubes into the stigma. The composition of the stigmatic exudate is such that it is permeable enough for pollen hydration to occur but not so permeable for water supply to become non-directional. Thus pollen tube penetration into the stigma is more a hydrotropic response than a chemotropic one.

The situation prevailing in tobacco and *Petunia*, both of which produce lipidic exudate on their stigma, does not hold good for species with a dry stigma or those which produce aqueous exudate (e.g. *Lilium*). Many studies carried out on *Arabidopsis* and *Brassica* (both characterized by dry stigma) have indicated that the function of the stigmatic exudate has been taken over by the lipids present in the pollenkitt in species with a dry stigma. Subsequent to pollination, the pollenkitt lipids come into contact with the stigma surface, form a contact zone and provide a medium to establish directional water gradient for pollen hydration, germination and pollen tube guidance into the stigma. In *Brassica*, pollen grains from which the pollenkitt had been removed failed to hydrate and germinate (Doughty et al. 1993). Pollen grains of *Arabidopsis* mutants (*cer1, cer3, cer6* and *pop1*) defective in the synthesis of long-chain lipids failed to hydrate and germinate on the stigma (Preuss et al. 1993, Hulskamp et al. 1995a, Wolters-Arts et al. 1998). Interestingly, application of the triacylglyceride, trilinolein to pollen of all the four mutants of *Arabidopsis* restored their function; mutant pollen grains were able to hydrate and germinate and their tubes able to enter the stigma (Wolters-Arts et al. 1998).

Interestingly, when pollen grains of wild plants of both *Arabidopsis* and *Brassica* were placed on the stigma of transgenic tobacco, the pollen coat lipids formed a contact zone between pollen and the stigma; pollen hydration, germination and pollen tube entry into the transgenic stigma were achieved, although pollen tubes could grow only for a short distance in the stigma because of incompatibility. These studies support the concept that during evolution of the dry type of stigma the pollenkitt lipids took over the role of exudate of the wet stigma (Elleman and Dickinson 1996). The role of lipids in pollen tube guidance needs to be studied in other systems, particularly in those which produce aqueous exudate.

In Ovary

The pollen tube has to change direction of growth by almost 90° to enter the micropyle of the ovule. This is considered to be in response to a chemotropic stimulus localized at the micropyle. The nature of the chemotropic stimulus is not clearly understood. Much evidence suggests that the nucellar cells along the micropyle and/or the synergids are the source of the chemotropic substance (Russell 1992). Periodic acid-Schiffs positive substance and proteins apparently produced by the dissolution of integumentary and nucellar cells along the micropyle have been reported in some species of grasses (Chao 1977, Tilton 1981b, Reger et al. 1992). Recent studies have shown that in many species such as *Gasteria* and *Lilium* (Franssen-Verheijen and Willemse 1993) and *Arabidopsis* (Hulskamp et al. 1995b), the micropyle is covered with a drop of exudate. The exudate seems to be necessary for the entry of pollen tube into the ovule. Gentle washing of isolated ovules of *Lilium* with solvents that disrupt proteins affects pollen tube entry, indicating that the proteins of the exudate may have a role in pollen tube entry (Franssen-Verheijen and Willemse et al. 1993). In pearl millet, pollen tubes showed positive chemotropism in vitro to glucose, calcium as well as to a low molecular weight ovarian protein, suggesting a possible chemotropic role for the extracellular substances present in the micropyle (Reger et al. 1992). Calcium released from the degenerated synergid has been suggested to form a chemotropic gradient (Jensen and Ashton 1981, Tilton 1981b); reports of accumulation of a high level of calcium in the synergids of wheat and pearl millet (Chaubal and Reger 1990, 1992) accord with this suggestion.

Recent studies on *Arabidopsis* mutants defective in ovule development (Hulskamp et al. 1995a,b) have provided some definitive information on pollen tube guidance in the ovary. In the wild type, a substantial number of pollen tubes grew directly towards ovules soon after their emergence from the transmitting tissue and en-

tered the ovules (Fig. 8.7A). In the mutants, although growth of the pollen tubes in the stigma and the style was not affected, it was markedly affected in the ovary. Pollen tubes in the mutants lost direction in the ovary; instead of growing towards the ovules, they grew randomly on all available surfaces including the ovary wall. Detailed studies on one of the mutants of *A. thaliana* (54D12) in which the extent of embryo sac development varied (i.e., ovules either lacked an embryo sac or had a partial embryo sac or a normal embryo sac), have highlighted the importance of the embryo sac in pollen tube guidance. In this mutant over 90% of the ovules which had a normal embryo sac received pollen tubes, whereas only 28% of ovules which had a partial embryo sac and none of the ovules that lacked an embryo sac received pollen tubes (Fig. 8.7B). These results clearly show that development of the embryo sac is crucial for pollen tube guidance into the ovule.

Additional evidence for the role of embryo sac, particularly the synergids, has come from in-vitro studies on *Torenia fournieri*

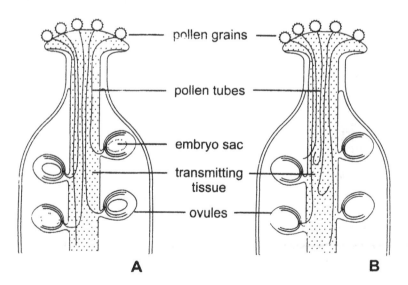

Fig. 8.7 Diagram of the pistil of *Arabidopsis thaliana*. A.Wild type. Pollen tubes grow through the transmitting tissue and enter the ovules (all with embryo sac). B. Mutant pistil in which many of the ovules lack embryo sac. The ovule with embryo sac has received pollen tube but those without embryo sac have not. Many of the pollen tubes have lost direction and grown randomly (redrawn from Shivanna and Rangaswamy 1999, based on Hulskamp et al. 1995b).

(Scrophulariaceae) in which the micropylar part of the embryo sac protrudes beyond the ovule (Higashiyama et al. 1997, 1998). In co-cultures of pollinated styles (following the semivivo method of pistil culture, see Shivanna and Rangaswamy 1992) and isolated ovules, the pollen tubes emerged through the cut end of the cultured style, grew towards the ovules and arrived at the filiform apparatus of the synergids (Fig. 8.8A, B). Pollen tubes failed to arrive at the embryo sacs that were immature, damaged or fertilized. These results clearly demonstrate that (i) pollen tubes are specifically attracted to the region of filiform apparatus in the micropylar part of the embryo sac and (ii) pollen tubes are not attracted to embryo sacs that have already received a pollen tube; this may be an adaptation to prevent polyspermy.

In a recent study, Higashiyama and co-authors (2001) were able to ablate each cell of the embryo sac by a UV laser beam before cultivation of the ovules and monitor the arrival of the pollen tube, using a set-up of semivivo pistil and ovule culture similar to that described above. Embryo sacs attracted pollen tubes even after 20–30 surrounding ovular cells were ablated, indicating that these cells are not required for generating the attraction signal. Pollen tube attraction was also not affected when the egg cell or central cell was ablated. When one of the synergids was ablated, the frequency of pollen tube attraction was reduced but did not cease; pollen tube attraction ceased completely in embryo sacs in which both the synergids were ablated. By a series of treatments in which different cells of the embryo sacs were ablated, Higashiyama and co-workers (2001) demonstrated that one synergid cell is necessary and sufficient to attract the pollen tube. These studies have opened up a new set of experiments for understanding further details of pollen tube guidance.

POLLEN TUBE ENTRY INTO OVULE AND EMBRYO SAC

Pollen tubes enter the ovary after growing through the transmitting tract of the style and continue to grow along the placenta. In peach (*Prunus persica*) (Herrero and Arbeloa 1989), pollen tubes come into contact with the obturator (a placental outgrowth in the micropylar region of the ovule) and cease further growth. The cells of the obturator are filled with starchy reserves by the time the pollen tubes arrive.

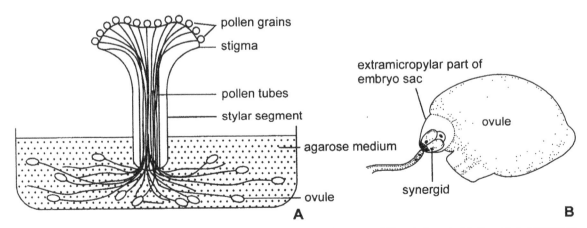

Fig. 8.8 Diagram of pollen tube guidance in vitro to partially naked embryo sac of *Torenia fournieri*. A. Semivivo pistil culture set-up. Pollen tubes have emerged from the cut end of the style and grown towards ovules. B. One of the ovules magnified to show extramicropylar part of the embryo sac and the arrival of the pollen tube at the filiform apparatus of the synergid (redrawn from Shivanna and Rangaswamy 1999, based on Higashiyama et al. 1998).

During the next 4–5 days the starch is hydrolyzed but a secretion that stains for carbohydrates and proteins is produced. Pollen tube growth is resumed only after production of this secretion, indicating that the secretion of the obturator is apparently necessary for the onset of the second phase of pollen tube growth. After the tubes have grown along the obturator into the ovule, callose accumulates in the cells of the obturator and the pollen tubes cannot grow across the obturator after callose deposition. Thus the obturator seems to play an important role in regulating pollen tube entry into the ovule. The obturator present in a number of taxa (Tilton and Horner 1980, Hill and Lord 1987, Scribailo and Barrett 1991)) may play a similar regulatory role.

DOUBLE FERTILIZATION

Basic understanding of the fusion events associated with double fertilization had been achieved by the end of the 19th century (see Maheshwari 1950). Subsequently, extensive studies carried out on a large number of species, confirmed double fertilization as a characteristic feature of flowering plants. Since the 1960s electron microscopy has added another dimension to studies on fertilization. Nevertheless, investigations on fertilization remained essentially structural and descriptive over the years. During the last decade recent techniques, particularly in-vitro culture and molecular biology, have been effectively used in studies on fertilization and a beginning has been made in subjecting the fertilization process to experimentation.

Synergids play an important role in receiving pollen tubes and in subsequent events of fertilization. As mentioned earlier, in many species one of the synergids degenerates after pollination but before the pollen tubes reach the ovule (Huang and Russell 1992). Degeneration results in increased electron opacity of the cytoplasm and disruption of its plasma membrane. The pollen tube enters the degenerated synergid. The factors responsible for the degenerated

synergid's receptivity are not understood. Lower resistance of the degenerated synergid to pollen tube penetration has been suggested as an important factor (Huang and Russell 1992). In many species such as *Petunia* (van Went 1970) and *Gasteria* (Willemse and Franssen-Verheijen 1988), however, both the synergids remain healthy at the time of pollen tube arrival. The pollen tube enters one of them and triggers degeneration. The other synergid remains intact for some time.

The filiform apparatus of the synergid seems to play a critical role in pollen tube entry as the tube always enters the embryo sac through this apparatus (Fig. 8.9A). Even in species such as *Plumbago*, which have no synergids, the egg cell itself develops a filiform apparatus at the micropylar region (Russell 1982). Soon after entering the synergid or after growing for a short distance in the synergid cytoplasm, the pollen tube releases its contents (Fig. 8.9A) through a pore or through rupture of the tube tip. During pollen tube discharge additional plasma membrane (of the vegetative cell) surrounding male gametes is disrupted. These gametes with their cytoplasm and plasma membrane intact, move into the chalazal end of the synergid. The mechanism of this movement is not understood; much evidence indicates the involvement of cytoskeleton elements, in particular the actin filaments in this movement (Russell 1993). Eventually one of the sperms comes into contact with the plasma membrane of the egg cell and the other with that of the central cell. Their plasma membranes fuse to form bridges; widening of the membrane bridges results in the entry of the contents of one of the male gametes into the egg and the other into the central cell (Fig. 8.9B). The nuclei of the male gametes migrate and align with the respective egg and polar nuclei. Nuclear fusion is also reported to be mediated through membrane fusion (Russell 1993).

Preferential Fertilization

The classical concept of fertilization considers that the two male gametes released into the

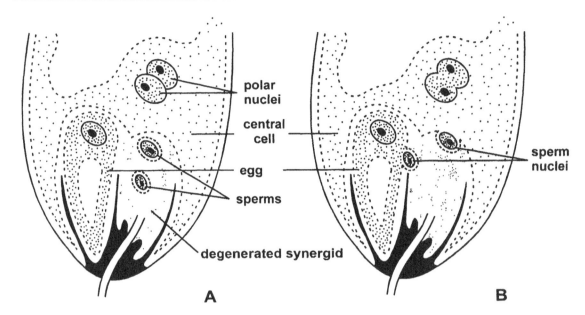

Fig. 8.9 Diagram of pollen tube entry into the embryo sac (A) and entry of sperm cells into the egg and central cell (B). Longisections of the micropylar part of the embryo sac in a plane passing through one of the synergids and the egg. The synergid that has received the pollen tube has degenerated and lacks intact plasma membrane; thus the sperms are able to come into direct contact with the membranes of the egg and central cell. Sperms enter the egg and the central cell through membrane contact and membrane fusion (after Shivanna et al. 1997).

synergids are identical and that fusion of one gamete with the egg and the other with the central cell is random. However, many studies have shown that in some species such as *Plumbago* (Russell 1984) and *Brassica* (Knox and Singh 1987, Mogensen 1992) the two male gametes formed by the generative cell are not identical (see Chapter 1); they show distinct differences in morphology and contents. Sperm dimorphism raises interesting questions about the possible existence of preferential fertilization, involving fusion of a specific sperm with the egg cell.

Detailed studies on *Plumbago* (Russell 1991, 1993) revealed fusion of the plastid-rich sperm with the egg in 94% cases (Russell 1985). In maize, nuclear differences in the two sperms due to failure of B-chromosomes to segregate have been reported (Carlson 1969); in such instances, the sperm cell with extra B-chromosome fused with the egg in up to 75% cases. Thus, at least in some of the aforementioned species, fusion of the gametes with the egg and the central cell

was not random. Preferential fertilization apparently involves gamete recognition in the embryo sac. The basis of such recognition is not yet known (Knox and Singh 1987).

Inheritance of Plastids

Details of fusion events described during fertilization suggest the entry of both cytoplasm and nucleus of the sperm cells into the egg/central cell. However, classical genetic studies on the inheritance of plastids and of cytoplasmic male sterility controlled by mitochondrial genome (see Chapter 5), clearly show that in the majority of species both plastids and mitochondria are maternally inherited. In several species, electron microscopic studies have shown different levels at which plastid elimination takes place in the pollen (Hagemann and Schroder 1989): (i) During asymmetric division of the microspore most of the plastids migrate to the cytoplasm away from the region of the nucleus. Because of this reorganization of the cytoplasm the resultant generative cell contains no plastids (*Lycopersicon*). (ii) The generative cell contains a few plastids soon after microspore mitosis, which are degraded during maturation of the pollen (Clauhs and Grun 1977) (*Solanum*). (iii) The generative cell as well as sperm cells contain plastids. After discharge of the sperm cell in the synergid, the plastids as well as mitochondria are stripped from the sperm cells and left in the synergid (*Triticum*). In some species, enucleated cytoplasmic bodies containing both plastids and mitochondria, apparently of the sperm cell(s), have been localized in the synergid (Jensen and Fischer 1968, Mogensen 1988). In a few species such as *Pelargonium* and *Plumbago*, the plastids show biparental inheritance (Russell 1984, 1985).

The structural aspects associated with mitochondrial elimination have yet to be clearly understood. Unlike plastids, which are present in the sperm cells of only a few species, sperm cells do contain mitochondria in all species. They may be eliminated at the time of fertilization (Mogensen 1988) or they may enter the egg and then be eliminated in the zygote (Yu et al. 1992).

Positional Effect of Ovules

In many multiovulate species, the number of seeds produced is less than the number of ovules present in the ovary. This is due either to failure of fertilization in a proportion of ovules or to the abortion of some of the young seeds (Bawa and Webb 1984, Nakamura 1988). Often the probability of an ovule getting fertilized and/or developing into a mature seed is related to its position in the ovary.

In some species which show a row of linearly arranged ovules, those situated at the stylar end develop into seeds in a significantly larger proportion of fruits than those situated at the pedicel end (*Raphanus*—Hill and Lord 1987, *Dalbergia*—Ganeshaiah and Shaanker 1988; *Phaseolus*—Nakamura 1988, Rocha and Stephenson 1990, 1991). The time of fertilization seems to influence the probability of an ovule producing a mature seed (Bawa and Webb 1984,

Ganeshaiah and Shaanker 1988, Rocha and Stephenson 1991). It is suggested that the ovules fertilized first have a temporal advantage in competing for maternal resources. In both *Dalbergia* (Ganeshaiah and Shaanker 1988) and *Phaseolus* (Rocha and Stephenson 1991), the ovules situated near the stylar end are the first to be fertilized and thus most likely to develop into mature seeds, while those located near the pedicel end are fertilized later and more likely to abort. In *Dalbergia*, seeds located at the pedicel end are aborted by the production of an abortive agent by the young seed developed at the stylar end and its diffusion toward the pedicel region (Ganeshaiah and Shaanker 1988). Selective abortion of some seeds in the developing fruits has been explained on the basis of parent-offspring conflict or sibling rivalry (Ganeshaiah and Shaanker 1988).

Studies on *Phaseolus* (Rocha and Stephenson 1990, 1991) have also shown that the ovules near the stylar end are more likely to produce heavier seeds and more vigorous progeny. When only the fastest growing pollen tubes are allowed to enter the ovary (through excision of the style after allowing only the fastest growing pollen tubes to pass through), only the ovules at the stylar end are fertilized. Development of heavier seeds, capable of producing more vigorous progeny, at the stylar end of the fruit seems to be due to the temporal advantage and/or superior paternal parentage they have over those situated at the pedicel end. Similarly, in *Cucurbita*, 25% of the ovules situated in the second quarter of the ovary from the stylar end are the first to get fertilized by the fastest growing pollen tubes and the seeds from this quarter are significantly larger than those produced in other parts of the fruit.

IN-VITRO POLLINATION AND FERTILIZATION

Since fertilization in flowering plants takes place deep inside the ovule, it imposes major technical problems for both descriptive and experimental studies. Because of this constraint, information on fertilization in flowering plants has lagged far behind that in animals and some of the lower plants, in which fertilization is external and can be observed under the microscope. In a few species of flowering plants the ovules are transparent (many orchids) or the micropylar part of the embryo sac grows out of the micropyle (species of *Torenia, Jasione*) (Poddubnaya-Arnoldi 1960, Erdelska 1974). Such species provide suitable systems for study of fertilization in the living embryo sac; so far only limited progress has been made on these aspects (Russell 1992). As pointed out earlier, in recent years one such species, *Torenia,* has been effectively used to understand the role of synergids in pollen tube guidance into the ovule (see page 130).

One of the main objectives of experimental embryologists has been to carry out fertilization in vitro by using isolated eggs and sperm cells, and development of the in vitro formed zygote into an embryo. This would enable effective experimentation on fertilization in flowering plants comparable to that in animals and lower plants and thus lead to rapid progress in both fundamental and applied aspects of fertilization. Although a beginning was made during the 1960s, it is only in the last decade that true in-vitro fertilization using isolated sperms and eggs has been achieved. The techniques of in-vitro pollination and fertilization provide powerful tools for studying fundamental aspects of pollen-pistil interaction and fertilization, and effective manipulation of these processes.

In-vitro Pollination

The technique of in-vitro pollination involves aseptic pollination of cultured pistils or ovules to achieve pollen germination, pollen tube growth and its entry into the embryo sac, double fertilization and subsequent development of embryo, endosperm and seed.

Pollination of cultured pistils

Pollination of cultured pistils and their growth in vitro into mature fruits provide a very convenient technique for experimental studies on pollen-pistil interaction, fertilization and development of fruits and seeds. Pollination of cultured pistils has been successfully achieved in species of *Nicotiana* (Dulieu 1966, Rao 1965), *Petunia* (Shivanna 1965) and *Antirrhinum* (Usha 1965). The pistils were surface-sterilized, cultured on a simple nutrient medium and pollinated with fresh pollen collected aseptically. Pollen grains in compatibly-pollinated pistils germinated, pollen tubes grew through the style, entered the ovary and effected fertilization. Such cultures developed into normal mature fruits in about 3 weeks, the same period required for fruit development in vivo. Pollination of cultured pistils with incompatible pollen did not result in fruit- and seed-set; incompatible pollen tubes were inhibited in the style (Shivanna 1965). Thus, pollination of cultured pistils was not effective in overcoming sexual incompatibility because the zone of inhibition, namely the stigma and style, was intact.

Pollination of cultured ovules

The technique of pollination of cultured ovules was standardized in the laboratory of the Late Panchanan Maheshwari, Department of Botany, University of Delhi, during the 1960s (see Rangaswamy 1977). The procedure for intraovarian pollination was the first technique standardized with members of Papavaraceae; it involves injecting the pollen grains (suspended in a suitable medium) directly into the ovary, achieving pollen germination, entry of pollen tubes into ovules and fertilization. Viable seeds consequent to intraovarian pollination have been obtained in *Papaver somniferum, P. rhoeas, Argemone mexicana* and *A. ochroleuca* (Maheshwari and Kanta 1961, Kanta and Maheshwari 1963a). The intraovarian pollination technique has also been applied to achieve interspecific hybridization between *A. mexicana* and *A. ochroleuca.*

There are many limitations in extending the technique of intraovarian pollination to other taxa. It is not suitable for species in which there is not enough space in the ovary to inject pollen sus-pension. Also, in species in which a sugar (usu-ally sucrose) is required for pollen germination, injection of pollen suspension in sugar solution may render the ovary prone to bacterial and fun-gal infections. These limitations have been over-come in techniques involving aseptic culture of ovules to achieve test-tube fertilization.

Kanta et al. (1962) cultured unpollinated ovules of *Papaver somniferum* on a nutrient medium and copiously dusted the ovules with fresh pollen grains. Following 'ovule pollination' (Fig 8.10A), the pollen grains germinated nor-mally, the pollen tubes entered the ovules and effected double fertilization. The fertilized ovules developed into mature seeds on the same cul-ture medium (Fig. 8.10B, C). After this initial suc-cess, the technique of test-tube fertilization was extended to *Argemone mexicana, Eschscholzia californica, Nicotiana tabacum, Dianthus caryophyllus* and *Dicranostigma franchetianum* (see Shivanna and Johri 1985, Zenkteler 1990).

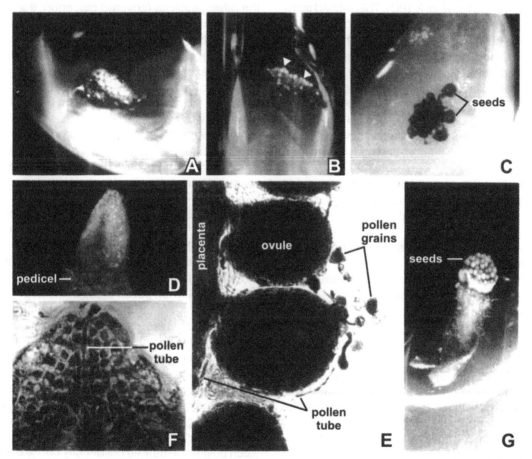

Fig. 8.10 In-vitro pollination of ovules. A, B. *Papaver somniferum* A—Group of ovules cultured on a nutrient medium and pollinated with pollen grains. B—A 7-day-old culture showing many developing seeds (arrowheads). C. *Argemone mexicana*. A 4-week-old culture showing development of seeds following ovule pollination. D-G. *Petunia axillaris*: Placental pollination. D—Placental explant made ready for culture. E—Free-hand transection through self-pollinated placenta 24 h after culture; only part of the placenta with two complete ovules shown. Pollen germination and profuse growth of pollen tubes are obvious. F—Longisection of micropylar part of an ovule to show pollen tube entry into the ovule. G—A 24-day-old placental self-pollinated culture. Mature seeds are apparent (A-C—after Kanta and Maheshwari 1963b, D-G—after Shivanna 1969).

Rangaswamy and Shivanna (1967, 1971a) attempted to overcome self-incompatibility in *Petunia* by using the technique of ovule pollination. The culture of isolated ovules or groups of ovules together with the pollen grains did not result in fertilization or seed development, even with compatible pollen. They modified the technique: instead of pollinating the isolated ovules, the entire mass of exposed ovules intact on the placenta together with a short length of pedicel was cultured on the medium and the ovule mass dusted with pollen (Fig. 8.10D). This refined technique, termed 'placental pollination', helped to bring the pollen into direct contact with the ovules without disturbing their original arrangement and thereby precluding any injury to them. Following placental pollination, pollen grains germinated readily on the ovules (Fig. 8.10E), pollen tubes entered the ovules (Fig. 8.10F), and effected fertilization. The fertilized ovules showed normal development of the embryo and endosperm, and mature viable seeds were obtained in 3 weeks after fertilization (Fig. 8.10G); the same period is required for seeds to mature in vivo. Unlike stigmatic pollination in which self-pollination is invariably a failure, in placental pollination both self- and cross-pollinations were equally effective in inducing seed-set.

The technique of placental pollination (in *Petunia*) was further modified to treat the ovules on the two placentae, of the same ovary, differently (differential pollination) (Shivanna 1971). The two placentae were separated mechanically by introducing a piece of cellophane between them. When one of the placentae was maintained as control and the other pollinated with self- or cross-pollen grains, the ovules on the control placenta invariably shrivelled, whereas the ovules on the pollinated placenta developed into seeds, irrespective of self- or cross-pollination. When one of the placentae was self-pollinated and the other cross-pollinated, seeds developed equally well on both placentae, thus demonstrating conclusively the equal efficacy of both self- and cross-pollinations of the placenta.

These experiments demonstrated that once the stigma and the style are eliminated, the ovules show no preferential receptivity to cross-pollen; this led to further modifications of the techniques to study the interaction between stigmatic pollination and placental pollination. The technique of two-site pollinations was devised with *P. axillaris* (Rangaswamy and Shivanna 1971b). The ovary wall was carefully peeled to expose the ovules on one of the placentae (exposed placenta) while the wall was retained on the other placenta (covered placenta); the style and the stigma were kept intact. Thus pollination could be carried out both on the stigma and on the exposed placenta in the same pistil. A series of experiments using two-site pollinations demonstrated that stigmatic self-pollination does not affect the response of placental self-pollination, and vice-versa.

Success with in-vitro pollination in several species attracted the attention of many investigators and attempts have been made to use the technique in many fundamental and applied aspects. It has also been successfully used to overcome self-incompatibility in *Petunia hybrida* (Niimi 1970, Wagner and Hess 1973). Balatkova and Tupy (1968) achieved in-vitro fertilization and seed formation in *Nicotiana tabacum* by using already germinated pollen grains. They further showed that when pollen tubes in which gamete formation had occurred were deposited on unpollinated ovules, successful fertilization was effected. These studies demonstrate the feasibility of treating either the male gametes or female gametes, without affecting the other.

Pollination with killed pollen, and with pollen of alien species, occasionally stimulated the unfertilized egg to develop parthenogenetically (Maheshwari and Rangaswamy 1965). Logically, such pollination carried out directly on the ovules would be more effective in inducing parthenogenesis than that carried out on the stigma, as the diffusible factors which stimulate the egg become available much closer to the embryo sac. The feasibility of this technique has been demonstrated by Hess and Wagner (1974). Their attempts to obtain androgenic haploids in *Mimulus luteus* by anther culture technique were unsuccessful. However, when the ovule mass of *M. luteus* was pollinated in vitro with the pollen of *Torenia fournieri*, 1% of the ovules developed parathenogenetically and gave rise to gynogenic haploid plantlets.

Zenkteler and his associates (Zenkteler 1967, 1970, 1990) successfully applied the technique of placental pollination to raise many interspecific and even intergeneric hybrids. Details are given in Chapter 13.

In-vitro Fertilization

Further progress during the 1960s to accomplish in-vitro fertilization with isolated eggs and sperm cells was hampered for want of techniques to isolate viable male gametes (sperms) and eggs and to fuse them in vitro to achieve fertilization. Rapid advances in protoplast technology and somatic hybridization during the 1970s and the 1980s gave new impetus to studies on in-vitro fertilization (Yang and Zhou 1992, Shivanna and Rangaswamy 1997). Studies were initiated in many laboratories on isolation of sperms and embryo sacs in the early 1980s and progress astounding. Basically two procedures were used to isolate the sperms from pollen grains/tubes: mechanical method and osmotic shock method (Russell 1991, Chaboud and Perez 1992, Cass 1997). In the first method, the pollen grains/tubes are ruptured mechanically by grinding them in an isolation medium; in the second, the pollen grains/tubes are allowed to rupture in a hypotonic solution. The debris is removed through filtration and the sperms then separated through density gradient centrifugation. Embryo sacs are generally isolated by combining enzymic maceration of ovules with mi-

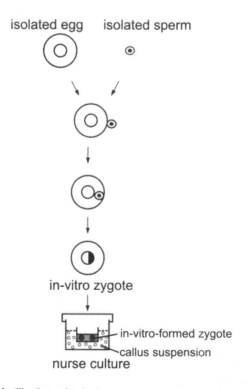

Fig. 8.11 Diagram of in-vitro fertilization using isolated egg and sperm cells (modified from Kranz et al. 1991).

crodissection (Theunis et al. 1991, Matthys-Rochon et al. 1997). Subsequently the technique was further refined to isolate protoplasts from constituent cells of the embryo sac (egg, synergids and central cell). Thus, by 1990 protocols were available for isolation of viable protoplasts from sperms and eggs of many plant species and the stage set for serious attempts at achieving in-vitro fertilization.

The pioneering success of fusing isolated egg and sperm cells in vitro was accomplished in maize (Fig. 8.11) by Kranz and his associates (Kranz 1997, Kranz and Lorz 1993, 1994, Kranz and Dresselhaus 1996) at the University of Hamburg. Fusion of gametes was performed under the microscope in microdroplets (2 µl) of fusion medium overlaid with mineral oil. Isolated egg and sperm cells held in microcapillaries were transferred to the fusion droplets by using a computer-controlled dispenser. The egg and sperm cells were aligned electrophoretically or mechanically with microneedles and fusion induced by giving one or a few short pulses of direct current. Fusion was completed in less than 1 s and karyogamy occurred in less than 1 h from fusion (Faure et al. 1993). As in somatic protoplast fusion, in gametic fusion also the composition and osmolarity of the fusion medium were critical.

The in-vitro-formed zygotes were cultured on a semipermeable membrane placed on fast-growing non-morphogenetic cell suspension cultures derived from maize embryos or microspores. In these nurse cultures the zygotes established polarity and as many as 85% zygotes divided, giving rise to globular structures, proembryos and transition phase embryos comparable to seed embryos in vivo, and eventually to fertile plants.

Subsequently, in-vitro fusion of isolated gametes in maize was achieved not with electric pulse but in the presence of high calcium and high alkalinity (pH 11) (Kranz and Lorz 1993); microinjection of sperm nuclei into isolated embryo sacs was also essayed (Matthys-Rochon

et al. 1997). The sperm cell was recently fused with the central cell also; the resultant fusion product gave rise to a callus comparable to that of the endosperm (Kranz and Dresselhaus 1996). Further studies have shown that the fusion product of maize egg and sperm of related plants such as wheat and sorghum underwent divisions, while that of maize egg and sperm of unrelated plant, *Brassica*, did not. The fusion product of two maize eggs failed to undergo divisions (Kranz and Dresselhaus 1996). Most of the achievements of the past decade on in-vitro fertilization have been with maize, except for a report of electrofusion mediated fertilization of isolated egg and sperm cells in wheat (Kovacs et al. 1995). However, in wheat, in-vitro fertilized zygotes have given rise to multicellular structures and not embryos. The technique is bound to be extended to other species in the coming years.

Applications of in-vitro Pollination and Fertilization

In-vitro pollination and fertilization provide very effective techniques in fundamental and applied areas of fertilization and seed development. Pollination of cultured pistils is simple and can be used for a number of experiments on seed and fruit development. The in-vitro system with its controlled conditions provides a much more convenient method for understanding many fundamental aspects in particular of ovule physiology before, during and after fertilization than in-vivo systems.

Potential applications of pollination of cultured ovules, especially of placental pollination, are many and the techniques have already been used in a number of applied areas. The placental pollination method, for example, should prove promising in studies of induced parthenogenesis and in mutation research. Placental pollination provides a better means of treating only the female partner (ovules) with any chemical or physical factor. Likewise, use of pollen tube cultures for in-vitro pollination offers a convenient method of treating only the male partner (gametes in pollen tubes) with physical and chemical mutagens.

Advances in in-vitro fertilization using isolated gametes and embryogenesis of in-vitro-formed zygotes have opened a vista of experimental problems on fertilization, viz: (a) mechanisms of recognition, adhesion and fusion of gametes; (b) parthenogenesis; (c) polyspermy; (d) establishment of polarity in the zygote and (e) genetic regulation of early embryogenesis. Research on fertilization in flowering plants has entered an exciting phase; rapid progress in understanding the process of fertilization is expected in the coming years.

9

Self-incompatibility

Self-incompatibility (SI) is an important outbreeding mechanism (see Chapter 7) and is widespread in flowering plants. About half the known species of flowering plants are estimated to be self-incompatible (Brewbaker 1959, de Nettancourt 1977, 2001). Although SI was known since the second half of the 18th century, it was Darwin (1876, 1877) who carried out systematic studies on SI and emphasized its significance as an outbreeding mechanism. SI is defined as the inability of a hermaphrodite plant producing functional gametes to effect fertilization upon self-pollination. In SI species, self-pollen grains are actively recognized and inhibited on the surface of the stigma before germination or during the growth of pollen tubes through the pistil before reaching the ovule; only cross-pollen grains effect fertilization. SI is therefore a prefertilization barrier. Self-pollination does not result in the loss of ovule fertility; even after self-pollination ovules remain available for fertilization by cross-pollen (Crowe 1964, Heslop-Harrison 1978). There are many reports of post-fertilization operation of self-incompatibility due to abortion of the embryo (Dulberger 1964, Crowe 1971, Seavey and Bawa 1986). The genetics and physiological mechanism of post-fertilization incompatibility are obviously different from SI and should be considered different from SI (Mulcahy and Mulcahy 1983).

The genetics of SI is well investigated in at least a few species. SI represents a model sys-

tem for investigating cell recognition events in plants and is uniquely accessible for molecular analysis. SI is also important for plant breeders as it provides an efficient pollination control system for commercial production of hybrid seeds. Thus, apart from reproductive biologists, SI has attracted the attention of population geneticists, plant breeders, biochemists and molecular biologists. The literature has been reviewed from time to time: Lewis (1949, 1954, 1979), Brewbaker (1957), Crowe (1964), Linskens (1965, 1975a), Lundqvist (1965), Vuilleumier (1967), Arasu (1968a), de Nettancourt (1977), Heslop-Harrison (1978, 1983), Shivanna (1982), Shivanna and Johri (1985), Richards (1986), Hodgkin et al. (1988), Bell (1995), and Raghavan (1999). Advances in the molecular biology of SI have been adequately covered by Hinata et al. (1993), Dickinson (1995), Nasrallah and Nasrallah (1993), Sims (1993), Franklin et al. (1995), Brugiere et al. (2000), Franklin-Tong and Franklin (2000), Wheeler et al. (2001) and Stone and Goring (2001).

EVOLUTION OF SI

SI is generally considered a primitive condition and self-compatibility a derived condition (Brewbaker 1957, de Nettancourt 1977). However, there is no unanimity as to whether SI arose once or several times independently in different groups of flowering plants. According to

Whitehouse (1950) and Crowe (1964), SI arose once and different types of SI have evolved from a primitive, multiallelic and gametophytic type. Nevertheless, many recent investigations, especially those on the homology of S-genes in different species and the mechanism of action of S-genes, support polyphyletic origin of SI (Read et al. 1995).

On the basis of floral morphology, SI has been divided into two broad categories—homomorphic and heteromorphic. In homomorphic SI, there are no morphological differences in the flowers produced between plants of the same species. In heteromorphic SI, flowers produced on different plants can be grouped into two or three morphological types.

HOMOMORPHIC SI

Homomorphic SI has been reported in over 250 species of 600 genera belonging to at least 71 families (Brewbaker 1959, Crowe 1964). However, detailed studies have been conducted on just a limited number of species belonging to Solanaceae, Cruciferae, Compositae, Liliaceae and Poaceae; *Nicotiana, Petunia, Papaver* and *Brassica* are the most intensively investigated systems, particularly with reference to aspects of molecular biology.

Genetics

The first satisfactory genetic interpretation of homomorphic SI was given by East and Mangelsdorf (1925) for *Nicotiana.* SI is governed by a single locus termed S-locus with multiple alleles designated as S_1, S_2, ... S_n. Pollen grain carrying an S allele identical to one or both the alleles present in the pistil is inhibited (Fig. 9.1). Subsequent genetic studies (Crowe 1954, Bateman 1955) showed that on the basis of genetic control of incompatibility in the pollen, homomorphic SI can be broadly classified into two types: gametophytic and sporophytic. In the gametophytic type, the incompatibility phenotype of the pollen is based on the haploid genotype of the individual pollen grain, while in the sporo-

phytic type the incompatibility phenotype depends on the diploid genotype of the pollen-producing parent (Fig. 9.1). Gametophytic (GSI) and sporophytic (SSI) types of self-incompatibility are associated with many other differences (Table 9.1). Members of Solanaceae, Leguminosae, Rosaceae, Onagraceae, Papaveraceae, Liliaceae, Commelinaceae and Poaceae show GSI whereas those of Cruciferae, Compositae and Convolvulaceae exhibit SSI.

Table 9.1 Correlations between the genetics of self-incompatibility and some other features

Features	Gametophytic	Sporophytic
Pollen cytology	2-celled	3-celled
Zone of inhibition	style	stigma surface
Nature of stigma	wet/dry	dry
Development of callose plug in stigmatic papilla	absent	present

Lewis (1949, 1956) and Pandey (1958, 1970) explained the two types of SI on the basis of the differences in the time of S-gene action (Fig. 9.2). In sporophytic systems, the S-gene is activated before completion of meiosis, with the result that products of both the alleles are distributed in all the four microspores and thus in all the pollen grains produced by the plant. In gametophytic systems, on the other hand, S-gene activation is delayed until completion of meiosis, with the result that two of the microspores of each tetrad receive the products of one of the S-alleles, and the other two the products of the other S-allele; thus, each pollen will have the products of only one of the S-alleles.

Much evidence indicates (see page 144) that in SSI, S-allele specific products are synthesized in the tapetum and incorporated into the pollen exine following breakdown of the tapetum (Fig. 9.3) (Dickinson and Lewis 1973a, Howlett et al. 1975, Heslop-Harrison 1978).

It is quite likely that S-allele products in sporophytic systems are synthesized by the pollen grain as well as the tapetum, and are present in the pollen cytoplasm as well as pollen exine. Until

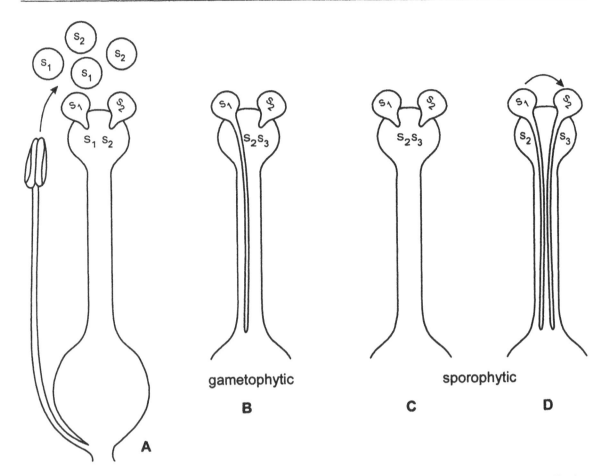

Fig. 9.1 Diagram of operation of gametophytic and sporophytic self-incompatibility in homomorphic systems. Self-pollination of a plant with $S_1 S_2$ alleles results in inhibition of both S_1 and S_2 pollen (A). B-D. Cross-pollination of $S_2 S_3$ pistil with pollen of $S_1 S_2$ plant. Only S_2 pollen is inhibited in gametophytic system because of the presence of the matching allele in the pistil; S_1 pollen will function as it does not have a matching allele (B). In sporophytic systems S alleles may act independently (C) or show dominance of one over the other (D, dominance of S_1 over S_2 is indicated by arrow)). When S_2 and S_3 alleles have independent action, both are inhibited, when S_1 is dominant over S_2, both are functional (redrawn from Shivanna and Johri 1985).

recently, there was no information on the nature of S-allele components in the pollen of any SI species. This was a major limitation in understanding the time of expression of S-genes or the location of S-allele products in the pollen. Only recently has the S-allele product been identified in the pollen of *Brassica* (Schopfer 1999). These studies confirmed that the pollen S-allele product is synthesized in the pollen cytoplasm as well as the tapetum, but the involve-

ment of the tapetum and/or the cytoplasmic products in SI response is yet to be clearly understood.

Although in the majority of gametophytic systems SI is controlled by a single locus with multiple alleles, there are some variations. In members of Poaceae, SI is controlled by multiple alleles at two loci, S and Z, which are inherited independently (Lundqvist 1956, Hayman 1956, Stone and Goring 2001). Allelic matching of both

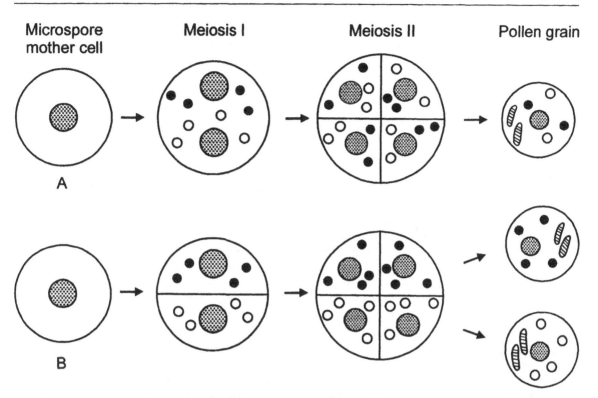

| Microspore mother cell | Meiosis I | Meiosis II | Pollen grain |

Fig. 9.2 Proposed model to explain sporophytic (A) and gametophytic (B) self- incompatibility based on the time of S-gene activation. In sporophytic SI, S-allele products are synthesized before meiosis with the result that all microspores of the tetrad receive products of the S-alleles. In gametophytic SI, S-allele specific products are synthesized after completion of meiosis; two of the microspores receive products of one of the S-alleles and the other two the products of the other S-allele (modified from Pandey 1979).

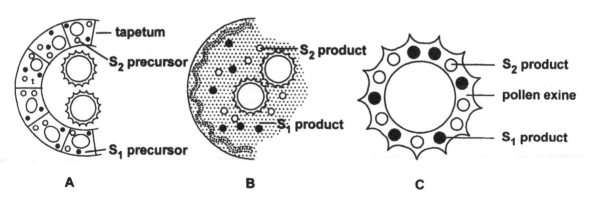

Fig. 9.3 Tapetal origin of S-allele products in sporophytic systems and their incorporation into pollen exine (modified from Pandey 1979).

genes is required for SI. In members of Ranunculaceae and Chenopodiaceae, SI is reported to be controlled by four multiallelic genes that are linked in their inheritance (Lundqvist 1975, 1990, Osterbye 1975, Larsen 1977). According to Lundqvist (1975) SI controlled by many genes is the primitive one and that controlled by a single gene the derived condition.

Even in members of Cruciferae (*Eruca sativa, Raphanus sativus* and *Brassica campestris*) SI has been reported to be determined by multiple genes (Verma et al. 1977, Lewis 1979). Lewis and associates (1988) and Zuberi and Lewis (1988) explained that in *R. sativus* and *B. campestris,* SI is controlled by two linked genes, S and G; the former has sporophytic control and the latter gametophytic control in the pollen. Matching of alleles at both S and G loci is required for SI. These researchers consider the G locus the ancestor of the S-gene in the gametophytic system that has been retained in the sporophytic system.

Determination of the genetics of SI, in particular of SSI, is time consuming and laborious; it involves a large number of controlled pollinations and analyses of the progeny often extending to many generations (Wallace 1979). Therefore, only a few species have been genetically analysed. However, studies of other correlated features (Table 9.1) give an indication of the genetic basis of SI.

The number of alleles in a population is potentially large. Over 40 S-alleles have been reported in populations of *Oenothera organensis* (Emerson 1939), *Papaver rhoeas* (Lawrence et al. 1993) and *Lolium perenne* (Fearon et al. 1994). In *Sinapis arvensis*, Stevens and Kay (1989) reported 35 different S-alleles from 35 plants raised from seeds collected randomly from a population. In *Brassica oleracea* (Ockendon 1975), 49 alleles have been identified in a wide range of cultivars.

Cytology

Cytological details of pollen inhibition have been investigated in a number of species. In SSI inhi-

bition is strictly confined to the stigma surface. Self-pollen grains either fail to germinate or the small tubes that emerge, at most, penetrate the cuticle (Kanno and Hinata 1969, Dickinson and Lewis 1973a, b). Inhibition of pollen is generally associated with the deposition of callose in the germ-pore or at the tube tip (Figs. 9.4, 9.5).

Interestingly, stigmatic papillae in contact with incompatible pollen, also develop a lenticular plug of callose at the tip, between the cell wall and the plasma membrane (Dickinson and Lewis 1973a, b, Heslop-Harrison et al. 1974, Howlett et al. 1975, Shivanna et al. 1978a, Kerhoas et al. 1983). This reaction is termed the rejection reaction (Fig. 9.4A, B, 9.5). No such plugs develop following compatible pollination. The deposition of callose plug in the papillae is rapid and is often apparent within 10 min of pollination. Although it is a clear manifestation of self-pollination in SSI systems, it does not seem to play a direct role in pollen tube inhibition. Inhibition of callose deposition by treating the stigmas with 2-deoxy-D-glucose before self-pollination (Singh and Paolillo 1989) showed that callose deposition in the stigmatic papillae is not essential for pollen inhibition (see also Sulaman et al. 1997).

Rejection reaction on the stigmatic papilla has been used as an assay to analyse the location of pollen components involved in SSI in *Raphanus* and *Iberis* (Dickinson and Lewis 1973a, Heslop-Harrison et al. 1974). Deposition of exine material (isolated through high-speed centrifugation of pollen or leached onto an agarose film) from the self-pollen onto the stigmatic papillae stimulated development of the rejection reaction; the pollen exine material from cross-pollen was not effective. Thus the exine components are involved in at least one phenotypic manifestation of SI, development of the rejection reaction.

In gametophytic species in general, pollen germination is not affected but pollen tubes are inhibited in the style (de Nettancourt 1977). However there are variations. In members of Papaveraceae, the zone of inhibition is generally the stigma. In some species such as

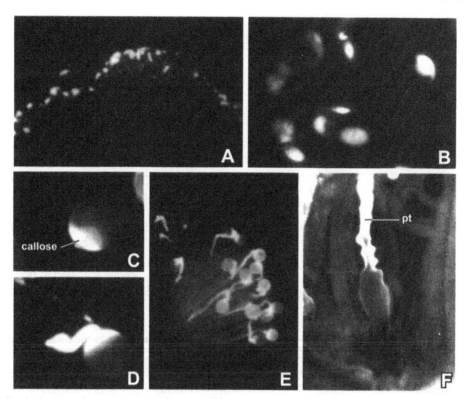

Fig. 9.4 Pollen inhibition in self-pollinated pistils. Fluorescence micrographs of the stigma 6 h after self-pollination stained with decolourized aniline blue. A, B. *Brassica campestris*. Most of the papillae show rejection reaction (development of callose plug at the tip). A few of the papillae are magnified in B. C-E. *Saccharum bengalensis*. Pollen grains are inhibited before germination (C), after germination but before pollen tube entry (D), or after growing a short distance in the stigma (E); callose has accumulated at the germ pore and pollen tubes. E. *Lycopersicon peruvianum*. Pollen tube (pt) is swollen at the tip and has burst in the transmitting tissue of the style (A-E—after Shivanna 1982, F—after de Nettancourt et al. 1974).

Oenothera (Dickinson and Lawson 1975a) and members of Commelinaceae (Owens 1981), pollen tubes are inhibited in the tissues of the stigma itself. In many species of grasses (Heslop-Harrison et al. 1974, Sastri and Shivanna 1979) the zone of inhibition is variable often within the same species (Fig. 9.4C-E). Inhibition may take place at the levels of pollen germination, pollen tube entry into the papillae, and pollen tube growth at different levels in the stigmatic tissue. In some species, inhibition is delayed until the tubes reach the main axis of the pistil (Thomas and Murray 1975). Pollen tube inhibition is associated with deposition of an excessive amount of callose in the tube, particularly at the tip (Fig. 9.4C-E).

The tips of incompatible tubes often show swelling and/or bursting in the stylar region. In *Lycopersicon peruvianum* (de Nettancourt et al. 1974), compatible tubes have typical two-layered wall—an outer pectocellulosic wall of loose fibrils and an inner homogeneous callose wall. Incompatible tubes display similar wall structure during initial growth. As they grow down one-third of the style, the inner layer gradually becomes thin and numerous particles (ca 0.2 μm diameter) accumulate in the tube cytoplasm. Also, the endoplasmic reticulum in the incompatible tubes is altered into a whorl of concentric layers. Eventually the inner wall disappears completely and the tube tip bursts releasing these particles into the intercellular spaces of the

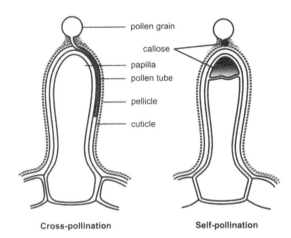

Cross-pollination   Self-pollination

Fig. 9.5 Diagram of early events of pollen-pistil interaction in sporophytic SI systems. A. Cross-pollination. B. Self-pollination. Note development of callose plug (rejection reaction) in the stigmatic papilla following self-pollination (after Shivanna 1982).

transmitting tissue (Fig. 9.4F). On the basis of such evidence de Nettancourt and colleagues (1975) suggested that SI is an active process resulting in the destruction of the pollen tube wall, and inhibition of protein synthesis a result of altered endoplasmic reticulum.

Although excessive deposition of callose in the pollen tube is a characteristic feature of self-pollen tubes (Fig. 9.4C-F), detailed studies of Shivanna and co-authors (1978b, 1982) on SI response of grasses have shown that callose deposition is not the primary event and therefore not the cause of inhibition. Deposition of pectic material at the tube tip is the earliest deviation observed in selfed tubes. Callose is initially deposited behind the tip region and later may extend to the tip.

Physiological and Biochemical Studies

A number of physiological and biochemical differences in the pollen and/or pistil between self- and cross-pollinations have been reported in both GSI and SSI systems. In *Brassica*, differences in pollen adhesion and hydration, and mobility of pollen wall components between self- and cross-pollen have been reported (Stead et

al. 1979, Roberts et al. 1980). The extent of adhesion and hydration (Fig. 9.6) was much less in self-pollen than in cross-pollen.

There are many differences in the metabolism between self- and cross-pollinated pistils. Detailed studies have been carried out on many enzymes, in particular peroxidases (Pandey 1967, Bredemeijer 1974, 1982) and esterases (Pandey 1973), in self- and cross-pollinated pistils mainly to analyse their role in the inhibition of self-pollen tubes. Changes have also been shown in the synthesis of RNA and proteins in the style and ovary in response to self- and cross-pollination (van der Donk 1974, Deurenberg 1976, 1977b). However, to date no evidence has been adduced that any of these differences is directly involved in pollen tube inhibition (see Shivanna and Johri 1985).

One of the effective approaches to studying the basis of SI recognition and inhibition is through the use of an in-vitro assay in which self-pollen grains are selectively recognized and inhibited (see Shivanna and Johri 1985). Many such assays have been reported from time to

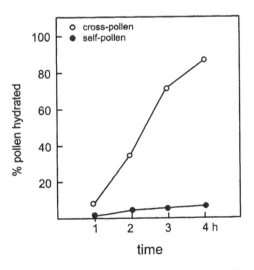

Fig. 9.6 Differences in pollen hydration following self- and cross-pollinations in *Brassica oleracea*. Pollen hydration is expressed as % of the total number of pollen grains that have become spheroidal (after Roberts et al. 1980).

time using crude/dialyzed extracts (see Jackson and Linskens 1990). Although Ferrari and Wallace (1975) reported selective inhibition of in-vitro pollen germination by self-stigma leachate in *Brassica oleracea*, subsequent studies failed to record such a differential effect (Roberts et al. 1983). One of the major difficulties in establishing such an assay in *Brassica* was lack of a suitable medium to achieve satisfactory in-vitro germination and tube growth. Later many media were formulated to achieve good in-vitro germination (Roberts et al. 1983, Hodgkin and Lyon 1983) as well as tube growth (Shivanna and Sawhney 1995). Using one of these media Singh and Paolillo (1989) reported differential effects of stigmatic eluates on self- and cross-pollen.

The progress made through use of in-vitro assay has been more substantial in gametophytic systems. In *Petunia hybrida* (Sharma and Shivanna 1982, 1983a) as well as *Nicotiana alata* (Sharma 1986), both crude as well as dialyzed extracts from unpollinated pistils inhibited in-vitro germination and tube growth of self-pollen but not of cross-pollen (Fig. 9.7A, B, 9.8A-

C). Pistil extract from the bud, which permits incompatible pollen to grow through, did not show inhibition of self-pollen.

Incorporation of a transcription inhibitor (actinomycin D/cordycepin) in the culture medium (containing pistil extract) was effective in overcoming inhibition of self-pollen (Fig. 9.8D) (Shivanna and Sharma 1984). These studies clearly established that (i) the S-allele components are synthesized in the pistil before pollination and (ii) inhibition of self-pollen is dependent on fresh transcription in the pollen. Subsequently an in-vitro assay was established by incorporating isolated S-glycoproteins (Jahnen et al. 1989).

In many animal systems and lower plants (see Marx 1978, Shapiro and Eddy 1980), it has long been known that gamete recognition is established as a result of complementation between surface saccharides of one gamete and saccharide-specific receptors (lectins) on the other gamete. This is the result of in vitro experiments in which gamete recognition is inhibited when the recognition molecules are blocked by specific sugars or lectins.

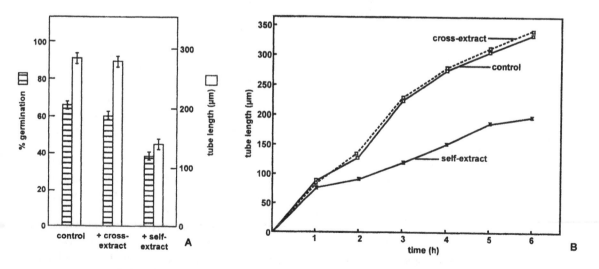

Fig. 9.7 In-vitro assay for self-incompatibility inhibition in *Petunia hybrida*. A. Effects of self- and cross-pistil extracts on pollen germination and pollen tube growth. Cross-extract had no effect while self-extract inhibited both pollen germination and tube growth. B. Temporal details of pollen tube growth in control and presence of self- and cross-pistil extract (after Sharma 1986).

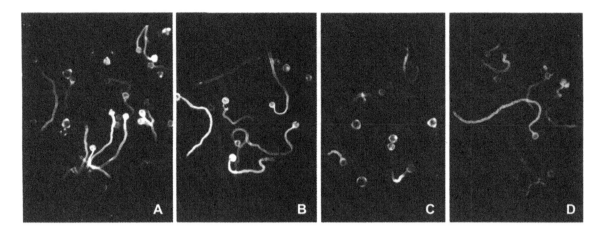

Fig. 9.8 In-vitro assay for self-incompatibility in *Petunia hybrida*. Fluorescence micrographs of pollen cultures grown in vitro in the presence of pistil extract. A. Control medium (no pistil extract), B. Cross-pistil extract, C. Self-pistil extract, D. Self-pistil extract + actinomycin D. Actinomycin D overcomes the inhibition of self-pistil extract (after Sharma 1986)

Using in vitro pollen germination assay, Sharma and Shivanna (1983a) and Shivanna and Sharma (1984) carried out experiments in *Petunia* comparable to those in animal systems. Pollen grains were treated with lectins or sugars before culturing in a medium containing pistil extract. Inhibition of self-pollen was overcome when pollen grains were treated with glucose or lactose but not with lectins. Similarly, incorporation of specific lectins into the medium containing pistil extract was also effective in overcoming inhibition of self-pollen. However, neither the sugars nor the lectins showed S-allele specificity. Although these results indicate that recognition of self-pollen is established as a result of complementation between lectin-like components of the pollen and specific sugar moiety of the pistil, presumably of S-allele specific glycoproteins, recent studies (see page 149–154) on characterization of S-allele components have not indicated a role for carbohydrates of S-allele glycoproteins in pollen recognition or inhibition.

In *Papaver rhoeas* also a consistent in vitro assay has been established and used effectively to understand the details of inhibition (Franklin-Tong et al. 1988, 1989, Foote et al. 1994). Addition of crude stigma extract and, more impor-

tantly, of purified stigmatic S-protein or recombinant S-gene product, was effective in inhibition of tube growth of self-pollen. As in *Petunia* (Shivanna and Sharma 1984), in *Papaver* also incorporation of actinomycin D in the in-vitro assay could overcome inhibition of self-pollen. These results confirm that SI inhibition requires de novo pollen gene expression (Franklin-Tong et al. 1990).

Temporal Expression of SI

In both GSI and SSI, expression of the self-incompatibility response is developmentally regulated. The SI is weak or even absent in immature flower buds, while mature flowers are strongly incompatible (Shivanna and Johri 1985). For example, in *Petunia* as well as *Brassica* self-pollination of buds 2–4 days before anthesis results in fruit- and seed-set. This has been explained on the basis that immature buds do not contain S-allele products or only a low concentration; S-allele components reach effective concentration only towards maturity. This knowledge of bud compatibility has been effectively used to produce homozygous lines for breeding purposes. More importantly, bud compatibility has proved invaluable in isolation and characterization of S-genes.

Operation of SI

SI involves highly specific cellular and molecular recognition. For effective operation of SI, the following are the essential requirements:

- Production of S-allele specific products in the pistil and the pollen.
- Interaction of S-allele specific products of the pistil with those of the pollen at some stage of pollen-pistil interaction to establish pollen recognition.
- Inhibition of incompatible pollen following pollen recognition.

An understanding of the operation of SI therefore involves isolation and characterization of S-alleles and their products in the pistil and pollen, and elucidation of the basis of pollen recognition and the mechanism of pollen inhibition. Remarkable progress has been made during the last two decades in our understanding of the details of S-genes and their products and, to a lesser extent, the details of pollen inhibition. The basis of pollen recognition is yet to be clearly elucidated, however.

Characterization of S-allele Products and S-alleles in Pistil

Sporophytic systems

Initial progress in understanding S-allele products was made by Nasrallah and associates in *Brassica oleracea* using immunodiffusion and electrophoretic studies (Nasrallah and Wallace 1967, Nasrallah et al. 1970, 1972). They showed that each S-allele produces a specific protein and each S-allele specific protein has a different electrophoretic mobility; each could therefore be localized to a specific band in the gel. These proteins were heritable, as evidenced by the presence in the heterozygous plant, of both the S-allele specific bands of the parents. Segregation of S-allele specific bands in F_2 progeny correlated precisely with segregation of the S-allele phenotype. Acquisition of self-incompatibility in the pistil coincided with the appearance of S-allele-specific proteins. In *B. oleracea* for

example, S-allele-specific proteins in the stigma increased in concentration with the onset of SI about 2 days prior to anthesis; immature buds about 5 days before anthesis, which are self-compatible, contained only trace levels of S-allele proteins (Nasrallah 1974, Nasrallah et al. 1985, Roberts et al. 1979).

Subsequently Hinata and associates (Nishio and Hinata 1977, 1978, 1979, 1980, 1982, Hinata and Nishio 1981) demonstrated the presence of S-allele specific band(s) following isoelectric focusing in *B. oleracea* as well as *B. campestris*. These proteins were shown to be glycoproteins and many were purified from stigma extracts by gel filtration followed by affinity chromatography on concanavalin A-sepharose column (Nishio and Hinata 1979). All S-allele glycoproteins showed binding to a lectin, concanavalin A (Nishio and Hinata 1980, Hinata and Nishio 1981). The molecular weight of all the S-allele glycoproteins in *Brassica* has been estimated to be around 55 kD; they usually have high isoelectric points (pI 10.3–11.1). The amino acid sequences of many S-glycoproteins have been determined (Nasrallah et al. 1987, Takayama et al. 1987). These studies apart from establishing S-allele glycoprotein-SI phenotype relationship, indicated that the gene(s) encoding these glycoproteins represent either the S-genes or sequences tightly linked to S-genes.

All S-allele specific glycoproteins of *Brassica* have 12 conserved cysteine residues located in the C-terminal region of the proteins (Nasrallah et al. 1987). These glycoproteins show 13 potential N-glycosylation sites; of these. 4 are conserved among all the alleles (see Dickinson et al. 1992). Analyses of carbohydrate chains of three S-allele specific glycoproteins from *B. campestris* have shown that the major N-glycosidic oligosaccharide chains A and B are identical and are commonly found in other plant glycoproteins (Takayama et al. 1986), rather than being specific to S-gene products of *Brassica*. These studies indicate that oligosaccharide moiety may not play a role in S-allele specificity (Takayama et al. 1986). Nevertheless, glycosylation seems to be required for SI; treatment of stigma with tunicamycin, an inhibitor of glycosylation, overcomes inhibition of self-pollen and results in normal germination and tube growth (Sarker et al. 1988).

Recent studies using molecular approaches have shown that the S-gene in *Brassica* is a complex multigene family and contains a secreted S-locus specific glycoprotein gene (SLG) and an S-locus receptor kinase gene (SRK) (Lalonde et al. 1989, Nasrallah and Nasrallah 1993, Hinata et al. 1993, Sims 1993, Franklin et al. 1995). The SLG protein is around 57 kD and the SRK protein is around 120 kD. Besides these, another group of the S-gene family, S-locus related genes (SLR-1 and SLR-2), has also been identified.

S-locus-specific glycoprotein gene (SLG): Nasrallah et al. (1985) were the first to isolate a cDNA clone that contained sequences encoding part of the S-glycoprotein corresponding to S_6 allele of *B. oleracea* by differential colony hybridization. Since then cDNA and genomic clones have been isolated for many other S-alleles (Chen and Nasrallah 1990, Toriyama et al. 1991, Nasrallah et al. 1987, 1988, Trick and Flavell 1989, Nasrallah and Nasrallah 1993, Wheeler et al 2001). Analysis of RFLPs from parents, F_1 and F_2 populations have shown exact co-segregation of RFLP bands with specific S-alleles, providing evidence for the identification of S-allele specific genes. Studies using Northern hybridization and in-situ hybridization (Nasrallah et al. 1985, 1988, Kandasamy et al. 1989) have revealed that the S-gene is highly

expressed in the stigmatic papillae, the site of pollen inhibition, and its glycosylation products accumulate in high levels in the wall of the papillae. S-allele specific genes can be divided into two classes based on intensity of incompatibility (Nasrallah et al. 1991). Class I alleles exhibit a strong SI reaction and Class II alleles show a weaker SI and are recessive in pollen.

Identification and characterization of S-receptor kinase gene (SRK): Nucleotide sequencing of the flanking regions of a number of SLG homologous genomic clones from *B. oleracea* resulted in identification of the SRK gene (Stein et al. 1991). Studies on the structural details of the SRK gene suggest that SRK protein is an integral membrane protein with an external SLG-like receptor domain (which shares extensive sequence similarity with SLG), a transmembrane domain and a cytoplasmic kinase domain (Fig 9.9). The SLG-like domain of SRK_6 shares 94% nucleotide sequence identity with SLG_6 (Nasrallah et al. 1991). SRK_2 and SLG_2 similarly share over 90% identity. The identity between SLG_2 and SLG_6 and between SRK_2 and SRK_6 is only about 68% (Stein et al. 1991). The sequence of the cytoplasmic domain of SRK is very similar to the serine/threonine kinase. SRK exhibits the same developmental and spatial pattern of expression as the SLG gene (Stein et al. 1991). Analysis of RFLPs in F_2 populations segregating for different SI genotypes has shown that SLG and SRK genes are genetically inseparable from each other and from the SI phenotype; they behave as components of a single locus (Fig. 9.9). On the basis of this data Nasrallah and associates (Nasrallah and Nasrallah 1993)

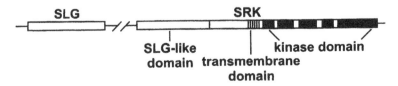

Fig. 9.9 Diagram of SLG and SRK genes in S locus of *Brassica* (after Nasrallah and Nasrallah 1993).

adapted the term 'S-haplotype' instead of 'S-allele' for SI genotypes.

Isolation and characterization of S-locus related genes: The pattern of expression of SLR genes (SLR-1 and SLR-2), both spatially and temporally, is identical to the SLG gene; their expression is also confined to stigmatic papillae and the products reach maximum accumulation towards maturity. However, unlike the SLG gene, SLR genes are not linked to the S-locus and therefore do not contribute to S-allele specificity. Although SLR-1 and SLR-2 loci are distinct from SLG, SLR-1 and SLR-2 are linked (Boyes et al. 1997). Analysis of SLR-1 and SLR-2 genes from different *Brassica* genotypes revealed an extremely low level of allelic variation (99–100% homology) (Trick 1990, Scutt et al. 1990). Detection of SLR genes in both self-incompatible (*Brassica* and *Raphanus*) and self-compatible (*Brassica, Arabidopsis*) spp. and their high level of expression in both self-incompatible and self-compatible plants indicate that SLR genes have no role in SI but may have a role in other post-pollination processes. Recent studies have indicated that SLR products may play a role in pollen adhesion (Luu et al 1999).

Requirement of SLG and/or SRK in SI Response: Initial analyses of some of the self-compatible mutants of *B. campestris* and *B. oleracea* showed drastic reduction in the level of stigma SLG products (Nasrallah et al. 1972, Nasrallah 1974) but not of SRK products. Other self-compatible mutants likewise showed SLG products comparable to SI lines but aberrant SRK transcription (Nasrallah and Nasrallah 1993). These studies indicated involvement of both SLG and SRK genes in SI response in *Brassica*. Attempts to modify SI specificity in the stigma through transgenic expression of SLG/SRK alleles isolated from plants of one S haplotype in plants of another S haplotype were not successful. High sequence similarity between SLG and SRK resulted in homology-dependent gene silencing and reduced the expression of both transgene

and the endogenous gene pair; this resulted in self-fertile transgenic plants (Nasrallah 2000). It was thus not possible to attribute loss of SI to SLG or SRK.

Subsequently two genomic clones that share SLG_{910} and SRK_{910} were transformed into self-compatible *B. napus* line Westar. The transgenic lines expressed both SLG_{910} and SRK_{910} at levels comparable to that found in the wild line (Cui et al. 2000). The transgenic stigmas expressed the SI response in a haplotype specific manner but the pollen phenotype remained unchanged (see Brugiere et al. 2000).

Recent transformation studies, however, have clearly shown that SRK but not SLG plays a critical role in SI responses (Takasaki et al. 2000). Either SRK_{28} or SLG_{28} was introduced into a homozygous S_{60} of *B. campestris*. Transgenic plants with only SRK_{28} introduced rejected both S_{28} and S_{60} pollen indicating the gain of S_{28} specificity in the transgenic stigma; the pollen phenotype did not change. These experiments indicated that SRK is the female S-specificity determinant. However, further analysis indicated that SLG might enhance the SI recognition process. Plants expressing both SLG_{28} and SRK_{28} transgenes exhibited a stronger SI than those with only SRK_{28}. It was suggested that SLG may play a role in stabilizing SRK or in enhancing diffusion of the pollen factor determining SI specificity (Brugiere et al. 2000, Franklin-Tong and Franklin 2000, Wheeler et al. 2001).

Subsequent studies (Silva et al. 2001) have confirmed that SRK is the primary determinant of SI in the pistil. The SRK_{910} and SLG_{910} cDNAs isolated from SI *B. napus* W1 line were individually transformed into SC *B. napus.* cv Westar. The SRK_{910} transgenic Westar plants assumed the SI function and rejected W1 pollen but SLG_{910} transgenic Westar plants were fully compatible to W1 pollen. Unlike earlier studies, transgenic plants expressing both SLG_{910} and SRK_{910} showed no enhancement of SI reaction.

Limited studies have been carried out to understand SI genes in *Ipomoea trifida*

(Convolvulaceae), which also has SSI. Results have suggested the operation of an SI mechanism in this species different from the one found in *Brassica* (see Wheeler et al. 2001).

Gametophytic systems

S-allele specific glycoproteins from the styles which co-segregated with specific S-alleles have been identified in several gametophytic systems belonging to Solanaceae, namely *Nicotiana, Lycopersicon, Petunia* and *Solanum* (Bredemeijer and Blaas 1981, Clarke et al. 1985, Kheyr-Pour and Pernes 1985, Mau et al. 1986, Kamboj and Jackson 1986, Kirch et al. 1989, Xu et al. 1990a). All the S-proteins isolated to date from solanaceous members are glycosylated. They show considerable variability in carbohydrate chains. The glycoproteins are basic (pI 7.5–9.5) and their molecular weights range from 27–33 kD. The putative S-allele specific proteins accumulate late in pistil development. All the S-allele specific proteins have 15 highly conserved N-terminal amino acids (Mau et al. 1986).

Most of the initial studies on cloning of S-genes were carried out by Clarke and her associates on *N. alata* (Anderson et al. 1986, 1989, Cornish et al. 1987, Dodds et al. 1993). A cDNA clone encoding a stylar glycoprotein of *N. alata* which co-segregated with S_2-allele was the first to be isolated (Anderson et al. 1986). Subsequently, cDNAs for several other S-glycoproteins of *N. alata* were cloned. ^{32}P-labelled cDNA encoding S_2 glycoproteins was used to detect S_2 mRNA by in-situ hybridization using sections of the pistil; expression of the S-gene was shown to occur in the stigma and throughout the transmitting tissue. Similar localization of S_2 specific-glycoproteins was also shown through immunocytochemistry using antibody raised against specific S-glycoprotein (Cornish et al. 1987, Anderson et al. 1989).

cDNA clones corresponding to several S-alleles of *P. hybrida* (Clark et al. 1990) and *P. inflata* (Ai et al. 1990) and two of *Solanum chacoense*

(Xu et al. 1990b) have also been obtained. Amino acid sequence alignments of different S-glycoproteins have shown five highly conserved regions and several hypervariable regions. S-locus is highly polymorphic and shows a high degree of diversity between S-allele glycoproteins in hypervariable regions (Ioerger et al. 1990, Kheyr-Pour et al. 1990). For example, S_1, S_2, S_3 and S_6 glycoproteins from *N. alata* share only 51.5% overall sequence identity (Haring et al. 1990, Clark and Kao 1991).

One of the important findings with respect to the solanaceous S-gene is that the amino acid sequences of the S-gene product in two of the conserved regions are homologous to the regions conserved in the ribonucleases (RNases) (McClure et al. 1989); they are close to the catalytic domain of RNases and include histidine residues which are critical for enzyme activity. Further studies showed that all S-glycoproteins from solanaceous species are functional RNases (Clark et al. 1990, Singh et al. 1991, Xu et al. 1990b) and are referred to as S-RNase proteins.

Later studies revealed that S-RNase entered the pollen tubes and degraded r-RNA in incompatible tubes but not in compatible tubes indicating that S-RNase functioned enzymatically in the SI reaction (McClure et al. 1990). However, there was no S-allele specificity in the activity of S-RNase in vitro grown pollen tubes (Gray et al. 1991). S-gene products in *Pyrus*, another gametophytic system, belonging to Rosaceae, have also been shown to be RNases (Sassa et al. 1992, Certal et al. 1999). Subsequent studies have shown that in members of Scrophulariaceae also S-gene products of pistil are RNases (Xu et al. 1996)

Studies on genetic transformation in *Petunia* and *Nicotiana* have provided direct evidence for the involvement of putative S-genes in SI response. Transformation of S_1S_2 plant with an S_3 gene resulted in transgenic plants which rejected S_3 pollen; such plants had S_3 proteins in addition to S_1 and S_2 proteins at comparable levels.

Other transgenic plants which displayed partial seed-set with pollen had smaller amounts of S_3 proteins (Lee et al. 1994).

Ribonuclease activity is essential for rejection of self-pollen (Huang et al. 1994). S_3 gene of *P. inflata* was mutated by replacing the codon for His-93 (which is essential for ribonuclease activity) with a codon for asparagine, and the mutant gene was transferred to the plants of S_1S_2 genotype. Two transgenic plants produced a mutant S_3 protein at a level comparable to that of S_3 protein in SI plants, but did not exhibit RNase activity. These two transgenic plants failed to reject S_3 pollen.

In *Nicotiana* also transformed plants with SA_2 allele showed high levels of SA_2 RNase and rejected SA_2 pollen (Murfett et al. 1994). In *Lycopersicon peruvianum* (Royo et al. 1994), loss of histidine residue at the active site of S-locus ribonuclease has been shown to be associated with self-compatibility.

Papaver system

Papaver rhoeas, a gametophytic SI system, shows distinct differences from other gametophytic systems. Inhibition takes place on the stigma, similar to sporophytic systems. Using in vitro assay, stigmatic S-gene products have been isolated and characterized (Franklin-Tong et al. 1989, Franklin-Tong and Franklin 1992). The proteins which co-segregate with the S alleles have been purified (Foote et al. 1994). The putative S_1 products are small extracellular molecules (ca. 15 kD, Foote et al. 1994). These S_1 products also correlated developmentally with the expression of SI. Unlike other gametophytic systems in which S-glycoproteins form a major component of pistil proteins, in *P. rhoeas* they form only ca 0.5–1.0% of the total proteins of stigmatic papillae. More importantly the S-gene products in *P. rhoeas* are not ribonucleases, indicating that the mechanism of SI operation is different in this system from other gametophytic systems (Franklin-Tong et al. 1991).

Several alleles of *P. rhoeas* and *P. nudicaule* have been cloned (Foote et al. 1994). The sequence predicts the position of a single potential glycosylation site. Expression of S allele is confined to the stigmatic tissue and is largely in the mature stage.

S-allele-specific components in pollen

In contrast to the progress made in understanding the details of S-alleles and their products in the pistil, our knowledge of S-allele components in the pollen is very meagre. Attempts to identify pollen proteins which co-segregate with S-alleles have not been successful. In *Brassica*, a low level of expression (several hundred fold lower than in stigmas) of SLG product was detected in post-meiotic anthers but not in mature pollen (Nasrallah and Nasrallah 1989, Guilluy et al. 1991). Also, transgene-induced silencing of SLG (Toriyama et al. 1991) or SLG and SRK (Conner et al. 1997, Stahl et al. 1998) that resulted in the breakdown of SI in the stigma, did not affect the pollen phenotype, indicating that SLG and SRK do not function as S-genes in the pollen. Doughty et al. (1998) reported a 7 kD pollen coat protein that forms a complex with stigmatic extracts. However, further studies indicated that this interaction was not S-allele specific.

Studies on *Nicotiana alata* (Dodds et al. 1993) have shown the expression of a transcript homologous to the cDNA of the S_2 gene in developing pollen. The level of expression was only ca 1% that found in the style. Evidence has also been obtained for the presence of S-proteins in the intine of mature pollen. None of these studies on *Nicotiana*, however, have categorically demonstrated the involvement of any of these pollen components in SI.

Recently a pollen S-gene from *Brassica oleracea* was identified and designated as SCR (S-locus cysteine-rich protein) (Schopfer et al. 1999). SCR is expressed exclusively in anthers during pollen development and encodes a short open reading frame with an unusually high frequency of cysteine residues; it accumulates after meiosis. SCR product is a hydrophilic protein of 8.4–8.6 kD. So far over 20 SCR alleles have been identified. Takayama et al. (2000) also isolated an analogous gene in *B. campestris* and designated it as SP 11 (S-locus protein 11).

The following data have confirmed that SCR is indeed the pollen S-gene:

(i) A self-compatible mutant of *B. oleracea* in which SI breakdown was confined to the male determinant did not have a detectable amount of SCR expression. (ii) Deduced amino acid sequence analyses of SCR proteins from S_6, S_8, and S_{13} haplotypes showed a high degree of polymorphism giving an overall similarity of only 30–42%. (iii) Transformation studies involving transfer of SCR_6 coding region to S_2S_2 homozygous *B. oleracea* have also confirmed the role of SCR in SI. The pollen from $S_2S_2/SCR_6{}^+$ transformants acquired S_6 specificity.

SCR is expressed gametophytically in microspores. It is suggested that the proteins of the two SCR alleles are synthesized during pollen development and are incorporated into the pollen coat as a mixture when pollen grains get mixed within the anther locule. In-situ hybridization studies have indicated that SCR is expressed in the tapetal cells also, suggesting that this gene is expressed both sporophytically and gametophytically (Takayama et al. 2000). Apparently, the pollen coat substances, derived from the tapetum, would contain the products of both the S-alleles (see Nasrallah 2000).

Basis of S-allele Specificity

In spite of the progress made in understanding the details of S-genes in *Brassica* and members of Solanaceae, the basis of S-allele specificity is not clear. In *Brassica*, SLG protein has two highly conserved and two highly polymorphic regions. In *Nicotiana*, there are five conserved regions which are important for S-RNase activity and several hypervariable regions. The variable regions are likely to be involved in allelic specificity. Although S-allele proteins of both *Brassica* and *Nicotiana* are glycosylated, there is no evidence to suggest that carbohydrate chains have any role in S-allele specificity (Broothaerts et al. 1991, Karunanandaa et al. 1994).

Mechanism of Pollen Recognition and Inhibition

There is no complete understanding either on the basis of recognition or on the mechanism of inhibition in any SI species. However, on the basis of available evidence, three different models have been proposed: (i) *Brassica* and other sporophytic systems, (ii) solanaceous gametophytic systems, and (iii) members of Papaveraceae. A summary of the three models suggested (Dickinson 1994, Wheeler et al. 2001, Stone and Goring 2001) follows.

In *Brassica* studies described earlier indicated that SI response requires a functional interaction between SLG and SRK products. According to Nasrallah and Nasrallah (1993), there are two possibilities for such an interaction: (i) the SLG may become competent for binding with SRK only after it is modified by pollen-borne S-allele product. In this model, pollen S-product acts as a modifying factor for SLG and (ii) the S-allele product acts as an extracellular ligand to bind SLG and SRK. The binding of SLG and SRK activates the receptor kinase in the stigma papillae which then initiates phosphorylation of one or more intermediates that ultimately produce a localized response within the stigmatic papillae that inhibits the pollen. This model envisages the primary site of inhibition reaction to be the papillae, unlike in gametophytic systems in which the pollen seems to be the site of inhibition reaction. Also, the model explains neither the basis of S-allele specificity nor the mechanism of inhibition.

Recent evidence indicating that only SRK is involved in SI response in the pistil and SCR is the pollen S-gene has led to reinterpretation of SI inhibition in *Brassica* (Brugiere et al. 2000, Franklin-Tong and Franklin 2000). Following pollination, pollen coat substances which include SCR products (and many others involved in other aspects of post-pollination events such as pollen adhesion) flow from the pollen, form a meniscus and come into contact with SLG and/or SRK products. The SCR and extracellular domain of SRK protein encoded by the same haplotype interact and establish recognition (Fig. 9.10). This interaction is likely to result in activation of the intracellular serine-threonine protein kinase domain of SRK. Activated kinase domain phosphorylates ARC 1 (arm-repeat containing protein 1, a stigma specific molecule that is phosphorylated by SRK in vitro), which in turn initiates an intracellular signalling cascade in the papillar cell and eventually inhibits pollen germination or pollen tube growth. ARC 1 has been shown to be required for the SI response in *Bras-*

sica. Transgenic plants of *B. napus* in which ARC 1 expression was down-regulated by an antisense approach showed reduced ability to reject self-pollen (Stone et al. 1999). On the basis of indirect evidence it has been suggested that signal transduction pathway may finally regulate the activity of aquaporins, water channel proteins, in the stigmatic papilla to limit the availability of water to self-pollen. As pointed out earlier, the role of SLG protein in SI response is not yet clear; it may play a role in stabilizing the SRK or in enhancing diffusion of pollen SRC

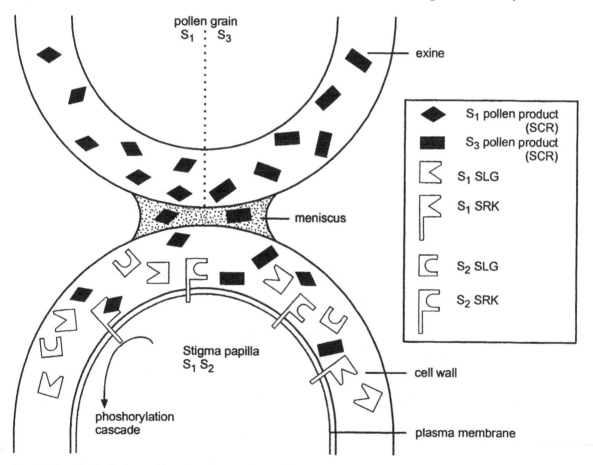

Fig. 9.10 Proposed model for operation of SI in sporophytic system (*Brassica*). In pollen grain, the products of two S-alleles, S_1 (matching allele with that of the stigma) and S_3 (non-matching allele with that of the stigma) are shown. S_1-pollen product (SCR) binds to the S_1-SRK (and possibly SLG) product on the stigmatic papilla. This recognition leads to phosphorylation cascade that eventually inhibits pollen germination. S_3-SCR does not match with the S-allele products of the stigma and thus does not activate phosphorylation cascade.

products or may act as accessory receptor or have a more general role in pollen-pistil interaction.

The model suggested for members of Solanaceae, Rosaceae and Sterculiaceae, is based on the fact that S-allele products in the pistil are functional RNases. According to this model, the S-RNase of the transmitting tissue enters the self-pollen tubes (Fig. 9.11). The internalized S-RNase then degrades the rRNA in the pollen tube, leading to cessation of protein synthesis and consequent arrest of pollen tube growth (Newbigin et al. 1993, Dickinson 1994). Although all the data including transformation experiments support the model on the mechanism of pollen tube inhibition, there is no evidence to explain

the basis of specificity. It is suggested that either RNase enter only the self-pollen tubes as a result of recognition and not the cross-pollen tubes, or the cross-pollen tubes somehow inactivate RNase.

Recent immunological studies (Luu et al. 2000) indicate that uptake of S-RNase is independent of S-genotype. S_{12} pollen was used to pollinate pistils of $S_{11}S_{12}$ genotype. S_{11} antibodies detected markedly higher amounts of S_{11} RNase within the pollen tubes compared to the extracellular matrix of the pistil.

The evidence in *Papaver rhoeas*, in which S-allele products are not RNases, suggest that the operation of SI in this species differs compared to other gametophytic systems (Franklin et al.

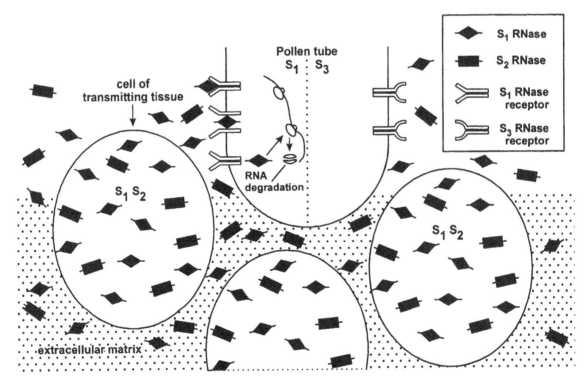

Fig. 9.11 Proposed model for SI in solanaceous species such as *Nicotiana* and *Petunia* in which S-allele product in the pistil is RNase. Receptors of both S_1 and S_3 alleles are depicted in the pollen tube. S_1-RNase released into the extracellular matrix of the pistil is recognized and internalized by the matching S_1-receptors on the pollen membrane; S_1-RNase degrade the r-RNA in the pollen tube leading to cessation of protein synthesis and arrest of pollen tube growth. According to the model depicted, when S-RNase (S_2) of the ECM does not match with the S-allele receptor (S_3) of the pollen, the S-RNase fails to enter the pollen tube.

1995). According to the model for *Papaver* (Franklin-Tong et al. 1993, Dickinson 1994, Franklin et al. 1995), stigmatic S-protein acts as a signal molecule and binds with homologous S-receptors of the pollen, believed to be located in the plasma membrane (Fig. 9.12). Inhibition of self-pollen is rapid and is brought about within minutes. The signal molecule (S-allele product of the stigma) is not internalized but the binding triggers a calcium-mediated signal transduction pathway inside the pollen, leading to the release of Ca^{2+} from internal stores within a few seconds

of SI interaction. This transient increase of intracellular Ca^{2+} is the initial step and leads to inhibition of pollen tube growth.

Evidence for the operation of Ca^{2+}-mediated signalling during SI response in *P. rhoeas* has come from studies using Ca^{2+}-selective dye, Calcium Green-1, which has high affinity and selectivity for free calcium, in the in vitro assay (Franklin-Tong et al. 1993). Calcium Green undergoes a marked change in fluorescence on binding to cytoplasmic free Ca^{2+}. Pollen tubes growing in vitro were microinjected with the dye

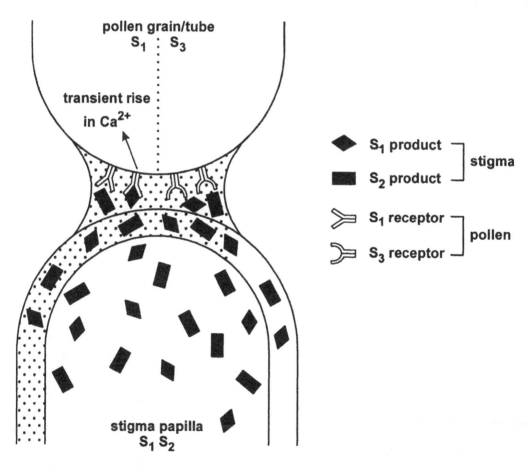

Fig. 9.12 Proposed model for operation of SI in *Papaver* system. Both S_1 and S_3 receptors are depicted in the pollen. Binding of S_1-allele product of the stigma to the S_1-allele receptor present on the pollen membrane results in the activation of calcium-mediated signal transduction pathway leading to transient release of calcium from internal store. Increased levels of intracellular calcium inhibit pollen germination/pollen tube growth.

and cytosolic free Ca^{2+} was monitored using laser scanning confocal microscopy. Addition of S-glycoproteins from self-stigma in the medium induced a transient increase in the level of cytosolic free Ca^{2+} in the pollen tubes and resulted in the arrest of their growth. No rise in Ca^{2+} levels was detected after addition of protein from compatible pistil or heat denatured S-glycoproteins. The photoactivation of bound Ca^{2+} to artificially elevate cytosolic free Ca^{2+} also resulted in the inhibition of pollen tube growth, similar to SI response. These studies clearly establish a link between transient rise in cytosolic free calcium and pollen tube inhibition and the mediation of SI inhibition by Ca^{2+}. Calcium is important for the regulation of pollen tube growth (see Chapter 4) and is known to act as a second messenger in plant cells.

Further studies have indicated that the initial Ca^{2+} signal results in phosphorylation of several proteins, some Ca^{2+}-dependent, and others Ca^{2+}-independent (Rudd et al. 1997). SI response at a later stage seems to target the actin cytoskeleton of pollen tubes (Geitmann et al 2000). In actively growing pollen tubes, F-actin bundles are generally absent at the tip region; within 1–2 min of SI reaction, there is a reorganization of F-actin bundles at the tip of self-pollen tubes. In another 8–10 min, fine fragments of actin are seen throughout the cytoplasm indicating its disorganization. Nuclear fragmentation, a feature of programmed cell death, has been detected 4–12 h after SI response (Wheeler et al. 2001).

HETEROMORPHIC SI

In heteromorphic SI, plants of the same species produce more than one type of flower. The most prominent difference between floral morphs is the relative position of the anthers and the stigma. Floral morphs may be of two (dimorphic/distylous) or three (trimorphic/tristylous) types in a given species but each plant produces only one type of flowers. According to de Nettancourt (2001) heteromorphic incompati-

bility is recorded in 22 families of flowering plants.

Evolution of heterostyly has been discussed by many investigators. According to Baker (1966), Charlesworth and Charlesworth (1979) and Ganders (1979), dillelic SI evolved first and the morphological differences were superimposed on pre-existing SI. According to Lloyd and Webb (1986) and Barrett (1990), however, morphological differences preceded physiological SI, and the latter evolved by gradual adjustment of pollen tube growth to different morphs. On the basis of its taxonomic distribution as well as the prevailing diversity of structural and physiological mechanisms, most of the investigators support polyphyletic origin of heterostyly.

Dimorphic Systems

The two floral forms of dimorphic species are referred to as thrum/short-styled and pin/long-styled morphs. The anthers in one morph correspond to the level of the stigma in the other morph. Pollination between flowers of different morphs is compatible (intermorph compatibility) while those within the morph are incompatible (intramorph incompatibility). Intramorph incompatibility and intermorph compatibility transgress species limits. In *Linum grandiflorum,* intramorph inter-specific pollinations are invariably incompatible while intermorph pollinations are compatible (Ghosh and Shivanna 1984b). The floral morphs of the two types and their compatibility relationships are diagrammatically presented in Figure 9.13A. Dimorphic condition is common in members of Rubiaceae, Plumbaginaceae, Linaceae and Boraginaceae (de Nettancourt 1977, Ganders 1979).

Distyly is controlled by a single gene complex, S, with two alleles S and s. The allele for short-style (S) is dominant over that of long-style (s). Long-styled morphs are homozygous recessive (ss) while short-styled morphs are heterozygous (Ss). Thus, intermorph compatible pollinations (ss x Ss) produce progeny of approximately equal number of short- and long-styled morphs. Pollen with recessive s allele from a short-styled

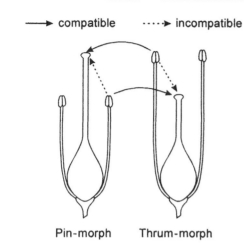

→ compatible   ····▶ incompatible

Pin-morph   Thrum-morph

A

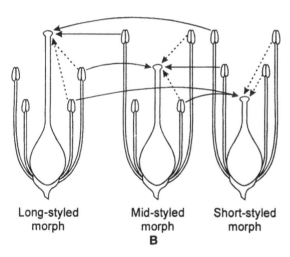

Long-styled   Mid-styled   Short-styled
morph   morph   morph

B

Fig. 9.13 Diagram of heteromorphic incompatibility. A. Dimorphic system. B. Trimorphic system. Compatible pollinations are depicted by solid arrows and incompatible pollinations by dashed arrows (after Shivanna 1982).

morph is compatible on a long-styled morph (ss) but incompatible on a short-styled one (Ss). Thus, incompatibility in the pollen is sporophytically determined.

    The style length ratios of pin and thrum morphs are variable; for example, 4:1 in *Primula auricula* and 2:1 in *Limonium vulgare*. Differences in the length of the style and stamens in the two morphs are associated with other morphological differences such as size and/or or-

namentation of the pollen and morphology of the stigmatic papillae (Figs. 9.14 and 9.15, Table 9.2). All these differences may not be present in all the dimorphic species.

Table 9.2 Morphological differences between thrum and pin morphs in some dimorphic species (based on data presented in Dulberger 1992; see Dulberger 1992 for details of references)

Character	Differences	Some examples
Relative lengths of style and stigma	Longer style and shorter stamen in pin, shorter style and longer stamen in thrum	All dimorphic species
Pollen size	Thrum pollen larger than pin pollen	*Armeria maritima* *Amsinckia* spp. *Oldenlandia* spp. *Primula* spp.
	Pin pollen larger than thrum pollen	*Linum* spp. *Limonium* spp.
Exine sculpture	Variable in different species	*Armeria maritima* *Limonium vulgare* *Linum* spp.
Pollen production (No. of pollen grains per flower)	More in pin morph	*Amsinckia grandiflora* *Primula* spp.
Anther size	Thrum anthers larger	*Linum* spp.
	Pin anthers larger	*Amsinckia* spp.
Pollen colour	Thrum pollen dark blue, pin pollen dark grey	*Linum grandiflorum*
	Pin pollen pale brick-red and thrum pollen yellow	*Linum tenuifolium*
Stigma surface	Thrum stigma wet, pin stigma dry	*Linum grandiflorum*
	Pin stigma wet, thrum stigma dry	*Primula* spp.
	Pin stigma non-papillate, thrum stigma papillate	*Limonium* sp.
Stigmatic papillae	Papillae of pin morph longer/larger	*Linum* spp., *Primula* spp.
Wall of papillae	Thrum papillae with subcuticular pectin cap, absent in pin papillae	*Linum pubescens*

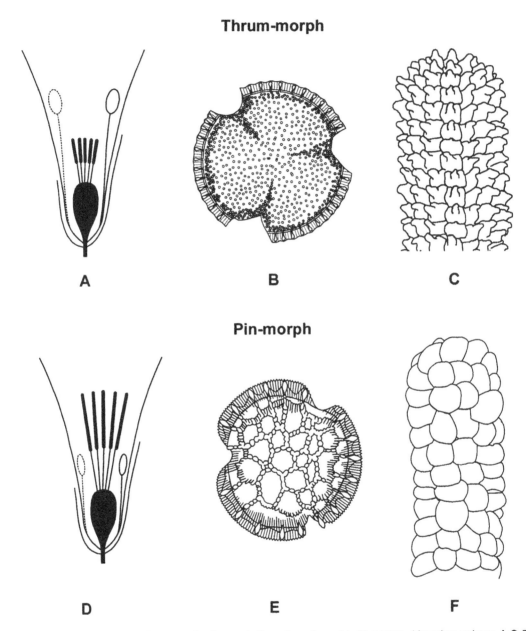

Fig. 9.14 Differences in structural features of pollen and stigma in a dimorphic SI system, *Limonium vulgare*. A-C. Thrum morph (A) with non-reticulate pollen exine (B) and papillate stigma (C). D-F. Pin morph (D) with reticulate pollen exine (E) and non-papillate stigma (F) (after Baker 1966).

The S-gene in heteromorphic species has been traditionally considered a supergene made up of several closely linked genes. According to Lewis (1954) the S-gene complex comprises six linked genes: G, S, I′, I″, P and A (G = style length, S = stigma surface, I′ = pollen

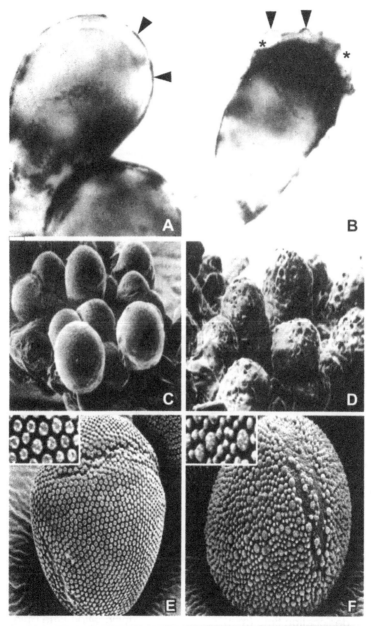

Fig. 9.15 Stigmatic papilla of pin and thrum morphs in *Linum grandiflorum*. A, B. Cytochemical localization of non-specific esterases on the stigma surface of pin (A) and thrum (B) papillae. In A the pellicle (arrowheads) is smooth and continuous while in B, it is raised and ruptured at places because of the accumulation of secretion product just below the pellicle (asterisk). C, D. Scanning electron micrographs of stigmatic papillae of pin (C) and thrum (D) stigmas. The pin stigma is of the dry type and that of the thrum stigma, the wet type. Scanning electron micrographs of pollen grains of thrum (E) and pin (F) morphs of *Linum austriacum*. The exine processes are large and uniform in thrum pollen while they are of two sizes in pin pollen. Insets show magnified portions of exine of respective pollen grains (after Ghosh 1981).

incompatibility, I″ = stylar incompatibility, P = pollen size and A = stamen height). Incompatibility genes are closely linked to genes determining floral polymorphism and thus are inherited together. Several studies have indicated physiological and biochemical differences between the pollen of pin and thrum morphs (Fig. 9.16, Table 9.3). Compared to the number of studies on the structural differences between thrum and pin morphs, relatively few studies have been carried out on physiological and biochemical aspects. In the light of these physiological differ-

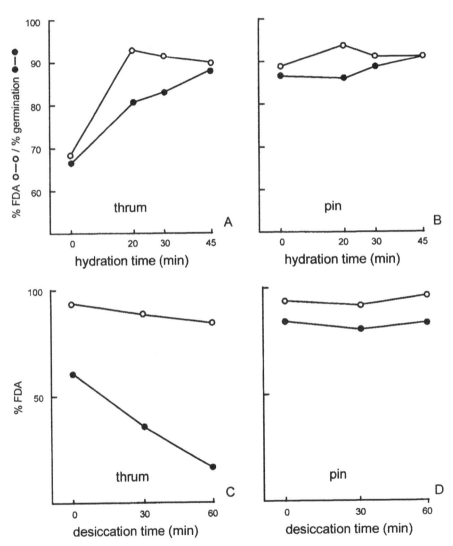

Fig. 9.16 *Primula vulgaris*. Physiological differences between pollen grains of thrum and pin morphs. A, B. Effect of controlled hydration on FDA and germination responses of thrum (A) and pin (B) pollen. Controlled hydration improved the responses of thrum pollen but not of pin pollen. C, D. Effect of desiccation (for up to 60 min) on FDA responses of pollen of the two morphs without controlled hydration (•) and after exposing to controlled hydration for 30 min (o). Desiccation reduced the FDA response of thrum pollen but not of pin pollen. Similarly, controlled hydration markedly improves the response of thrum pollen but not of pin pollen (after Shivanna et al. 1983).

Table 9.3 Physiological differences between pin and thrum morphs in some dimorphic species

Pollen germination	Desiccation reduces FDA response in thrum but not in pin pollen; controlled hydration improves in-vitro germination in thrum but not in pin pollen; pollen coat plays a role in controlled hydration in thrum but not in pin pollen	*Primula vulgaris* (Shivanna et al. 1983)
Protein profiles	Differences in banding pattern of proteins in pollen leachates of pin and thrum morphs	*Linum grandiflorum* (Ghosh 1981)
	Differences in protein profiles of the leachates of pin and thrum stigma	*L. grandiflorum* (Ghosh and Shivanna 1980)
	Differences in protein profiles of pin and thrum stamen and style in 2-D electrophoresis; a 72 kD polypeptide detected exclusively in pin style, a 45 kD polypeptide detected exclusively in thrum stamen	*Averrhoa carambola* (Wong et al. 1994)
	Three stylar proteins and two pollen proteins specific to thrum morph	*Turnera* sp. (Anthansiou and Shore 1997)
In-vitro assay	Stylar extracts inhibit in-vitro pollen tube growth of the same morph more than that of the other morph	*Primula vulgaris* (Shivanna et al. 1981) *P. abconica* (Golynskaya et al. 1976)

ences more genes are likely to be associated with the S-gene complex.

Trimorphic Systems

Trimorphic plants produce three floral morphs: long-styled, mid-styled and short-styled; each morph produces anthers at two levels corresponding to the levels of the stigma in the other two morphs (Fig.9.13B). For compatibility, pollen grains have to come not only from flowers of others morphs, but also from anthers corresponding to the same level as the stigma. For example, in the short-styled morph, pollen from long stamens are compatible on pistils of the long-styled morph (and not on the mid-styled morph) and pollen from mid-stamens are compatible on pistils of the mid-styled morph (and not on the long-styled). Tristyly is reported to occur in 10 genera of 5 unrelated families—Lythraceae, Oxalidaceae, Pontederiaceae, Amaryllidaceae and Connaraceae (Ganders 1979, Barrett 1998). Trimorphic species also show one or more of the other structural differences described for dimorphic species particularly in pollen size (Ganders 1979, Dulberger 1992, Barrett 1998). Trimorphic species have been investigated far less than dimorphic species and further studies are likely to reveal more differences between the morphs.

Genetic studies have indicated that tristyly is controlled by two genes, S and M, with two alleles each; S is epistatic over M. Long-styled morphs are homozygous recessive for both the genes (ssmm), mid-styled morphs are homozygous recessive for S and heterozygous or homozygous dominant for M (ssMM/ssMm) and short-styled morphs are heterozygous for S, and M may be in any form (Ssmm/SsMM/SsMm). As in dimorphic species, in trimorphic species also, incompatibility in the pollen is sporophytically controlled.

Zone of Inhibition

Variation in zone of inhibition is marked not only between species, but also between different morphs of the same species (Dulberger 1992). In many dimorphic species, particularly in members of Plumbaginaceae, inhibition occurs on the stigma surface in both morphs (pin × pin, thrum × thrum); pollen grains either fail to germinate or pollen tubes fail to enter the stigma (Baker 1966). In the few species pollen tubes enter the stigma but are inhibited in the tissues of the stigma or at the junction of the stigma and style. In the majority of heterostylous species the zone of inhibition in the two forms differs. In *Fagopyrum*, inhibition in thrum × thrum pollination occurs inside the stigma while in pin × pin pollination, inhibition occurs in the middle part of the style (Fig. 9.17 A, B). In *Linum grandiflorum* pin pollen fails to adhere and hydrate on the pin stigma (pin × pin), while thrum pollen shows adhesion, hydration and germination on the thrum stigma (thrum × thrum); however, thrum pollen

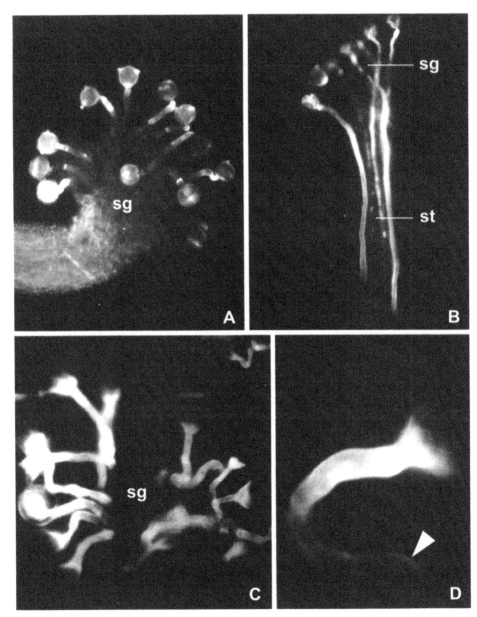

Fig. 9.17 Inhibition of pollen tube growth in intramorph pollinations. A, B. *Fagopyrum esculentum*. A. Pollen tubes are inhibited soon after entering the stigma (sg) in thrum morph (A), and after growing about half the distance in the style (st) in pin morph (B). C, D. *Linum grandiflorum*. Pollen tubes are inhibited soon after entry into the stigma in thrum morph (C); D. One of the pollen tubes similar to those shown in C magnified to show swelling (arrowhead) and rupture of tube tip. (A, B—after Sunanda Ghosh, unpublished; C, D—after Ghosh and Shivanna 1982b).

tubes are effectively inhibited in the thrum stigma (Ghosh and Shivanna 1982 b) (Fig. 9.17 C,D). Similarly, in trimorphic *Lythrum*, inhibition occurs at the base of the stigma (in short-styled morph) or in the upper part of the style (in mid- and long-styled morphs) (Dulberger 1970). In some species such as *Oxalis alpina* and *Linum* spp. (see Dulberger 1992), the zone of inhibition is the same in both illegitimate combinations.

In a few species such as *Primula vulgaris* (Shivanna et al. 1981, Heslop-Harrison et al. 1981) there is no localized zone of inhibition. Incompatible pollen grains are inhibited at several sites—on the surface of the stigma due to the failure of pollen germination or pollen tube penetration, in the stigma head or in the stylar region due to inhibition of pollen tube growth. Therefore, the incompatibility in *P. vulgaris* is the result of cumulative effects of inhibition at different levels.

Mechanism of Inhibition

Unlike homomorphic species in which dramatic progress has been made in understanding the details of pollen inhibition, very little information is available on the mechanism of inhibition in heteromorphic species. No molecular biology approaches have been carried out thus far in heteromorphic systems. Two concepts have been posited to explain the mechanism of pollen inhibition in heteromorphic species.

Passive inhibition

Lewis (1949), in his classical paper on *Linum grandiflorum,* reported that pin x pin pollination does not support pollen adhesion; even those pollen grains which remain on the stigma fail to hydrate (see also Dickinson and Lewis 1974). Thrum x thrum pollinations support pollen adhesion, hydration and germination but pollen tubes show swelling and bursting soon after entering the stigma. Comparative studies on the osmotic potential of the stigma and pollen of the two morphs (by fixing a reference value of 1 to thrum styles) showed clear differences: thrum morph—style 1.0, pollen 7.0; pin morph—style

1.75, pollen 4.0. Thus, in both compatible pollinations (thrum x pin, pin x thrum), the osmotic potential ratio of pollen:style was 4:1. In pin x pin incompatible pollinations (in which pollen grains fail to take up water from the stigma), osmotic potential ratio was 5:2 and in thrum x thrum pollinations (in which pollen tubes burst soon after entering the stigma) it was 7:1. On the basis of these studies, Lewis (1949, 1952) suggested that in this species, incompatibility is determined by mismatching of osmotic potential of the pollen and the stigma, and the differences in the nature of cell colloids. The osmotic potential ratio of 4:1 between pollen and style is optimal for pollen hydration, germination and pollen tube growth. If the ratio is too high (T x T), the pollen tube bursts and if too low (P x P) pollen hydration itself is prevented.

Moewus (1950) likewise studied the distribution of some flavonols in pollen of the two morphs and reported the presence of quercitrin in pin pollen and rutin in thrum pollen. Both flavonols were inhibitory for pollen germination. He suggested that these inhibitors are inactivated by specific enzymes present only in compatible stigmas. Incompatible pollinations do not result in the breakdown of inhibitors and therefore pollen is inhibited. However, Lewis (1954) could not confirm Moewus' results. Also, the recent demonstration that some of these flavonols are essential for pollen germination and tube growth (see Chapter 8) does not accord with this hypothesis.

The hypotheses of both Lewis (1949) and Moewus (1950) explain heteromorphic incompatibility as a passive mechanism but ascribe no role to the morphological differences of the two morphs. Studies of Barrett and his associates on *Pontederia* spp. (see Barrett 1990), have shown positive correlation between pollen size and pollen tube length in illegitimate pollinations. It has been suggested that pollen inhibition in some of the incompatible pollinations is passive and associated with storage products. According to this hypothesis, although inhibition is passive, pollen polymorphism plays a direct role in incompatibility.

Combination of passive and active inhibition

Based on their studies on *Linum grandiflorum*, Ghosh and Shivanna (1982b) posited an alternate hypothesis. According to them, self-incompatibility in heteromorphic species is a combination of passive and active inhibition. In *Linum grandiflorum,* as mentioned earlier, most of the pollen grains in pin × pin pollinations fail to adhere. However, contrary to earlier reports, Ghosh and Shivanna (1982b) showed that a few pollen do adhere and such pollen hydrate and germinate; pollen tubes enter the stigma but their growth is inhibited as in thrum × thrum pollinations. The number of pin pollen that adhere and germinate on the pin stigma can be greatly increased by treating the pollen with organic solvents such as hexane or smearing of pin stigma with lipoidal exudate from *Petunia* stigma. Pin × pin pollinations in such treatments were comparable to thrum × thrum pollinations. However, pollen tubes were invariably inhibited soon after entering the stigma. These studies clearly show that lack of adhesion and hydration of pin pollen on pin stigma is passive and is due to differences in morphological or physiological features. They are not the only manifestation of incompatibility, however. Even when these barriers are overcome by suitable treatments, pistils are endowed with the capacity to inhibit incompatible pollen tube growth.

Further, treatment of the pistil of *L. grandiflorum* with a protein synthesis inhibitor (cyclohexamide) before incompatible pollination, permitted continued pollen tube growth for a longer distance before they were inhibited, indicating the requirement of protein synthesis for pollen tube inhibition (Ghosh and Shivanna 1982b).

These studies of Ghosh and Shivanna (1982b) showed that in this species intramorph incompatibility operates at at least three levels: pollen adhesion, pollen hydration and pollen tube growth in the stigma. The first two are passive due to lack of morphological and/or physiological complementation and the third is active as a result of pollen recognition similar to homomorphic systems. All three operate in pin × pin pollinations but only the third is involved in thrum × thrum pollinations. According to this hypothesis, differences in osmotic potentials of the pollen and the stigma have no role in incompatibility; physiological incompatibility is superimposed by morphological heteromorphism and the two supplement each other in preventing illegitimate pollinations. Dulberger (1974, 1992) has given a detailed analysis of possible involvement of various differences between morphs in physiological incompatibility.

Subsequent studies of Murray (1986) on *L. grandiflorum* also did not support differences in the osmotic potential hypothesis of Lewis (1949). Murray (1986) determined the osmotic potential (OP) of stigmatic cells of many *Linum* species. The responses of the pollen in interspecific pollinations did not correlate with OPs of the stigmatic papillae. Pollen of all the species hydrated and germinated on stigmas with a wide range of OPs, indicating that the ability of the pollen to hydrate is independent of OP of the stigma cell.

Factors involved in active recognition and/or inhibition in heteromorphic species are not clearly understood. In *Primula obconica* (Golynskaya et al 1976), aqueous extracts of pistils inhibited tube growth of pin and thrum pollen differentially; pin extract inhibited growth of pin pollen tubes while thrum extract inhibited growth of thrum pollen tubes. The authors also reported that the effect of pistil extract was due to water-soluble proteins which are adsorbed onto erythrocyte membranes, suggesting their lectin-like nature. In *Primula vulgaris* also, studies of Shivanna and co-authors (1981) implicated stylar proteins in the incompatibility responses. Stylar dialysates inhibited tube growth of pollen of the same morph more strongly than that of the other morph. Unfortunately these in-vitro studies have not been extended.

10

Interspecific Incompatibility

Interspecific incompatibility is a reproductive barrier resulting in failure of seed-set between two reproductively isolated species. Table 10.1 presents details of interspecific barriers caused by the inability of male and female gametes from plants of different species/genera to effect fertilization and/or the resultant fertilized ovule to develop into a viable seed. Interspecific incompatibility may therefore operate before fertilization (prefertilization) and/or after fertilization (post-fertilization). Interspecific incompatibility prevents gene flow between species and maintains species identity. Although interspecific incompatibility has been reported in a large number of crosses, most of them are based on seed-set data; only limited studies have been done to understand the details of incompatibility or its mechanism. Stebbins (1958) gave the first comprehensive account of interspecific incompatibility. Since then the topic has been reviewed from time to time (Maheshwari and Rangaswamy 1965, de Nettancourt 1977, 2001, Shivanna and Johri 1985, Raghavan 1986, Hadley and Openshaw 1980, Khush and Brar 1992, Shivanna 1997).

PREFERTILIZATION BARRIERS

Prefertilization barriers prevent fertilization by arresting post-pollination events at one or many levels: pollen hydration, pollen germination, pollen tube entry into the stigma and pollen tube growth through the stigma and style.

Unilateral Incompatibility

In many interspecific crosses, incompatibility operates in one direction only; the reciprocal cross is successful. Such incompatibility is referred to as unilateral incompatibility (UI) (Lewis and Crowe 1958). Most authors restrict the term UI to incompatibility operating before fertilization (Dhaliwal 1992). UI has been reported in a large number of interspecific crosses (Lewis and Crowe 1958, Abdalla 1974, de Nettancourt 1977, Dhaliwal 1992, Harder et al. 1993). UI is more common when one of the parental species is self-incompatible (SI). Generally, the cross is successful when an SI species is used as the male parent (SC × SI) but unsuccessful when

Table 10.1 Major interspecific crossability barriers

I. Temporal and spatial isolation of parent species
II. Prefertilization barriers
On the surface of the stigma before pollen tube entry
Inside the tissues of the stigma and style
Inside the ovary and embryo sac
III. Post-fertilization barriers
Non-viability of hybrid embryos
Failure of hybrid to flower
Hybrid sterility
Lack of recombination
Hybrid breakdown in F_2 or later generations

an SI species is used as the female parent (SI × SC). However, there are reports of UI in other combinations as well (SC × SC, SC × SI, SI × SI). Two hypotheses have been put forward to explain UI. According to the first, UI is controlled by SI genes (Lewis and Crowe 1958). The SI gene has a dual function; SI allele inhibits pollen grains carrying SC allele from a self-compatible plant, in addition to pollen grains carrying matching SI allele. Lewis and Crowe (1958) considered SC species to have evolved from SI species through mutation of SI alleles in stages: SI → Sc → Sc' → SC. They explained the growth of pollen tubes of SC species in the pistil of SI species reported in some crosses on the basis that the pollen has an Sc allele and has not yet stabilized to an SC allele.

Pandey (1968, 1969, 1970) elaborated the dual function hypothesis further to accommodate all UI irrespective of the presence of SI in one of the parents. According to him, the S-gene complex has two specificities—primary specificity and secondary specificity; the former controls interspecific incompatibility and the latter, SI. S-gene-mediated responses of UI are often modified by other genes (Pandey 1968, de Nettancourt 1977).

According to the second hypothesis, SI genes play no role in UI (Grun and Radlow 1961, Grun and Aubertin 1966, Abdalla 1974). Inhibition of the pollen carrying the SC allele is mediated by specific genes different from the S-locus. According to Abdalla (1974), there are specific UI genes that inhibit fertilization by pollen grains carrying specific SC alleles. In the absence of matching UI genes, the cross between SI × SC plants will be successful in both directions. In the presence of matching UI genes, the cross is successful in one direction only (SC × SI). Crosses between distantly related species will fail in both directions, irrespective of the nature of incompatibility (SC × SI, SI × SI, SC × SC) because of other barriers.

There are hardly any studies on the physiological aspects of UI. A few studies carried out on the cytological details of pollen tube inhibition have shown that the inhibition of pollen in UI is stronger than that in SI (*Nicotiana*—Pandey 1964, *Solanum*—Grun and Aubertin 1966; *Lycopersicon*—Lewis and Crowe 1958, Hogenboom 1972). In the pistil of *N. alata*, for example, selfed pollen tubes are inhibited after growing through 10–27% of the style length, whereas pollen tubes of *N. longsdorfii*, a self-compatible species, are inhibited in the stigma (of *N. alata*) before entering the style (Pandey 1964).

Incongruity/Passive Inhibition

As discussed earlier (Chapter 9), SI inhibition is the result of active recognition. Self-pollen is actively recognized as a result of interaction of S-allele products; positive recognition results in activation of metabolic processes in the pollen and/or the pistil and brings about pollen inhibition. Compatible pollination activates many physiological processes in the pistil, as a result of co-adaptation between the pollen and the pistil, which facilitates sequential completion of all post-pollination events.

In most interspecific crosses, however, the arrest of post-pollination events seems to be passive as a result of lack of co-adaptation between the pollen and the pistil. Such an incompatibility which is passive (and not the result of active recognition) has been termed incongruity (Hogenboom 1975). Compatibility requires matching of the genetic system between the pistil and pollen; for each barrier gene or gene complex active in the pistil, there must be a penetration gene or gene complex in the pollen (Table 10.2). Incongruity is due to the absence of genetic information in one partner for some relevant character of the other and the details vary in different crosses.

Incongruity is a by-product of evolutionary divergence. Among a freely interbreeding population, a subpopulation may differentiate as a result of changed environment. This may bring in an extra barrier in the pistil of one of the population for the pollen of the other, resulting in incongruity between the two populations. The

Table 10.2 Compatibility relationships based on incongruity concepts of Hogenboom (1975). Presence/absence of penetration capacity of pollen to the barriers present in the pistil determines whether the cross is compatible (+) or incompatible (-)

Barriers in pistil	Penetration capacity of the pollen for			
	A	AB	ABC	ABCD
A	+	+	+	+
AB	-	+	+	+
ABC	-	-	+	+
ABCD	-	-	-	+

barriers increase as the divergence of the two populations increases. The incongruity operates not only during pollen-pistil interaction, but also during post-fertilization stages. Thus, incongruity is comparable to the lock and key mechanism; absence of a suitable key(s) with the pollen for the lock(s) present in the pistil results in incongruity. Incongruity is a complex phenomenon and may involve many genes depending on the extent of genetic divergence. The greater the divergence, the earlier the barrier.

The hypotheses put forward to explain interspecific incompatibility are based mostly on genetic studies. There is very little information on physiological and biochemical aspects. In the cross *Petunia hybrida × Salpiglossis sinuata*, the morphological abnormalities of incompatible pollen tubes (branching, swelling of tube tip, increased callose deposition etc.) and the pattern of respiration in pollinated pistils were similar to those following SI pollinations (Roggen and Linskens 1967). On the basis of these results, Roggen and Linskens (1967) suggested that the mechanism of pollen tube inhibition is similar in both SI and interspecific incompatibility. In *Lilium longiflorum*, however, unlike SI which can be overcome by giving hot water treatment to the pistil or exposing the plants to elevated temperatures, interspecific incompatibility could not be overcome by either treatment (Ashcer and Peloquin 1967, Ascher 1975, 1979).

The evidence available so far suggests that interspecific incompatibility between closely re-

lated species involving one or both SI parents; pre-fertilization inhibition is likely to be mediated through S-genes. This is well illustrated in heteromorphic species of *Linum*. Studies conducted by Ghosh and Shivanna (1984b) and Murray (1986) on different species of *Linum* have shown that intermorph pollinations (pin × thrum, thrum × pin), both within as well as between species, are compatible whereas intramorph (pin × pin, thrum × thrum) pollinations are incompatible. Thus, intermorph compatibility and intramorph incompatibility operate within as well as between species. Obviously pollen tube inhibition in such interspecific crosses is mediated through the S-gene complex. Subsequent post-fertilization barriers, such as embryo and/or endosperm breakdown are, apparently, controlled by other genes.

In distantly related species, SI genes are not involved in incongruity even when one or both the parents are self-incompatible. In such crosses all individuals of a species regardless of the S-genotype of the plants behave similarly for incongruity. Incongruity operates even when SC accessions of SI species are used in the crosses. Many recent studies on wide hybridization, particularly on *Brassica*, have shown that UI operates more often in crosses between the cultivars (female) and wild species (male) than in the reciprocal combination, irrespective of the presence or absence of SI in one of the species (see Shivanna 2000). Such UI in which S-genes do not have a role has been termed unilateral incongruity (Liedl et al. 1996).

According to Sampson (1962), successful pollen-pistil interaction depends on the degree of complementation between stigmatic and pollen molecules. There are two areas—'species area' and 'S-allele area'—in which complementation can occur. Complementation at the S-allele area results in SI due to S-allele identity. Interspecific crosses are compatible only when there is complementation at the species area, but not at the S-allele area. Thus complementation at both areas or none of the areas results in incompatibility (see also Heslop-Harrison 1975).

De Nettancourt and colleagues (1973a, b, 1974, 1975) studied ultrastructural details of pollen tube inhibition in SI (*Lycopersicon peruvianum*) and UI (*L. peruvianum* × *L. esculentum*). Pollen tubes in both types of pollination accumulated a large number of bipartite particles at the tip and formed characteristic concentric endoplasmic reticulum. Consistent differences were observed between the pollen tubes inhibited in SI and those in UI only in details of the outer wall; it was thick in the former but rather thin in the latter. Based on these and other genetic studies de Nettancourt (1977) suggested that SI and UI are two distinct but closely related rejection processes, and UI results from the interaction of S-elements in the pollen grains with products of one or several unidentified stylar genes.

Mechanism of Passive Inhibition

Figure 10.1 summarizes different levels at which passive inhibition may operate during pollen-pistil interaction and possible mechanisms. As pointed out earlier (Chapter 8), all post-pollination steps require co-adaptation between the pollen and the pistil for successful completion of pollen-pistil interaction. Lack of such a co-adaptation at any stage of pollen-pistil interaction would result in inability of the pollen to complete post-pollination events (Fig 10.2A-C).

Pollen adhesion and hydration depend on morphological and physiological co-adaptation between the pollen and the stigma. The amount and nature of secretion products on the stigma and pollen coat play an important role in pollen adhesion. Absence or insufficient amount of these components in one of the partners will result in failure of pollen adhesion. For example, *Arabidopsis thaliana* stigma selectively binds *A. thaliana* pollen with higher affinity than pollen from related species (Zinkl et al. 1999). Pollen grains from other dicotyledonous species could be washed off by detergent treatment of the stigma; such treatment has little effect on the binding capacity of the pollen of the same species (Zinkl and Preuss 2000). This is obviously

due to lack of specific components on the stigma/pollen required for adhesion of pollen of other taxa. In the absence of effective pollen adhesion, pollen hydration and, consequently, pollen germination would be prevented.

Pollen hydration is the result of water uptake from the stigma mediated through differences in the osmotic potential of the pollen and the stigma. Mismatch in the optimal osmotic potential between pollen and stigma results in lack of hydration or insufficient hydration. Lack of co-adaptation between the supply and demand position would also result in insufficient hydration. This would happen particularly in crosses involving the male parent with large pollen and the female parent with small pollen; the stigmatic papillae may not be able to provide sufficient water for full hydration of larger pollen. Alternatively the limited number of germ pores of the large pollen may not establish effective contact with the stigmatic papillae (Heslop-Harrison 1979d). Recent studies have given a clear indication that some of the components of the pollen coat are required for pollen adhesion (see Chapter 8).

Failure of pollen to germinate on the stigma or the entry of pollen tubes into the stigma is one of the most common barriers observed, particularly in wide crosses (Martin 1970, Knox et al. 1976, Stettler et al. 1980, Batra et al. 1990, Barone et al. 1992, Gundimeda et al. 1992). The causative factors have not been worked out in most of the crosses. However, many studies on pollen-pistil interaction have indicated some of the possible causes for lack of pollen germination and pollen tube entry.

As pointed out earlier, lack of pollen adhesion or hydration would result in the failure of pollen germination. Even in those crosses in which adhesion and hydration are normal, pollen grains require suitable conditions/factors for germination. Pollen grains of many species are known to require boron and/or calcium for germination and the stigma provides these components to the pollen (Bednarska 1991). In *Vitis vinifera* boron-deficient stigmas fail to support

Pollen lands on the stigma

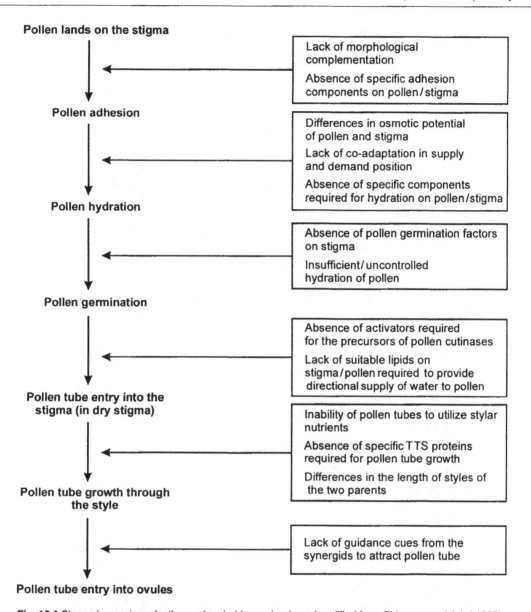

Fig. 10.1 Stages in passive rejection and probable mechanisms (modified from Shivanna and Johri 1985).

pollen germination (see Chapter 4). Thus, lack of the required amount of boron and/or calcium on the stigma may be a cause for failure of pollen germination. Suitable pH is also an important requirement for pollen germination (Roberts et al. 1983, Ganeshaiah and Shaankar 1988). Stigmatic exudates of different species often show marked variation in pH. Species of *Rosa*, for example, show a pH variation of 5 to 9 (Gudin and Arena 1991). Such pH differences may act as important barriers.

Many flavonols are required for pollen germination and pollen tube growth (see Chapter 8). This requirement is met either by the pollen

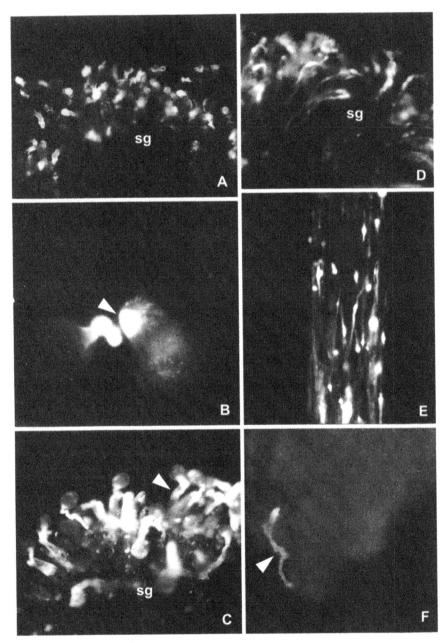

Fig. 10.2 Fluorescence micrographs of pistils stained with decolourized aniline blue following intergeneric crosses. A, B. *Brassica campestris* × *Diplotaxis berthautii*. Pollen grains have germinated on the stigma (sg) but pollen tubes have failed to enter the papilla. B is a magnified view of one of the pollen grains that has induced the development of callose in the stigmatic papilla (arrowhead). C. *Brassica carinata* × *Sinapis puoescence*. Some of the pollen tubes have entered the papillae (arrowhead) but are inhibited soon after. D-F. *Erucastrum abyssinicum* × *B. juncea*. Pollen germination (D), pollen tube growth in style (E) and pollen tube (arrow head) entry into ovule (F) are normal. Thus this cross has no prefertilization barriers (A, B—after Vyas 1993; C—after Ranbir Singh, unpublished, D-F—after Rao 1995).

or the stigma. Flavonol requirement would not be met in a cross in which pollen and stigma are both deficient in flavonols. Some components of the stigma surface proteins/glycoproteins are also required for pollen germination in some species (Heslop-Harrison and Heslop-Harrison 1975). Lack of such proteins would inhibit pollen germination.

Many studies have shown that phenolic compounds of the stigma may play an important role in passive inhibition; they selectively inhibit or promote germination (Martin 1969, 1970, 1972, Martin and Ruberte 1972, Sedgley 1975, Tara and Namboodiri 1976). Of the three male sterile mutants of *Impatiens sultani*, orange, crimson and pink, only orange and crimson set seeds upon pollination, but not the pink (Tara and Namboodiri 1976). This was due to the failure of the pink stigma to support pollen germination. Chromatographic studies showed that the two phenolic spots present in the stigmas of orange and crimson were absent in the pink (Tara and Namboodiri 1976, Bhaskar and Namboodiri 1978). The authors suggested that absence of these phenolics in the stigma of the pink mutant was responsible for its failure to support pollen germination.

Uncontrolled hydration, as happens on a wet stigma with aqueous exudate, may be another cause for failure of pollen germination. Several in-vitro germination studies have shown the importance of controlled hydration for successful germination of pollen (see Chapter 4). This is particularly true for those species characterized by dry stigma (Heslop-Harrison and Shivanna 1977, Shivanna and Heslop-Harrison 1981). Pollen germination will fail in crosses involving pollen requiring controlled hydration and the stigma lacking adaptation for controlled hydration.

Recent studies have shown that pollen adhesion, hydration and germination are more complex and a number of components on the surface of both pollen and the pistil play critical roles in these processes (see Chapter 8). Of these, the proteins and lipids present on the pollen as part of the pollen coat substances and/or the stigma surface are required for these processes, especially in species with dry-type stigma. Several mutants defective in one of these components have been isolated in *Arabidopsis* (see Chapter 8); these mutations act as barriers to pollen function.

Entry of the pollen tube into the stigma is critical. In species with a dry stigma covered with a layer of cuticle, pollen tubes enter the latter by activation of pollen cutinases. Pollen cutinases seem to require activators of the stigma surface. In some species, treatment of the stigma with proteases (Heslop-Harrison and Heslop-Harrison 1975) or concanavalin A (Knox et al. 1976) does not affect pollen germination but totally inhibits pollen tube entry, indicating that some components of the pellicle are required for effective operation of the pollen cutinases.

Results of recent studies have demonstrated that several species with wet as well as dry type stigma require long-chain lipids, which facilitate directional supply of water to the pollen and thus guide the resultant pollen tubes into the stigma (see Chapter 8). In wet stigma such as *Nicotiana*, these lipids are present in the stigmatic exudate; in dry stigma such as *Brassica*, they are present in the pollen coat. Lack of suitable lipids on the surface of the stigma or pollen coat may act as a barrier for pollen tube entry.

The stage at which the barriers operate seems to be related to the evolutionary divergence of the parental species. When the stigma of *Gladiolus* (Iridaceae) was pollinated with the pollen of *Crocosmia,* belonging to the same family, pollen hydration and germination were normal but pollen tubes failed to enter the stigma (Knox et al. 1976). However, when the pollen of *Gloriosa,* belonging to a different family (Liliaceae) was used for pollination, pollen hydration itself was inhibited.

Failure of pollen tubes to reach the ovary (after entering the stigma) is another most frequent prefertilization barrier. This is due to cessation of pollen tubes in the stigma, at various levels in the style or slow growth of pollen tubes, with the result that the pistil commences senescence before the pollen tubes reach the ovary. Such arrested tubes show many abnormalities, e.g. thicker walls, excessive deposition of callose, swollen tips and branching of the tube (de Nettancourt 1977, Dumas and Knox 1983, Fritz and Hannaman Jr 1989, Barone et al. 1992). It is now well established that pollen tubes growing in the pistil utilize stylar nutrients (see Chapter 8). Inability of pollen tubes to utilize stylar nutrients may be a cause for pollen tube arrest. This may be due to the absence of suitable nutrients in the pistil or of suitable enzyme in the pollen to utilize the available stylar nutrients. Recent studies have shown that extracellular matrix of the transmitting tract plays a crucial role in pollen-pistil interaction (see Chapter 8). Considerable data indicate that some of the proteins synthesized exclusively in the transmitting tissue (TTS proteins) participate in pollen tube growth, nutrition and their guidance. Absence of specific TTS proteins required to facilitate pollen tube growth may act as an important barrier.

Many crosses involve parents that show significant difference in length of style; in some, such as species of *Nicotiana, Datura* and *Lilium*, this difference may be up to several centimetres. Crosses in which a short-styled type was used as the female parent and long-styled as the male parent were often successful while reciprocal crosses were not (Avery et al. 1959, Maheshwari and Rangaswamy 1965, Gopinathan et al. 1986, Potts et al. 1987). In maize × sorghum cross, pollen tubes do not reach the ovary because of the unusual length of maize silk. However, when pollen grains were deposited on the basal parts of the silk, pollen tubes readily reached the ovary (Heslop-Harrison et al. 1985a, b). In such crosses failure of the pollen tube to complete growth through the style appears to be due to the intrinsic inability of pollen tubes of the male parent to grow beyond the length of its own pistil, rather than to active inhibition.

Examples of interspecific crosses showing inhibition of pollen tubes in the ovary are lim-

ited. In crosses involving many species of Poaceae, pollen tubes grow normally and reach the ovary. In maize × sorghum cross, no barriers were observed until pollen tubes entered the ovary; however, the pollen tube failed to enter the ovule (Heslop-Harrison et al. 1984, 1985a, b). As entry of the pollen tube into the ovule is considered to be mediated through secretion of a chemotropic substance in the micropyle, absence of such an attraction may act as a barrier. Recent studies have shown that the synergids provide guidance cues for pollen tubes to change the direction towards the ovule (see Chapter 8). Absence of such cues may form a barrier preventing entry of the pollen tube into the ovules.

Even after entry of the pollen tubes into the embryo sac, many crosses show disturbances in double fertilization. In the wheat × maize cross, only 12% of the ovules showed development of both the embryo and endosperm, 80% of the ovules showed only embryo development but not the endosperm, and the remaining 8% showed only endosperm development but not the embryo (Laurie and Bennett 1988). Obviously, this is due to some limitation(s) for effective syngamy and triple fusion; the controlling factors after discharge of male gametes in the embryo sac are yet to be understood.

In many crosses, prefertilization barriers are not restricted to a particular level or event, but are active over most of the post-pollination events. Thus the proportion of pollen grains that completes sequential post-pollination events is reduced at each level (Fritz and Hannaman 1989) with the result that very few or no pollen tubes reach the embryo sac. In the cross *Vigna unguiculata × V. vexillata* (Barone et al. 1992), 6.1% of the pistils showed inhibition of pollen germination, 5.3% showed pollen tube inhibition on the stigma surface, 32.6% showed inhibition inside the stigma and only in 8.2% of the pistils did the pollen tubes reached the base of the style. As against 76.6% of the ovules showing fertilization following intraspecific compatible pollinations, only 18.3% ovules showed fertilization in this cross. Similarly, in the cross *Brassica fruticulosa* × *B. campestris*, only 30% of the ovules showed pollen tube entry in contrast to 80% of the ovules in compatible pollinations (Nanda Kumar et al. 1988).

In many species different accessions show variation in intensity of prefertilization barriers (Snape et al. 1979, Zeven and Keijzer 1980). In the cross *Vigna vexillata* × *V. unguiculata*, for example, acc. TVnu 72 as the male parent resulted in fertilization of 18.6% of the ovules, while acc. TVnu 73 resulted in fertilization of 31.7% of the ovules (Barone et al. 1992). Of over a dozen accessions of *Tripsacum dactyloides* used in crosses with maize, only one proved effective in achieving fertilization (Harlan and de Wet 1977).

POST-FERTILIZATION BARRIERS

In crosses showing post-fertilization barriers, pollen germination, pollen tube entry into the stigma, pollen tube growth through the style and pollen tube entry into the ovule proceed normally (Fig. 10.2D-F). Post-fertilization barriers result in the abortion of fertilized ovules at different stages of development. This is due to the breakdown of the endosperm and/or embryo. Barriers may operate even during germination of the hybrid seed and subsequent growth of the F_1 hybrid (Zhou et al. 1991, Marubashi et al. 1988). Post-fertilization barriers are more prevalent than prefertilization barriers. Most crosses that show

prefertilization barriers show post-fertilization barriers also. Details of post-fertilization barriers are much less understood. On the basis of indirect evidence many causative factors have been suggested. Some of these are: presence of lethal genes, genic disharmony in the embryo, failure of the endosperm to develop or its early breakdown, and unfavourable interactions between the embryo sac and the surrounding ovular tissue (Khush and Brar 1992). In several crosses embryo abortion is associated with proliferation of nucellar or integumentary cells (Cooper and Brink 1940).

Chromosome elimination, in which chromosomes of one of the parents get eliminated during embryo development, is another manifestation of post-fertilization incompatibility, particularly in cereals (see Chapter 14). Chromosome elimination leads to the development of haploids; in most of the crosses embryo rescue is needed to recover haploids (see Palmer and Keller 1997). Many instances of haploid production in interspecific crosses, which are generally considered to be due to parthenogenesis, may be the result of chromosome elimination. Detailed cytological studies are needed to understand post-fertilization barriers in such crosses. Detailed discussions on post-fertilization barriers are presented by Stebbins (1958), Maheshwari and Rangaswamy (1965), de Nettancourt (1977, 2001), Hadley and Openshaw (1980), Raghavan (1986), and Khush and Brar (1992).

Part II
POLLEN BIOTECHNOLOGY

Pollen biotechnology is the management or manipulation of pollen grains for production and improvement of crops and other related economic products. Pollen grains can be manipulated during all phases of pollen biology—development, free-dispersed phase, pollination, pollen-pistil interaction and fertilization. Impressive progress made in recent years on pollen biology, covered in earlier chapters, has made the applications of pollen biotechnology more diverse, effective and viable.

A steady and significant increment in crop productivity has become imperative to feed the growing millions. There are two possible strategies to enhance crop productivity: (i) realization of potential yield by providing optimal conditions for plant growth and (ii) enhancement of genetic potential of crop species through transfer of agronomic or other useful traits from other species/genera. Increase in productivity through both these strategies has to be achieved under many constraints, some of which are:

- non-availability of additional land for cultivation,
- steady decrease in area of agricultural land due to increasing salinity, alkalinity, waterlogging, soil erosion and industrial contamination and
- need to drastically reduce the application of environment-unfriendly agrochemicals such as herbicides and pesticides.

Future plant breeders have therefore to develop varieties that give not only higher yield with better quality, but are also able to grow on marginal lands with minimum use of environment-unfriendly chemicals.

Realization of these objectives is very challenging and traditional methods of crop production and improvement are inadequate for achieving them. Some of the major challenges that cannot be effectively tackled by traditional methods are:

- insufficient pollination, particularly in many cross-pollinated species, due to lack of adequate number of pollinators;
- inability to exploit hybrid vigour in a number of crops due to non-availability of effective pollination control systems,
- narrow genetic base of crop species which has greatly increased their susceptibility to biotic and abiotic stresses such as diseases, pests, drought and salinity; and
- need to extend breeding programmes beyond species limits since the required genes are no longer available in the cultivated species.

Constant infiltration of new adaptive genes into cultivars is necessary to sustain and further improve the yield. Many wild and weedy relatives of crop species provide a vast source of such genes. However, strong crossability barriers between wild and cultivated species make gene transfer almost impossible using traditional methods.

Integration of pollen biotechnology into the traditional breeding programme offers effective approaches in a number of areas of production and improvement of crops and other products. Like any other technology, pollen biotechnology also decreases the time and cost of crop improvement and increases its efficiency. The following are the potential areas of pollen biotechnology presently available for effective application:

- overcoming pollination constraints,
- developing effective pollination control systems to exploit hybrid vigour,
- overcoming crossability barriers and many other constraints to transfer genes across species barriers and
- developing other economic products related to pollen.

The chapters on pollen biotechnology give a concise and integrated account on the progress made in these areas.

11

Optimization of Crop Yield

Fruits and seeds are the economic products of most crop plants. Effective pollination is a prerequisite for fruit- and seed-set. Therefore, successful pollination is of vital importance to realize optimal yield. In self-incompatible species, yield is dependent on adequate cross-pollination. Even in self-compatible species, pollination is largely dependent on pollinating agents as automatic selfing seldom occurs or is insufficient in most of the self-compatible crop species. The majority of our crop plants, except cereals (which are anemophilous), are pollinated by insects particularly bees. Adequate pollination is often a major constraint in many crop species due to one or more of the following reasons:

1. Drastic reduction in native pollinator populations because of the steady disappearance of natural habitats of insects, marked increase in level of pollutants and extensive use of environment-unfriendly agrochemicals, in particular pesticides and herbicides.
2. Lack of sufficient number of native pollinators due to enormous increase in area covered by the same crop species (often extending to hundreds of hectares) in the present-day monoculture cropping system.
3. Absence of natural pollinators for crops introduced from other regions.

Over 60 years ago it became apparent in the USA that production in many of the fruit, seed and nut crops could be increased substantially by careful management of pollination (Roubik 1995). This led to the initiation of extensive studies to increase pollination efficiency in crop species. Increased pollination efficiency can lead to increase in crop value by increasing crop yield, uniformity, quality and decreasing the time of crop maturity. Several effective approaches are available to overcome pollination constraints (Jay 1986, Torchio 1990, Free 1993, Currie 1997). However, progress in pollination management is largely confined to developed countries. In developing countries, although the yield in a number of crop species has been identified to be pollination-limited, pollination management has not yet been integrated into regular crop production practices (Savoor 1998), largely because of lack of data on pollination biology, pollination efficiency and details of pollinators of various crops (Roubik 1995).

ENHANCING POPULATIONS OF NATIVE POLLINATORS THROUGH HABITAT MANAGEMENT

Increasing local populations of native pollinator species through habitat management is one approach to removing pollination constraint. This is particularly useful when availability of nest site is a limiting factor. Habitat management can be achieved by maintaining uncultivated strips along field margins and providing permanent nest boxes (Pomeroy 1981, Osborne et al. 1991). Such strategies should also ensure availability of forage sources when the target crop is not in bloom. This approach also requires management of cropping practices in such a way that the flowering period of the target crop coincides with the peak populations of the pollinator (Free 1993). However, management of habitat is more expensive, particularly in areas of intensive agronomic practices (Torchio 1990).

USE OF COMMERCIALLY MANAGED POLLINATORS

So far the most economically viable and effective approach to overcome pollination constraints has been use of commercially managed pollinators, in particular honey-bees (*Apis* spp.), for pollination services. Honey-bees are the most effective pollinators in a range of crop species (legumes, fruits, vegetables and nuts) (Robinson 1979, Schmidt 1982). Management of honey-bees is convenient because of their large foraging populations, year-round availability and easy transportation. Methods for their management and control of their diseases, predators and parasites are well established (van Heemert et al. 1990, Free 1993). Increasing pollination efficiency through management of pollinators is warranted only when crops are pollen-limited and the cost involved is lower than the value realized through increase in crop production (Currie 1997).

The acreage of bee-pollinated crops has increased enormously in recent years. It is estimated that the value of bee-pollinated crops in the USA alone is ca US $9.3 billion. The demand for pollinators is steadily increasing (Torchio 1990). In India also many studies have indicated significant increase in yield of several oil-seed crops, especially sunflower, as a result of introduction of honey-bee colonies (Anonymous 1996, Savoor 1998).

Introduction of beehives in crop fields to enhance pollination efficiency and thus crop yield was known as early as 1892 (Savoor 1998). This technology was introduced on a commercial scale in the USA during the 1940s. Use of honeybees for pollination services has grown steadily over the years and it is presently estimated that in the USA alone over one million honey-bee colonies are rented every year for pollination services (Torchio 1990). These colonies are moved from one crop to another coinciding with the flowering season. Hives rentals in the last 2–3 years in the USA is reported to be ca $45 per colony/crop season (Gadagkar 1998). In recent years, maintenance of honey-bee colonies for pollination services has become more profitable and honey and wax have become byproducts.

Until recently, management of honey-bees for pollination services was confined largely to field-grown crops. Lately honey-bees have been increasingly used to achieve pollination in glasshouse crops, particularly tomatoes. Apart from bees, many other pollinators have shown potential for use in pollination services (Torchio 1987). For some crops, bumble-bees (*Bombus* spp.) are better pollinators (Holm 1966) than honey-bees because of their larger size, long tongue, and ability to vibrate flowers and fly at relatively low temperatures (Currie 1997). However, the high cost associated with bumble-bee management has so far restricted their use largely to red clover (Macfarlane et al. 1983) and to high-value glasshouse crops (Banda and Paxton 1991, Straver and Plowright 1991, Kearns and Inouye 1993). A commercial company in Europe is reported to rear >1000 bumble bee colonies per year for pollination services of glasshouse-grown tomato (Kearns and Inouye 1993).

Pollen collectors are generally more effective pollinators than honey collectors. Because pollen is primarily required for providing food for larvae and young bees, the proportion of pollen collecting bees is correlated with the level of egg-laying by the queen and the amount of brood present in the colony. Supply of extra frames of brood to colonies generally increases pollen collection.

Pollen transfer can be made more efficient by using 'pollen dispensers' mounted in the hive. The foragers are forced to pass through these dispensers which contain pollen of a compatible cultivar and thus pollen grains get deposited on the body parts of the foragers (Legge 1976, Ferrari 1990). Some of this pollen is eventually transferred to the stigmas of the crop plants during forager visits (Griggs and Iwakiri 1960, Williams et al. 1979). Use of pollen dispensers may prove particularly valuable in the production of hybrid seeds since pollination of male sterile plants can be achieved even in the absence of male fertile plants (Farkas and Frank 1982). This technology is likewise useful for self-incompatible species, in particular orchard species for which compatible trees are absent or insufficient. Many private companies sell pollen of different crop species for use in pollination services (Kearns and Inouye 1993, Newman 1984). The pollen in dispensers has to be continually replaced.

The required number of honey-bee colonies is maintained close to the target crop. Timing of pollinator introduction should synchronize with flowering of the target crop; introduction of bees too early or too late would reduce overall pollination efficiency. During inclement weather, more colonies are required to achieve pollination. Information is available for a number of crops detailing the requirement for populations of honey-bees/bumble-bees to pollinate a given area and their management in the field (DeGrandi-Hoffman 1987, Free 1993).

Use of honey-bee colonies for pollination services poses some management problems; honey-bees forage on a wide range of host plants and thus maintaining them on the target crop often becomes difficult. Thorough knowledge is required about the behaviour of bees, pollination biology of crop species and competing crops, and the ecology of plant-pollinator relationship.

Effective management strategies have to be followed to retain pollinators on the target crop. Honey-bees locate flowers by sight and odour and their fidelity to the crop is determined by the quantity and quality of the reward the crop offers. In the absence of sufficient rewards, bees desert commercial target crops for other attractive pollen- and nectar-yielding species (competing species) (Jay 1986). When the target crop is sprayed with an insecticide, bee repellents are sprayed on the target crop to temporarily prevent bee visits to the crop species, thereby reducing bee mortality.

Bees generally expand their foraging range gradually after being released from colonies at a new sight. When their foraging extends beyond the target crop, the colonies need to be rotated or replaced with fresh ones. This management strategy is particularly effective when the flowers of the target crop are not very attractive to the bees.

Overlapping of the flowering period of the target crop with a competing crop can be avoided by changing the planting period of the target crop. When weeds divert the pollinators from the target crop, competition from weeds can be reduced by mowing them or treating them with herbicides. Spraying competing plants with bee repellents such as carbolic acid, acetic acid, benzaldehyde or calcium chloride is another approach to reducing bee visits to competing non-target crops (Jay 1986).

One of the most effective long-term strategies for improving pollination efficiency is rational breeding of crops as well as honey-bees. Breeders should be able to breed crop varieties to make them more attractive and to provide better rewards in the form of more nectar production (see Currie 1997). Similarly, breeding bees that have a limited or reduced flight range and show preferential attraction to specific crops would be a great advantage in retaining bees on target crops (see Jay 1986).

Spraying Pollinator Attractants on Target Crop

A number of investigations have shown the potential of sprays with various substances on the target crop to attract pollinators (see Jay 1986,

Currie 1997). Although spray with sugar syrup often increases the abundance of foragers on the target crop, it may not increase crop yield, as the foragers tend to collect the syrup rather than the pollen or nectar.

Sprays with dilute solutions of pheromones (chemicals used for communication between members of the same species) have shown considerable potential. Secretions from the Nasonov gland located on the dorsal side of the abdomen of worker bees consist of seven terpenoids. Of these, geraniol and citral have been shown to increase honey-bee foraging activity. Similarly, the queen honey-bee secretes a five-component pheromone from its mandibular glands. Sprays of synthetic mandibular pheromones have been shown to increase honey-bee foraging activity and crop yield under a wide range of conditions (Currie et al. 1992a,b; Winston and Slessor 1993). Application of pheromone seems to be particularly effective on crops with flowers relatively unattractive to bees or during inclement weather conditions.

Apart from pheromones, sprays of synthetic plant volatiles isolated from nectar or pollen are also effective in attracting honey-bees (see Dobson 1994). Sprays containing food supplement such as Beeline[R] have also been reported to act as bee attractant in some crop species (Margalith et al. 1984).

INTRODUCTION OF POLLINATORS

When crops are grown in areas where natural pollinators are absent, as often happens when a crop is introduced from one country to another, introduction of pollinators is one of effective approaches. This approach involves detailed studies on the biology of the pollinator and monitoring its establishment in the new area. Oil palm (*Elaeis guineensis*) is a native of Africa and Central South America. It was introduced to Malaysia and Indonesia at the beginning of this century, where it is grown extensively at present. In its native habitat, oil palm is pollinated by wind as well as many insects in particular weevils. In many parts of Malaysia, where pollinating insects are absent, natural pollination was inadequate. Introduction of the weevil *Elaeidobius kamerunicus*, an important pollinator of oil palm from Cameroon, to Malaysia has been a successful example of such an approach. Introduction of weevil during the 1980s has markedly increased the yield in these areas (Syed 1979). Weevil could establish well in oil-palm estates in most Malaysian plantations. Over the first seven months after weevil introduction, oil yield increased 20–53% (Syed 1979, Syed et al. 1982).

SUPPLEMENTARY POLLINATION/ ASSISTED POLLINATION

During prolonged inclement weather conditions insufficient pollination is common in many crops in spite of availability of a sufficient number of pollinators. Many orchard species, such as apple, pear and cherry, are self-incompatible and are clonally propagated. Fruit-set depends on availability of two or more intercompatible varieties. Even when such lines are available, the activity of bees is greatly reduced in cold and wet weather, resulting in insufficient pollination. Raising of clonally propagated orchards using self-compatible clones would not only overcome pollination constraints, but also permit raising orchards with a single elite variety (see Reimann-Phillipp 1965). It is possible to induce self-compatibility through induction of mutations in SI alleles by irradiation of flower buds (Lewis 1954). However, this approach cannot be used for the existing orchards. Assisted/supplementary pollination through pollen sprays (Brown and Perkins 1969, Williams and Legge 1979, Hopping and Jerram 1980a, b) or any other method is the most effective technique for sustaining crop yield. This is routinely carried out for small-scale production of a few crops such as passion-fruit (Roubik 1995).

In high-value plantation crops such as oil-palm, pollination is a major constraint even in the presence of weevils, especially in younger plantations. Although oil palm is monoecious, the male and female phases alternate, each extending for many months; thus at any given time the plant will be in either the male or female phase. Insufficient number of plants in the male phase in the plantation and unfavourable weather conditions (Hardon and Turner 1967, Veldhuis 1968) reduce pollination efficiency. Assisted pollination is a common practice in oil-palm plantations in Malaysia and Indonesia, particularly in younger plantations (Hartley 1988). Assisted pollination has been reported to increase yield 20–150%, depending on age of the plants and weather conditions (Hardon 1973).

Assisted pollination requires standardization of protocols for pollen collection, pollen storage and pollination. Different methods have been essayed for assisting pollination in oil-palm (Wong and Hardon 1971). For pollinating young palms a 'hand puffer' (a rubber bulb/plastic bottle with an extension tube) has been found convenient; pollen is applied directly on anthesizing female inflorescences. For taller palms, blanket dusting of crowns is practised using a portable mechanical blower or duster (Wong and Hardon 1971). Pollination is normally repeated at 3-day intervals. Mechanized pollination of fruit trees such as apple and peach by using sprays has been tried, but this approach has not yet reached a commercial scale.

Hand pollination is regularly practised for *Vanilla* orchid. This orchid is a native of southern Mexico and Central America where it is pollinated by the euglossine bee *Eulaema* (Roubik 1995). *Vanilla* is grown extensively in many parts of tropical Asia where the pollinator is absent; hand pollination is routinely carried out to induce fruit-set.

12

Commercial Production of Hybrid Seeds

Development of hybrid seed technology for improvement of crop productivity is one of the most important advances in agriculture. The presence of sufficient heterosis/hybrid vigour is the primary requirement for application of hybrid seed technology. Hybrid vigour refers to an increase in vigour and productivity of the hybrid compared to its parents. Hybrids also show greater uniformity. For commercial purposes the extent of hybrid vigour is assessed on the basis of high parent heterosis (yield of F_1 hybrid exceeding the better parent used in the cross) as well as standard heterosis (yield of F_1 hybrid exceeding the best non-hybrid cultivar).

The basis of the hybrid seed industry was established during the second half of the 19th century when several geneticists showed that varietal hybrids of corn were more vigorous than inbred varieties (see Allard 1960). It was Shull who suggested at the beginning of the 20th century, use of inbred lines obtained by continuous self-pollination for commercial production of hybrid seeds (single-cross hybrids). However, this method did not lead to commercial use of hybrid varieties due to lack of suitable inbred lines and many other limitations (see Allard 1960).

The double-cross method, suggested by Jones in 1918, made commercial production of hybrid maize economically feasible. Although the first commercial production of hybrid maize was initiated in 1921, its impact was felt in agriculture only from the 1930s onwards. In 1933 hybrid maize covered <1% acreage in USA; it was extended to over 50% by 1940 and over 80% by 1944; soon after open-pollinated varieties of maize virtually disappeared from cultivation. During this period the increase in annual yield was estimated to be over 20% (Allard 1960). Hybrid varieties are now grown in all corn-growing countries.

The following are the basic requirements for commercial production of hybrid seeds in any crop species:

1. Development and maintenance of inbred lines.
2. Prevention of self- and intraline pollinations in hybrid seed production plots.
3. Effective cross-pollination between the two inbred lines.

The success of hybrid seed technology depends on the cost of producing hybrid seeds. Prevention of self- and intraline pollination is most critical and the cost of hybrid seed production is largely determined by the technology used to achieve this objective. In maize, a cross-pollinated crop, the architecture of the plant producing male flowers in the tassel at the tip and female flowers in cobs in the axils of leaves, is conducive to prevention of self-pollination by simply removing the tassel of the female parent before pollen grains are shed (detasseling). In species with bisexual flowers, however, controlled pollinations require labour-intensive emasculation and/or manual pollination; this markedly increases the cost of hybrid seed production and so the cost of hybrid seed technology becomes prohibitive. Different types of male sterility (described in Chapter 5), and self-incompatibility (Chapter 9) systems provide effective pollination control systems for extending hybrid seed technology to species with bisexual flowers. Hybrid seed technology has now been extended to a number of seed crops, e.g. sorghum, pearl millet, sugar-beet and a range of vegetable crops, e.g. tomato, onion and brassicas.

Use of male sterility for hybrid seed production requires availability of three lines (Shivanna and Johri 1985): A-line (male sterile line, used as the female parent for hybrid seed production); B-line (maintainer line, male fertile but isogenic to A line in other traits, used to maintain A line) and C line (restorer line, male fertile with fertility restoration gene, used as male parent in hybrid seed production plots). Figure 12.1 presents details of the maintenance of male sterile lines and production of hybrid seeds using genic male

sterile (GMS) (Fig 12.1A) and cytoplasmic male sterile (CMS) (Fig. 12.1B) lines.

Effective cross-pollination between male fertile and sterile lines in seed production plots is another major problem, particularly in self-pollinated crops, in exploiting any pollination control system for commercial production of hybrid seeds. Even in cross-pollinated crops such as *Brassica*, pollinators tend to avoid male sterile plants because of their low reward value and thus pollination becomes a constraint. In many high value ornamental and vegetable crops such as tomato in which each fruit produces a large number of seeds, hand emasculation and/or pollination is feasible, especially in developing countries where manual labour is comparatively cheaper. The advantages and limitations of different pollination control systems for the production of hybrid seeds are briefly reviewed below.

USE OF GENIC MALE STERILITY

Use of GMS for hybrid seed production is more straightforward than CMS as it requires no special efforts for developing male lines and transfer

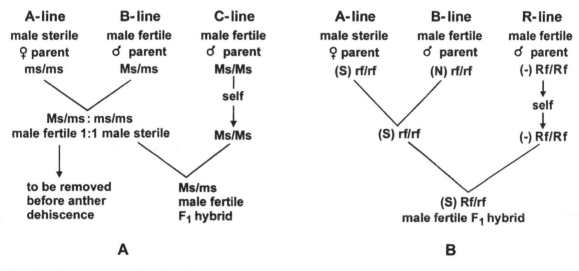

A

B

Fig. 12.1 Diagram of production of hybrid seeds using genic male sterility (GMS) (A) and cytoplasmic male sterility (CMS) (B). Maintenance of GMS line involves mechanical removal of male fertile plants in the progeny (Ms/ms—male sterile gene, Rf/rf—restorer gene, S/N—male sterile and normal cytoplasm).

of the cytoplasm to the good combiner. Since GMS in most species is expressed under a homozygous recessive condition, the restorer line has to be heterozygous. Thus, when A-line is crossed with B-line, only half the plants in the progeny are pollen sterile and the other half are pollen fertile; seeds are collected from male sterile plants to maintain the A-line. For production of hybrid seeds, the GMS line is planted in alternate rows with the male parent (C-line). The ratio of female and male lines used in seed production plots depends largely on the efficacy of cross-pollination and varies from 2:1 to 7:1 in different crops (Fig. 12.2). As 50% of the plants in female rows would be male fertile, the fertile plants have to be removed (rogued) before their pollen is shed (see Fig. 12.1A) (Frankel and Galun 1977). This is labour intensive and cumbersome as flowers of all the plants produced at the beginning have to be examined carefully to identify and remove the fertile plants; it may not always be possible to remove all male fertile plants before they shed their pollen. This is the major limitation of GMS for hybrid seed production. Various methods have been suggested to preclude/facilitate rogueing of fertile plants and thus to make GMS cost effective in hybrid seed production (Driscoll 1986, Rao et al. 1990). Some methods are described below.

Vegetative/Micropropagation of Female Lines

In some ornamental and vegetable crops, clonal multiplication of GMS lines through vegetative

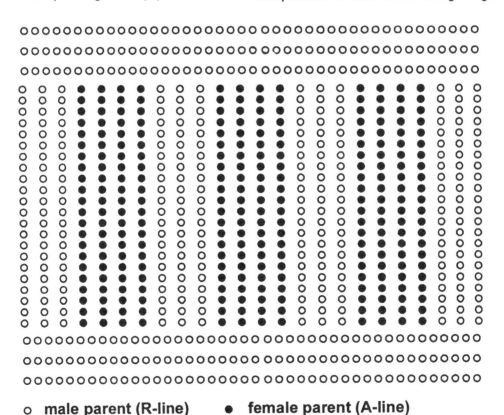

o **male parent (R-line)**   ● **female parent (A-line)**

Fig. 12.2 Arrangement of male and female lines in hybrid seed production plots. Ratio of male and female lines varies with the species (R-line—restorer line, A-line—pollen-sterile line).

propagation or micropropagation has been suggested. This approach may be feasible in high-value species in which vegetative propagation is profuse and/or cost effective protocols are available for micropropagation (Reimann-Phillipp 1983).

Use of Phenotypic Markers

A more effective method which can be used for a wide range of species is to identify male fertile or sterile plants through a marker phenotype controlled by pleiotropic effects of the GMS gene or a gene closely linked to it. A marker gene that is expressed in vegetative parts of the plant before flowering would greatly enhance the ease and effectiveness of rogueing of male fertile plants. The expression of markers such as stem pigmentation or nature of trichomes at the seedlings stage would be very convenient. Several such linkages have been reported in different GMS species (see Frankel 1973, Rao et al. 1990). A marker expressed in the seed stage itself would be most ideal as this would enable selection of seeds that produce only sterile plants for sowing in hybrid seed production plots.

Another effective approach is the linkage of male sterility gene with herbicide resistance. In barley, linkage of a GMS gene to herbicide resistance has been reported (Wiebe 1960). The advantage of herbicide resistance linked to male sterility is that the male fertile plants in hybrid seed production plots can be removed through spraying herbicide on the seed production plot at the seedling stage, thus eliminating the need for manual identification and rogueing of male fertile plants.

Environmental/Hormonal Induction of Male Fertility

Yet another method is based on the restoration of fertility in GMS plants through suitable environmental or hormonal treatments (Driscoll 1986, Kaul 1988, Sawhney 1994, 1997). This involves induction of pollen fertility in GMS plants by exposing them to suitable temperature/photoperiod, or through hormonal treatment (see Chapter 5); such plants selfed and/or their pollen used to pollinate pollen sterile plants to maintain the A-line (Fig. 12.3). The progeny raised from such GMS seeds would all be sterile and hence rogueing of plants in seed production plots unnecessary.

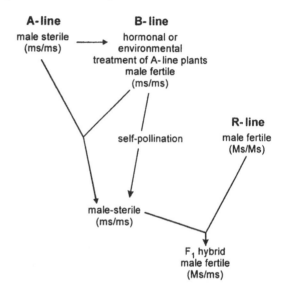

Fig. 12.3 Maintenance of GMS line through hormonal/environmental treatment and its use for hybrid seed production. GMS line is maintained through restoration of pollen fertility in GMS plants through environmental or hormonal treatment. Since the genotype of all the plants in the progeny remains ms/ms, all plants of the female line will be pollen sterile in hybrid seed production plots (adapted from Sawhney 1997).

USE OF CYTOPLASMIC MALE STERILITY

Cytoplasmic male sterility (CMS) is more convenient than GMS for hybrid seed production, as the progeny show no segregation into fertile and sterile plants and hence require no rogueing of fertile plants in seed production plots (Fig. 12.1B). Production of hybrid seeds using CMS was first reported for onion (Jones and Davis 1944). Since then it has been extended to many other crops, e.g. maize, sorghum, pearl millet, sunflower, carrot and *Petunia*.

Although CMS lines are available in many crop species, they are not used for hybrid seed production because of one or more limitations, such as enhanced susceptibility of CMS lines to certain diseases, environmentally related instability of male sterile and/or restored lines, reduced female fertility, leaf chlorosis particularly at seedling stage, pleiotropic negative effects of the cytoplasm on some agronomic traits, and insufficient nectar production that would affect pollination efficiency in insect-pollinated crops. Thus, availability of suitable CMS lines free from any of the above limitations is one of the primary requirements for commercial production of hybrid seeds using CMS.

Another constraint in the use of CMS in hybrid seed production is the availability of a suitable restorer line. Although CMS has been reported in over 250 species, restorer lines have been isolated in just a limited number of crop species. However, restorer lines are not needed in crops in which seed/fruit is not the commercial product, such as onion, vegetable brassicas and ornamental species.

In many crop species suitable CMS and restorer lines are not available and there is a need to develop new and better CMS and restorer systems. Even in those for which usable CMS and restorers are available, diversification of cytoplasmic sources is necessary to safeguard against the susceptibility of some CMS sources to pathogens. One of the effective methods for developing new CMS lines is the synthesis of alloplasmic lines (combining the cytoplasm of one species with the nucleus of another through sexual or somatic hybridization); many of the alloplasmic lines show CMS (see Chapter 5). CMS can also be transferred from one accession/species to another by repeated backcrossing of the CMS with pollen of the accession/species to which CMS is to be transferred.

CMS lines were extensively used in maize until 1970, when CMS-T, the most widely used cytoplasm, became highly susceptible to southern corn leaf blight (see Chapter 5). Although in recent years many other CMS and restorer systems have become available in maize, use of CMS has been replaced to a large extent by mechanical removal of the tassel of the female parent, the routine technique followed in earlier years.

In *Brassica napus*, although CMS and restorers are available, commercial hybrid seed production has not yet feasible due to insufficient pollination. As flowers of CMS plants do not produce pollen, bees tend to visit just male parent rows. To encourage bees to visit flowers of CMS plants, mixed sowing (as against sowing in rows) is being tried in Canada (Frauen 1987, McVetty 1997). Seeds of CMS and restorer plants are mixed 9:1 ratio and sown in seed production plots. As mixed sowing of such seeds results in pollen fertile plants being surrounded by sterile plants, bees are forced to visit sterile flowers after foraging fertile plants. Seeds harvested from such plots are contaminated with ca. 10% non-hybrid seeds. As 75% purity of the hybrid seeds is the regulation minimum in Canada, this fulfils the regulatory requirements.

USE OF SELF-INCOMPATIBILITY

Self-incompatibility (SI) is also an effective pollination control system to prevent self- and intra-line pollinations. Figure 12.4 presents broad procedures for production of hybrid seeds using SI. Use of SI has many advantages over GMS and CMS in production of hybrid seeds.

1. There is no need to develop maintainer or restorer lines for utilizing SI in hybrid seed production.
2. Hybrid seeds can be harvested from both parents (unlike GMS and CMS lines in which hybrid seeds can be harvested from only the female parent); thus use of SI reduces the cost of hybrid seed production.
3. There are no pollination constraints since both lines are pollen fertile and all SI species are naturally cross-pollinated.

In spite of these advantages, SI has not been widely used for hybrid seed production; its use is largely confined to vegetable crops, in

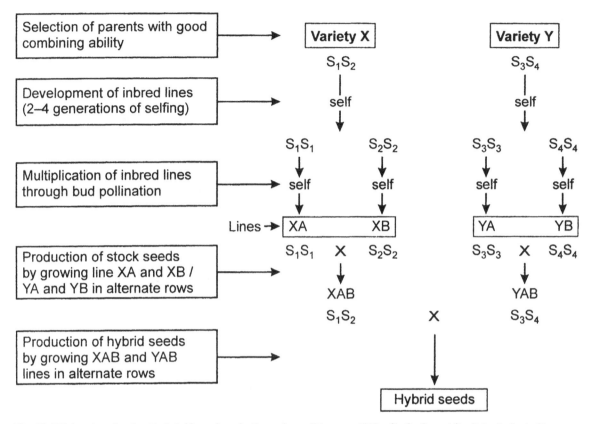

Fig. 12.4 Major steps involved in hybrid seed production using self-incompatibility. S_1, S_2, S_3 and S_4 alleles indicate SI genotypes.

particular brassicas—cabbage, cauliflower, kales and Brussels sprouts (Odland and Noll 1950, Thompson 1978, Gowers 1980, Pearson 1983). The following are some of the limitations for use of SI lines in hybrid seed production.

1. SI poses special problems in developing and maintaining pure lines (Fig. 12.4).

   Development of pure lines and their maintenance requires repeated self-pollination. As SI does not permit seed-set in self-pollinated flowers, some effective technique has to be used to overcome SI. Standardization of an effective and simple protocol to overcome SI is an essential requirement for developing and maintaining pure lines. This is cumbersome and expensive (see below), and often pure lines show inbreeding depression.

2. Susceptibility of SI to environmental variations, in particular temperature and humidity. This results in contamination of hybrid seeds with selfed seeds to different degrees depending on the prevailing environmental conditions.

3. SI is not present in many crop species.

Many approaches are in use or have the potential to circumvent these limitations.

Methods to Overcome SI

A range of methods is available to overcome SI (Table 12.1). However, most are cumbersome and can only be applied on a limited scale. This greatly increases the cost of hybrid seed production. Bud pollination has been the most common method used so far to overcome SI for development and maintenance of pure lines

Table 12.1 Some of the effective techniques to overcome self-incompatibility

Technique	Species	Reference
Induction of mutations	*Oenothera*	Lewis 1949
	Prunus	Lewis and Crowe 1953
	Petunia	Brewbaker and Natarajan 1960
	Trifolium	Denward 1963
	Nicotiana	Pandey 1965, van Gastel and de Nettancourt 1975
Induction of autotetraploidy	*Pyrus*	Crane and Lewis 1942
	Petunia	Stout and Chandler 1941
	Lolium	Ahloowalia 1973
Bud pollination	*Brassica*	Attia 1950, Odland and Noll 1950
	Raphanus	Kakizaki 1930
	Petunia	Eyster 1941, Shivanna and Rangaswamy 1969
Delayed pollination	*Brassica*	Kakizaki 1930
	Lilium	Ascher and Peloquin 1966a
Hot water/high temperature treatment	*Malus, Pyrus, Prunus*	Lewis 1942
	Oenothera	Hecht 1961
	Trifolium	Kendall and Taylor 1969
	Brassica	El Muraba 1957
	Lycopersicon	Hogenboom 1972
	Lilium	Ascher and Peloquin 1966b
	Raphanus	Matsubara 1980
Application of growth substances	*Lilium*	Emsweller and Staurt 1948
		Emsweller et al. 1960
		Matsubara 1973
	Petunia, Trifolium, Tagetes, Brassica	Eyster 1941
Use of mentor pollen	*Theobroma*	Opeke and Jacob 1969
	Cosmos	Howlett et al. 1975
	Malus	Dayton 1974
	Petunia	Sastri and Shivanna 1976a
	Nicotiana	Pandey 1979
Placental pollination	*Petunia*	Rangaswamy and Shivanna 1967
		Niimi 1970, Wagner and Hess 1973
High carbon dioxide concentration (3–6%)	*Brassica*	Nakanishi and Hinata 1975, Dhaliwal et al. 1981, Taylor 1982, O'Neill et al. 1984, Vitanova 1984, Tao and Yang 1986
Treatment of flowers with high CO_2 and humidity	*Brassica*	Gowers 1980
Treatment of stigma with NaCl (1.5–3%) before or after pollination	*Brassica*	Tao and Yang 1986, Monteiro et al. 1988, Fu et al. 1992
Treatment of stigma with lectins or pollen with sugars before pollination	*Petunia, Eruca*	Sharma et al. 1985

(Pearson 1983). It is effective in a range of species. However, it is labour intensive since it requires handling individual flowers. There is a need for development of simpler and more effective technique(s) to overcome SI.

Treatment with high CO_2 concentration or sodium chloride is very promising for overcoming SI in species of *Brassica*. CO_2 treatment can be given only in closed glasshouses and thus only a limited amount of seeds can be harvested.

Treatment with sodium chloride is more convenient and much less expensive. Since it can be applied under field conditions, a large amount of seeds can be harvested. However, the efficacy of both these treatments is variable between species and genotypes, and is affected by prevailing environmental conditions. Therefore, detailed studies are needed to standardize the protocol for different genotypes.

Apart from its use in the production of hybrid seeds, effective methods to overcome SI are necessary for other practical purposes. For example, selfing is the only alternative method to maintain hybrid characters in ornamental species which are difficult to propagate vegetatively or prone to viral infection. For genetic and breeding studies also it is desirable to obtain homozygous progeny. SI is a hindrance for such studies.

Many orchard species, e.g. apple, pear and cherry, are self-incompatible and must be clonally propagated. Fruit-set depends on the availability of two or more intercompatible lines. Even when such lines are available, bee activity is greatly reduced in cold and wet weather conditions, resulting in insufficient pollination. Raising clonally propagated orchards using self-compatible clones would not only overcome pollination constraints, but also permit raising orchards with a single elite variety (see Reimann-Phillipp 1965). It is possible to induce self-compatibility through induction of mutations in SI alleles through irradiation of flower buds (Lewis 1954). However, this approach cannot be used in the existing orchards.

Selection of Lines with Strong SI Alleles

To circumvent the problem of breakdown of SI under varied environmental conditions, it is necessary to select genotypes with strong SI alleles. This can be done either on the basis of seed-set in self-pollinated flowers or by scoring pollen tube penetration in selfed pistils through aniline blue fluorescence technique. A method which combines testing for the level of SI with production of selfed seeds through bud pollina-

tion, has been suggested for *Brassica* (Pearson 1983). Both open flowers and buds (3–4 days before anthesis) are self-pollinated in each inflorescence; open flowers are separated from pollinated buds by a tag tied to the inflorescence axis. After fruit maturity, the number of seeds from each part is determined. An effective SI allele is indicated by <1 seed/pod from open flowers; development of >50% of the ovules into seeds in self-pollinated buds (ca 10–12 seeds/pod) indicates the suitability of a line for pure line production through bud pollination.

Transfer of SI Alleles to Self-compatible Species

Most of the crop species are self-compatible (SC) although many of their wild relatives are SI. Transfer of SI (from SI species) to SC cultivars/species has not been feasible through conventional breeding programmes because of many practical difficulties (de Nettancourt 2001). Recent advances in cloning and characterization of S-genes (see Chapter 9) have opened up the possibility of transferring the S-gene through recombinant DNA technology.

HYBRID SEED PRODUCTION IN MONOECIOUS SPECIES

In many members of Cucurbitaceae and in spinach (*Spinacea oleracea*), which are monoecious, hybrid seeds are produced through manipulation of the development of male and female flowers (Pearson 1983). The proportion of male and female flowers on each plant is highly variable depending on the genotype of the plant and on prevailing environmental conditions; this proportion can also be altered by application of growth substances.

In *Cucurbita pepo*, male flowers which develop early in the flowering period, are manually removed from the seed parent (in seed production plots) at 2- or 3-day intervals to prevent self-pollination. This procedure though effective, increases the cost of hybrid seeds; it also often results in contamination of seeds since rogueing of all male flowers is difficult to achieve. In recent years a more convenient method based on application of plant growth regulator to induce development of only male or female flowers has been employed. Production of male flowers in the seed parent is suppressed for 2–3 weeks by spraying seedlings with ethylene. Female and male parents are usually planted in a 2:1 ratio in seed production plots, and the plants of the pollen parent are destroyed after the crop has set. This technology has markedly reduced the cost of hybrid seed production.

In many other cucurbits (e.g. *C. sativus* and *C. melo*) also, auxins and ethylene induce female flowers while gibberellic acid and antiethylene substances such as silver nitrate induce male flowers. In these crops, highly gynoecious lines have been isolated for use in hybrid seed production. Homozygous gynoecious lines are maintained by using gibberellic acid or silver nitrate to induce male flowers; self- and sib-pollinations in gynoecious plants are achieved through insects. Gynoecious and normal monoecious lines are raised in seed production plots and cross-pollination is achieved through insect activity.

Spinach (*Spinacea oleracea*) is monoecious and wind pollinated. The extent of female flowers on each plant ranges from 100% to 0%. Highly gynoecious female lines have been used for hybrid seed production. Unlike cucurbits, in which early-formed flowers are males, in spinach early-formed flowers are females. In highly gynoecious female lines, male flowers are formed only when female flowers do not set seeds and thus fail to act as metabolic sinks. If sufficient pollen is available in the field, as is the situation in seed production plots, early-formed female flowers set seeds which act as sinks; the plant matures with its seed crop and senesces without producing male flowers.

Manipulation of Sex-expression for Yield Improvement

In many dioecious (e.g. papaya) and monoecious (many cucurbits) crop species, productivity can be increased by increasing the number of female plants (in dioecious species) or female flowers (in monoecious species). Following the demonstration that many growth substances shift the sex-expression in favour of male flowers or female flowers such a technology has become feasible. Treatment of male plants with auxins or ethylene induces development of female flowers. Similarly treatment of monoecius plants with auxins or ethylene increases the number of female flowers. Such an increase in female flowers results in increased fruit- and seed-set.

USE OF POLLEN STERILITY INDUCED THROUGH r-DNA TECHNOLOGY

Details of development of pollen sterile and restorer systems using r-DNA technology have been described in Chapter 5. This pollination control system is being perfected for commercial production of hybrid seeds in *Brassica napus*. To overcome the problem of rogueing of male fertile plants in seed production plots, a herbicide resistant gene, *35S bar*, has been introduced into the female line along with *TA-29 barnase* (Fig. 12.5). Treatment of plants of the female parent with herbicide at the seedling stage eliminates male fertile plants, as they are susceptible to the herbicide. The transgenic-based pollination control system is undergoing field trials in *Brassica napus* and is likely to be released soon for commercial production of hybrid seeds.

USE OF POLLEN STERILITY INDUCED THROUGH CHEMICAL HYBRIDIZING AGENTS

Considerable progress has been made in recent years in development of effective chemical hybridizing agents (CHAs), particularly for wheat (see Chapter 5). A few have shown potential for commercial application (Cross and Schulz 1997). Limited studies have also been carried

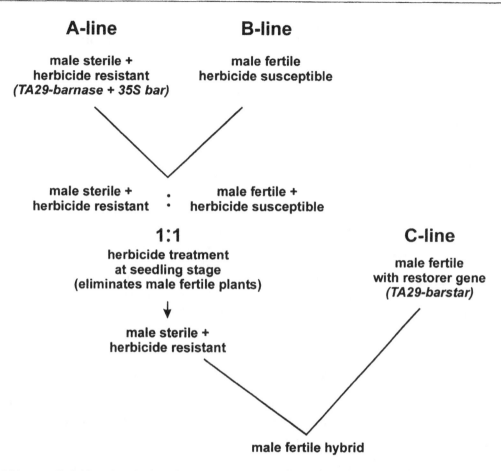

Fig. 12.5 Diagram of hybrid seed production using barnase-barstar approach by linking *TA-29 barnase* with selectable marker gene *35S bar* which confers tolerance to the herbicide glufosinate ammonium.

out to understand their mode of action. The following are some of the CHAs that have shown potential for application in hybrid seed production.

Proline Analogues

Proline is the most abundant free amino acid in pollen grains. Accumulation of high concentration of proline seems to be necessary for the production of fertile pollen. A few proline analogues such as methanoproline (*cis*-3, 4-methylene-S-proline), and azetidine-3-carboxylic acid (A3C) have been reported to be effective in inducing pollen sterility in wheat and barley (Devlin and Kerr 1979, Searle and Day 1980, Porter et

al. 1985, Johnson and Lucken 1986). Although pollen grains in A3C-treated plants appeared normal in morphology, they produced only small pollen tubes on the stigma, which eventually burst or failed to enter the stigma (Ladyman and Mogensen 1987, Mogensen and Ladyman 1989). However, these pollen grains surprisingly showed higher levels of proline accumulation indicating that A3C may not act as a proline analogue, but affect pollen development in some other way (Mogenson and Ladyman 1989).

Phenyl Pyridazones

Fenridazon (RH-0007), a phenyl-substituted pyridazone carboxylate developed by Rohm and

Hass, has been reported as an effective CHA, allowing commercially acceptable level of seed yield (Bucholtz 1988). This chemical did not affect meiosis in sporogenous cells, but affected development of microspores (Mizelle et al. 1989). The microspores showed irregular or reduced exine and eventually aborted. The tapetal cells enlarged and protruded into the locule; the orbicules deposited on the inner tapetal wall were greatly reduced. The pollen wall was 80% thinner in treated plants (El-Ghazaly and Jensen 1986). It was suggested that RH 0007 inhibited polymerization of sporopollenin precursors into exine wall and orbicules (Mizelle et al. 1989).

Phenylcinnoline Carboxylates

These CHAs have been developed by Sogetal Inc. (France) (Guilford et al. 1992). SC 1058 [1-(4 trifluoromethyl phenyl)-4-oxo-5-fluorocinnoline-3-carboxylic acid] and SC 1271 [1-(4'-chlorophenyl)-4-oxo-5-propoxy-cinnoline-3-carboxylic acid] induce complete pollen sterility in wheat when applied at the premeiotic stage of anther development (Cross et al., 1989, 1992). The effects of these CHAs are very similar to fenridazon (RH0007) (Schulz et al. 1993), indicating that SC 1058 and SC 1271 also mediate pollen sterility by interfering with the supply of sporopollenin wall materials (Schulz et al. 1993).

Subsequently, another compound, SC 2053 [1-(4'-chlorophenyl)-4-oxo-5-methoxyethoxy-cinnoline-3-carboxylic acid] was tested for CHA activity in wheat (Batreau et al. 1991). Treatment of plants with SC 2053 (700 and 1000 g/ha), when the spikes were 11–20 mm long, resulted in a high degree of pollen sterility (Batreau et al. 1991, Patterson et al. 1991, Cross and Schulz 1997). These CHAs have been reported to be effective in producing a usable level of male sterility with minimum phytotoxicity and loss of seed yield, when applied at the right stage and dosage.

LY 195259

LY 195259 [5-(aminocarbonyl)-1-(3-methylphenyl)-1H-pyrazole-4-carboxylic acid], a synthetic growth regulator, has been reported to be a commercially promising CHA for wheat (Tschabold et al. 1988). Even low dosage (1.12 kg/ha) produced >95% pollen sterility with a high degree of female fertility (seed-set as high as 80%). No other major cytotoxic effects have been reported with this CHA. Administration of the chemical to the soil was also effective in inducing pollen sterility (see Cross and Schulz 1997).

MON 21200

Monsanto reported the effects of a series of pyridine monocarboxylic and benzoic acid analogues on CHA activity in wheat (Ciha and Ruminski 1991). Currently a Monsanto-affiliated company (Hybri Tech) is developing a CHA compound MON 21200 for wheat (see Cross and Schulz 1997). MON 21200 is reported to provide good CHA activity over a wide range of genotypes, geographic regions and growth conditions. Outcrossing rates of around 80% have been reported for different wheat varieties over 6/7-year field trials.

POTENTIAL OF APOMIXIS IN HYBRID SEED TECHNOLOGY

One of the main constraints in hybrid seed technology at present is the need to produce hybrid seeds every year. Apomixis, an asexual reproductive trait, has the potential to overcome this major limitation (Khush 1994, Hanna and Bashaw 1987, Hanna 1991, Spillane et al. 2001). Broadly, apomixis includes all forms of asexual reproduction (Nogler 1984, Koltunow 1993). The type of apomixis relevant to hybrid seed technology is the one in which seeds develop without fertilization. Apomixis has been known for many decades but largely remained a trait of academic interest (Maheshwari 1950, Johri 1984). Apomixis has been reported in over 300 species belonging to about 40 families. It is more common in polyploid species, presumably as an adaptation to overcome meiotic abnormalities in seed development. Apomixis is prevalent in members of Asteraceae (*Hieracium*), Poaceae

(*Panicum, Pennisetum*) and a few other families. Several species are obligate apomicts (e.g. *Commiphora wightii*, Gupta et al. 1996) while many show facultative apomixis, producing seeds both apomictically and sexually.

In most plants apomictic embryos originate either from an unreduced egg or from the sporophytic cells surrounding the embryo sac. An unreduced egg develops as a result of the failure of female meiosis (*Antennaria*) or formation of restitution nucleus after the megaspore mother cell enters meiosis (*Taraxacum*); both events result in formation of a 2n embryo sac containing a 2n egg. In the other pathway the haploid egg of the reduced embryo sac degenerates but a sporophytic cell either from the nucellus or the endothelium (innermost layer of the integument) develops into the embryo. The former pathway is more prevalent. In many species, such as *Citrus*, both sexual and apomictic embryos develop in the seeds; the reduced egg gets fertilized and develops into the sexual embryo at the micropylar part of the embryo sac; apomictic embryos arise from the nucellar cells, protrude into the embryo sac and continue development. This results in polyembryony, one of the embryos being sexual and the remainder apomictic.

For full development of apomictic embryos, development of endosperm is necessary for nourishing the embryo. Endosperm development in many apomictic species is spontaneous (from fused polar nuclei) or pseudogamous (one of the sperms fuses with the secondary nucleus but the other sperm does not fuse with the egg). For the latter, pollination and pollen-pistil interaction have to be completed. Endosperm development is necessary not only for embryo nourishment, but also for human consumption in species with endospermous seeds. In many crop species, especially in all cereals, the economically important part of the seed is the endosperm as it stores most of the nutrients. In the absence of endosperm development, apomixis is of no commercial use in such species.

Since meiosis and fertilization are not involved in the formation of apomictic seeds, the progeny raised from such seeds will be identical to the parent. Transfer of the apomictic trait into the hybrid plants would result in fixation of hybridity; since the progeny do not show segregation for hybrid vigour, the farmer can use the seeds collected from hybrids generation after generation. This would enormously simplify hybrid seed technology and increase profitability to the farmer. Parthenogenesis (development of embryo from reduced egg without fertilization), another form of apomixis, is not useful for hybrid seed technology as it involves meiotic segregation and also results in sterile progeny.

Realization of the potential of apomixis in hybrid seed technology has stimulated great interest. Extensive studies are underway worldwide for screening a large germplasm for apomixis, understanding its genetics, and developing effective methods for its transfer to sexual crop species. Apomixis involves a series of events: failure of female meiosis, organization of the embryo sac, stimulation of the egg to develop into the embryo without fertilization and development of endosperm. In instances of adventive embryony, it also involves stimulation of the nucellar or integumentary cells to develop into embryos. Apomixis therefore seems to involve a number of genes controlling different aspects of meiosis and seed development; this makes transfer of the apomictic trait from one species to another more difficult. However, many studies have indicated that in several species, apomixis is controlled by a single gene or a few genes (Nogler 1984). The single locus is likely to be comprised of many tightly linked genes. Genes controlling apomixis may be dominant or recessive.

Although most of the crop species are sexual, many of their wild relatives show apomixis. Several attempts have been made to transfer apomixis from these sources to sexual species (Khush 1994) using traditional breeding approaches; for example, from *Tripsacum dactyloides* to maize, from *Pennisetum orientale* to pearl millet, from *Agropyron seabrum* to wheat (Asker and Jerling 1992). However, success so far achieved has been limited. The molecular approach seems to be a better option for this task. One limitation to developing effective technology for such a transfer is lack of sufficient basic information on the genetics of apomixis. Extensive studies are underway to identify and characterize the genes involved in meiosis and other events associated with apomixis in several species. These approaches have been elaborated by Maheshwari et al. (1998).

13

Transfer of Useful Genes to Crop Species

Transfer of useful genes to crop species from other accessions or species through hybridization has been the most effective approach in crop improvement programmes for many decades. Extensive breeding and selection have greatly reduced genetic variability of crop species. Dependence of agriculture on a narrow genetic base has markedly increased the vulnerability of present-day cultivars to a range of biotic and abiotic stresses. Although most cultivars have the genetic potential for high yield under optimal agronomic practices (with high input of fertilizers and other agrochemicals), their susceptibility to biotic and abiotic stresses reduces yield significantly, often up to 50% (Boyer 1982). Broadening the genetic base of cultivated species is an essential prebreeding requirement to maintain stability and further improve yield.

Until recently crop improvement programmes concentrated almost exclusively on increasing yield and improving quality and genetic uniformity. Remarkable success has been achieved over the years in fulfilling these limited objectives. In recent years, however, the objectives of the breeding programme have become more diverse; some of the additional breeding objectives apart from improvement in yield and quality are:

1. To develop varieties which are resistant/tolerant to biotic and abiotic stresses such as diseases, pests, temperature, salinity and drought.
2. To develop varieties which have low requirement for chemical fertilizers, herbicides, pesticides, and other environment-unfriendly chemicals. Excessive use of agrochemicals has been a major environmental problem of modern agriculture. This aspect has so far been ignored by farmers as well as breeders but has become increasingly important in recent years. The trend is to grow crops 'organically' without the use of pesticides, herbicides and chemical fertilizers.
3. To develop varieties that can be grown on marginal lands. A large area of agricultural land is being degraded by salinity, alkalinity, waterlogging, soil erosion and contamination with industrial chemicals. This is particularly severe in developing countries. As the recovery of these vast areas of land is not feasible because of the high cost involved and/or lack of effective technologies, the alternative strategy of breeding crops that can be grown on such marginal lands has to be adapted.

Realization of the aforesaid objectives is difficult and challenging. As many of the adaptive traits are controlled by multiple genes, transfer of such traits into a crop species by recombinant DNA technology is not feasible; a conventional breeding programme through the sexual cycle has to be followed. Because of the constant erosion of genetic diversity, genes imparting the required traits are no longer available within the cultivars of crop species. However, a large number of wild and weedy relatives of crop species form a good repository of such desirable traits (Harlan 1976). Unlike the early breeding programmes, which were largely confined to accessions within a species, present-day breeders have to extend the breeding programme across species limits, often to wild and weedy species (wide hybridization). A good number of examples of successful gene transfer through wide hybridization are already available (Table 13.1) (Hawkes 1977, Stalker 1980, Goodman et al. 1987, Hermsen 1992, Kalloo 1992, Chopra 1989, Chopra et al. 1996).

Apart from its use in introgression of desirable genes into cultivars, wide hybridization is also effective in: (a) production of haploids through chromosome elimination (see Chapter 14); (b) development of new cytoplasmic male sterile (CMS) lines through synthesis of alloplasmic lines (see Chapter 5) and (c) synthesis of new alloploid crops such as *Triticale* and *Raphanobrassica* (Sareen et al. 1992).

The following are the major steps involved in transfer of genes, especially from wild relatives through sexual means:

1. identification of species/accessions having desirable genes,
2. production of hybrids between the cultivar and the parents with desirable gene(s) and raising of successive backcross generations,
3. screening a large number of plants in each backcross generation for identification of the required recombinants, and
4. stabilization of the recombinants.

Figure 13.1 presents details of steps 2 and 3 using transfer of disease resistance as an example. The conventional breeding programme implemented so far is often ineffective because of the presence of strong crossability barriers. Even in the absence of such barriers, conventional breeding takes too much time. Integration of pollen biotechnological approaches into conventional breeding greatly improves the efficiency and reduces the cost and time of breeding. Table 13.2 lists appropriate pollen biotechnological approaches, which can be integrated at different levels of the breeding programme.

USE OF POLLEN FOR IDENTIFYING PLANTS WITH DESIRABLE GENES

A large number of accessions of related species and often wild species of cultivars have to be screened to identify those with desirable

Table 13.1 Some selected examples of transfer of important traits from wild species to crop species through wide hybridization (based on data presented in Goodman et al. 1987)

Crop species	Donor species	Trait
Beta vulgaris	*B. procumbens*	Sugar-beet nematode resistance
Brassica napus	*B. campestris*	Club-root resistance
Cucurbita pepo	*C. landelliana*	Mildew resistance
Gossypium hirsutum	*G. raimonddii*	Rust resistance
Lycopersicon esculentum	*L. peruvianum*	Nematode resistance
Nicotiana tabacum	*N. glutinosa*	TMV resistance
Oryza sativa	*O. nivara*	Grassy stunt virus resistance
Solanum tuberosum	*S. demissum*	Late blight resistance, leaf-roll resistance
Triticum aestivum	*Aegilops comosa*	Stripe rust resistance
Triticum aestivum	*Secale cereale*	Yellow rust resistance, leaf rust resistance, winter hardiness, stem rust resistance
Zea mays	*Tripsacum dactyloids*	Northern corn leaf blight resistance

Table 13.2 Integration of pollen biotechnology into conventional breeding programme

Major step/constraint	Pollen biotechnological approach
Identification of accessions with desirable genes	Use of pollen for screening
Spatial/temporal isolation of parental species	Pollen storage
Presence of pre- and post-fertilization barriers	Application of effective methods to overcome barriers
Multiplication of hybrids	Use of tissue culture technique
Hybrid sterility	Induction of amphiploidy
Handling of BC generations	Use of pollen for screening and application of selection pressure on pollen
Stabilization of recombinants	Induction of pollen embryos

genes before they can be transferred to the cultivated species. Screening accessions for desirable genes through the conventional method is laborious, time consuming and very expensive. For example, screening of an accession for disease resistance using the conventional method involves growing plants under glasshouse conditions suitable for growth of the pathogen, spraying the plants with spores of the pathogen and scoring for the symptoms after allowing sufficient time for pathogen establishment. It also restricts use of each plant for testing a single factor. Use of pollen for screening desirable genes offers many advantages; it is more effective and greatly reduces the time and cost of screening (Mulcahy 1984). As a single plant produces a large number of pollen grains, a range of factors can be tested against the pollen of the same plant.

Use of pollen for screening for desirable genes is dependent on the expression of the required gene(s) in the pollen grain. Recent studies have conclusively established that a large number of genes are expressed in the pollen (Tanksley et al. 1981, Mascarenhas 1993, Hamilton and Mascarenhas 1997) and about 60% of them are also expressed in the sporophyte (Ottaviano and Mulcahy 1989) (see Chapter 1). A simple method for testing the expression of gene(s) imparting resistance or tolerance to a particular stress in the pollen is to compare the responses of the pollen (Fig. 13.2) and the sporophyte under stress conditions. A number of investigations in several species have shown positive correlation between performance of the

pollen and that of the parent sporophyte under a range of stresses (Table 13.3), indicating that many of the adaptive genes are expressed in the pollen as well as the sporophyte. When such a correlation is established, plants resistant to a particular stress can be readily identified by studying the effect of stress on pollen germination and/or pollen tube growth under the stress condition (Hormaza and Herrero 1992, Sari-Gorla and Frova 1997). For example, pollen grains from a plant resistant to a disease would be able to germinate in the presence of specific pathotoxin, while those from a susceptible plant would fail to germinate.

Screening pollen for stress resistance/tolerance can be carried out using one of the following four methods:

i) in-vitro germination of pollen grains in the presence of stress,

ii) pollination of stress-treated pistils with untreated pollen grains,

iii) pollination of untreated pistils with pollen grains treated with stress,

iv) application of stress to pollinated pistils.

Although in-vitro germination is the most convenient and rapid method, it can only be used for those species in which satisfactory in-vitro pollen germination can be achieved. Pollen grains of many species, especially of 3-celled pollen species, are difficult to germinate in vitro (see Chapter 4). For such pollen other methods can be used. Pollen germination and pollen tube growth in the pistil can be conveniently studied through the use of aniline blue fluorescence (see Shivanna and Rangaswamy 1992). The number

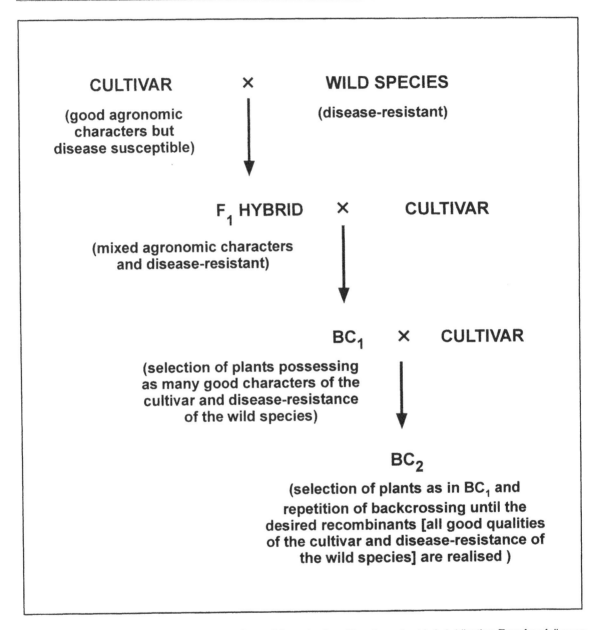

Fig. 13.1 Protocol for transfer of desirable gene from wild species to cultivar through wide hybridization. Transfer of disease resistance is used as an example (after Shivanna and Sawhney 1997).

of pollen tubes growing in the style compared to the controls (untreated pollen and pistil) indicate the sensitivity of pollen grains to the stress condition.

PRODUCTION OF HYBRIDS

As pointed out earlier, strong crossability barriers between parents which may operate before

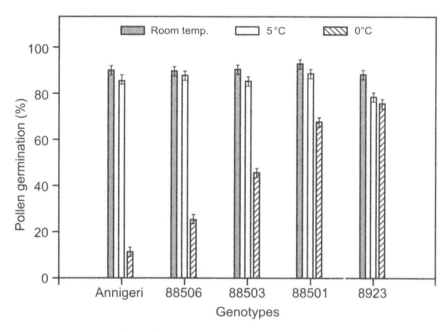

Fig. 13.2 Genotypic variation for low temperature tolerance using in-vitro pollen germination assay in *Cicer arietinum*. Different accessions show marked variation in in-vitro germination response at 0°C which broadly reflects the responses of parents (based on K R Shivanna, N P Saxena and N Seetharama, unpublished).

Table 13.3 Some of the agronomically important traits expressed in the pollen and the sporophytes (refer Sari-Gorla and Frova 1997 for additional references)

Character	Species	Reference
Tolerance to low temperature	Tomato	Zamir et al. 1982, Zamir and Gadish 1987, Zamir and Vallejos 1983
	Chickpea	Savithri et al. 1980
Tolerance to salinity	F_1 hybrids of *Lycopersicum* x *Solanum*	Sacher et al. 1983
Tolerance to heavy metals (zinc, copper)	*Silene dioica, Mimulus guttatus*	Searcy and Mulcahy 1985a, b
Tolerance to pathototoxins	Maize to *Helminthosporium*	Laughnan and Gabay 1973
	Brassica to *Alternaria*	Shivanna and Sawhney 1993
Tolerance to herbicides	Sugar-beet	Smith 1986
	Maize	Sari-Gorla et al. 1989, 1994

and/or after fertilization are the major constraints in any hybridization programme (see Chapter 10). Present-day plant breeders have a range of biotechnological methods available to overcome such barriers. For selection of the most suitable technique, it would be necessary to identify the barrier.

Effective Techniques to Overcome Physical Barriers

The most important physical barrier for wide hybridization is non-synchronous flowering or geographic isolation of the parent species. A conventional breeding programme is ineffective in such crosses. Pollen storage is one of the simple and effective methods to overcome this temporal and spacial isolation of the parent species.

Pollen storage

Pollen grains can be stored for extended periods under suitable conditions to maintain their viability and used when required. Pollen storage has important applications in both basic and applied areas of reproductive biology. These are some of them:

1. Overcoming crossability barriers imposed due to temporal and spatial isolation of the parent species.
2. Eliminating the need to grow pollen parents continuously in breeding programmes.
3. Implementing supplementary pollinations to sustain yield (see Chapter 11).
4. Preserving genetic resources and providing germplasm for international exchange.
5. Ensuring availability of pollen throughout the year for studies on various aspects of pollen biology and pollen allergy.

The need to establish 'pollen banks', which would ensure availability of pollen of the desired species at any time of the year and at any place, has long been emphasized. Pollen banks would greatly facilitate the breeding programme, particularly of tree species. Extensive studies have been carried out to assess various storage conditions for extending pollen viability and pollen grains of a large number of species have been stored successfully (Table 13.4) (Knowlton 1992, Holman and Brubaker 1926, Visser 1955, Johri and Vasil 1961, King 1961, 1965; Stanley and Linskens 1974, Shivanna and Johri 1985, Towill 1985, 1991, Hanna and Towill 1995, Barnabas and Kovacs 1997). The important methods used are briefly summarized below.

Storage under low temperature (+4 to –20°C) and low humidity (<10% RH): Pollen grains are taken in small unsealed vials and kept in a desiccator or a suitable airtight vial containing an appropriate dehydrating agent such as dry silica

to maintain low RH (Shivanna and Johri 1985). The sealed desiccators are then kept in a refrigerator or deep-freeze. This method is very convenient and effective for short-term storage (for a few weeks/months); it has been used extensively by amateur horticulturists. Storage under subfreezing temperatures (ca –20°C) is effective for storing pollen of several species for more than a year (Table 13.4). However, pollen grains of cereals in general cannot withstand desiccation and need to be stored under high RH in the refrigerator. Even under these conditions, viability of cereal pollen lasts for only a few days (Shivanna and Heslop-Harrison 1981).

Storage of freeze-dried/vacuum-dried pollen: Freeze-drying involves rapid freezing of pollen (–60 to –80°C) and gradual removal of water under sublimation. In vacuum drying, the pollen is exposed to simultaneous cooling and vacuum drying. Freeze-dried and vacuum-dried pollen grains exhibit no differences in responses to storage. The freeze/vacuum dried pollen is usually stored at subzero temperatures. For effective use of this storage method, optimum pollen water content, duration of drying and subsequent rehydration have to be optimized.

The freeze-drying method has been effective for long-term storage of pollen grains of a number of species (Table 13.4), except those of cereals (King 1965, Towill 1985). However, in recent years this method has not been very popular probably because of the success achieved by using a much simpler technique of cryopreservation.

Storage under ultralow temperature/ cryopreservation: In this method, pollen grains are dried to bring their water content below a threshold level and stored in liquid nitrogen. This has been the most effective method for long-term storage for a large number of species (Table 13.4) (Towill 1991, Bajaj 1987, Barnabas and Kovacs 1997).

Initial attempts to cryopreserve pollen grains of cereals were not successful largely because of their susceptibility to desiccation (which is

Table 13.4 Some examples of successful storage of pollen for up to one year or more (based on data given in Towill 1985, Hanna and Towill 1995, Barnabas and Kovacs 1997)

Taxa	Storage temperature, °C	Storage period
1. Under subfreezing temperature		
Citrus spp.	−20	1–3 years
Corylus spp.	−18	12 months
Juglans spp.	−19/−30	1–2 years
Linum regale	Deep-freeze	3 years
Lycopersicon esculentum	−20	1132 days
Olea europaea	−18	367 days
Phoenix spp.	−13	1 year
Pinus spp.	−20/−23	1–5 years
Poncirus trifolia	−18	57 weeks
Prunus spp.	−18	435–801 days
	−20	2.5 years
Pyrus	−18/−20	385–673 days
Rhododendron sp.	−20	662 days
Ricinus communis	−18	1 year
Solanum tuberosum	−34	1 year
Vitis vinifera	−12	4 years
Cocos nucifera	+5	1.5 years
Fragaria sp.	+2 to −4	3 years
Pistachia atlantica	0	550 days
2. After freeze/vacuum drying*		
Allium cepa	Room temperature	23 months
Amaryllis sp.	Room temperature	34 months
Antirrhinum majus	Room temperature	13 months
Beta vulgaris	+5	400 days
Betula verrucosa	+5	900 days
Citrus sp.	Room temperature	38 months
	−196	2 years
Elaeis guineensis	−10	1 year
Fragaria grandiflora	Room temperature	13 months
Gladiolus sp.	Room temperature	12 months
Larix sp.	−10	48 months
Lycopersicon esculentum	Room temperature	24 months
Medicago sativa	−21	11 years
Nicotiana glutinosa	+5	900 days
Picea sp.	−10	24 months
Pisum sativum	−25	1 year
	+5	700 days

Taxa	Storage temperature, °C	Storage period
Pinus spp.	Room temperature/ +5/−10	900 days
Populus hybrids	−18	4–5 years
Prunus spp.	Room temperature	21 months
	−10/−20	9 years
Pseudotsuga spp.	−10/−18	2–4 years
Pyrus spp.	Room temperature/−20	17 months– 6 years
Ricinus communis	−20	1 year
Solanum spp.	Room temperature	400 days
	+5	24 months
Trifolium pratense	+5	400 days
Tulipa spp.	+5	400 days
3. Under cryogenic temperatures		
Diospyros kaki	−80/−196	1 year
Juglans nigra	−196	2 years
Lycopersicon esculentum	−196	1062 days
Persea americana	−196	1 year
Pyrus spp.	−196	673–1062 days
Rhododendron catawbiense	−196	662 days
Beta vulgaris	−196	1 year
Brassica oleracea	−196	16 months
Capsicum annum	−196	42 months
Carya illinoensis	−196	1 year
Carica papaya	−196	485 days
Helianthus annus	−76/−196	4 years
Humulus lupulus	−196	2 years
Juglans nigra	−196	2 years
Pennisetum glaucum	−73	3 years
Pistacia sp.	−196	1 year
Prunus persica	−196	1 year
Secale cereale	−196	1–10 years
	−76	10 years
Solanum tuberosum	−196	24 months
Vitis vinifera	−196	64 weeks
Triticosecale	−196	10 years
Zea mays	−196	10 years

*Storage temperature of freeze-dried pollen

critical for cryopreservation). However, by using a 'pollen dryer' in which air of 20°C and 20–40% humidity is blown through pollen, Barnabas and associates (Barnabas and Rajki 1981, Barnabas 1985, 1994, Barnabas and Kovacs 1997) were able to successfully store pollen of many cereals for over 10 years (Table 13.4). Unlike earlier methods used for drying pollen grains of cereals over a desiccant, a rapid but gentle and uniform drying was achieved through a pollen dryer. At present cryopreservation seems to be the most frequently used and effective method for long-term pollen storage.

Storage in organic solvents: This is a very simple method of pollen storage, first reported by Iwanami (1972) and Iwanami and Nakamura (1972). Pollen grains are dried over silica and stored in organic solvents (taken in airtight vials) and maintained in a refrigerator or deep-freeze. After the storage period, pollen grains are recovered by filtration or evaporation of the solvent and used for pollination or for conducting any other viability test. The number of species in which this method has been effective is thus far limited (Iwanami 1972, 1975, Kobayashi et al. 1978, Mishra and Shivanna 1982, Jain and Shivanna 1987a, b, 1989, 1990). Extensive studies by Jain and Shivanna (1989, 1990) have shown that non-polar solvents such as hexane, cyclohexane and diethyl ether are more effective for pollen storage than polar solvents.

Pollen grains of *Chrysanthemum pacificum* (Iwanami 1975) lost viability within 60 min at 25°C over silica (Fig. 13.3A); pollen grains stored under such conditions for 30 min (viability reduced to 12%) when transferred to diethyl ether, maintained viability at the same level (12%) even after 20 days of storage (Fig. 13.3B). After removal from the organic solvent, viability was completely lost in another 30 min. Thus there was no loss of viability during storage in organic solvent; this has been termed absolute dormancy. Iwanami (1984) reported successful storage of *Lilium* pollen in organic solvents for up to 10 years; such pollen grains were effective in inducing seed-set. However, the efficacy of this method for long-term storage has not yet been shown in any other species. As the technique is very simple and convenient, it is worth investigating its suitability for long-term storage of pollen of other species.

Effective Techniques to Overcome Prefertilization Barriers

A number of techniques are available to overcome pre- and post-fertilization crossability barriers that operate after pollination. As genetic variability within the species affects interspecific crossability (see Chapter 10), it is worthwhile testing different accessions of both the parents

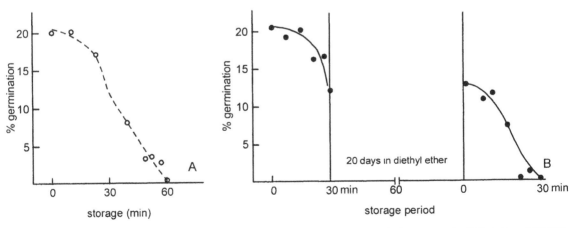

Fig. 13.3 Absolute dormancy in pollen grains of *Chrysanthemum pacificum* induced during storage in diethyl ether. A. Viability of pollen maintained under dry conditions at 25°C is lost in 60 min. B. Viability of pollen stored in diethyl ether after 30 min dry storage and tested after 20 days storage. There was no loss of viability during the period of storage in organic solvent (modified from Iwanami 1975).

in a hybridization programme. Also, it is necessary to ensure that the pollen grains are viable and the stigma of the female parent receptive at the time of pollination. The following are some of the methods used to overcome prefertilization barriers:

Application of growth substances and other chemicals

Application of growth substances (auxins, cytokinins and gibberellins) at the time of pollination or soon after pollination to the pedicel/ovary wound caused by severing one of the sepals or petals has been one of the common methods used to overcome prefertilization barriers (Table 13.5). Growth substances delay abscission of the style/flower so that slow-growing pollen tubes are able to enter the ovary and effect

fertilization. Growth substances may also promote post-fertilization development of the embryo up to a stage when hybrid embryos can be excised and cultured.

Immunosuppressors such as E-amino caproic acid, salicylic acid and acriflavin have been reported to be effective in overcoming interspecific incompatibility. According to Bates and Deyoe (1973), crossability barriers between distantly related species are mediated through inhibition reaction analogous to immunochemical reactions in animals. It affects both pre- and post-fertilization stages. Attempts have been made to overcome the barriers through the application of immunosuppressors. The female parent is treated with one of the immunosuppressors before and/or after pollination for a few days and the resultant embryo is cultured on a suitable

Table 13.5 Some interspecific hybrids obtained through the application of growth substances

Taxa	Growth substance	Reference
Agropyron junceum × *Triticum aestivum*	GA$_3$	Alonso and Kimber 1980
Arachis spp.	GA$_3$ followed by IAA/NAA+Kinetin	Sastri et al. 1983, 1982, 1983, Sastri and Moss 1982, Singh et al. 1980
Corchorus capsularis × *C. olitorius*	IAA	Islam 1964
Elymus canadensis × *Psathyrostachy juncea*	GA$_3$	Park and Walton 1990
Hibiscus cannabinus × *H. sabdariffa*	IAA	Kuwada and Mabuchi 1976
Hordeum spp.	GA$_3$	Subrahmanyam and Kasha 1973, Subrahmanyam 1978, 1999
Hordeum × *Triticum*	GA$_3$ + IAA	Larter and Chaubey 1965
Hordeum × *Alopecurus*	2,4-Dimethylamine	Kruse 1974
Hordeum × *Avena*		
Hordeum × *Dactylis*		
Hordeum × *Festuca*		
Hordeum × *Lolium*		
Hordeum × *Phleum*		
Hordeum vulgare × *Secale cereale*	GA$_3$+ Kinetin	Bajaj et al. 1980, Fedak 1978
Triticum timopheevii × *Secale cereale*	GA$_3$	Mujeeb-Kazi 1981
Lilium spp.	Naphthalene acetamide/Potassium gibberellate	Emsweller and Stuart 1948 Emsweller et al. 1960
Nicotiana repanda × *N. tabacum*	IAA	Pittarelli and Stavely 1975
Phaseolus vulgaris × *P. acutifolius*	α-naphthalene acetamide/ Postassium gibberellate	Al Yasiri and Coyne 1964
Pyrus × *Malus*	NAA	Crane and Marks 1952, Brock 1954
Solanum spp.	2,4-D	Dionne 1958
Trifolium spp.	ß-napththoxyacetic acid	Evans and Denward 1955
Triticum aestivum × *Elymus giganteus*	GA$_3$	Mujeeb-Kazi and Rodriguez 1980

medium. Success has been reported in crosses of many cereals: *Triticum* × *Hordeum*, *Hordeum* × *Secale* and *Zea* × *Sorghum* (Bates et al. 1979, Baker et al. 1975, Tiara and Larter 1977a,b, Mujeeb-Kazi 1981, Mujeeb-Kazi and Rodriguez 1980), and also *Vigna radiata* × *V. umbellata* (Baker et al. 1975, Chen et al. 1978). So far, success with immunosuppressors is limited to a few groups of plants and there is no evidence for the involvement of immunochemical reactions in interspecific crosses.

In many crosses of *Populus*, treatment of the stigma with organic solvents such as n-hexane or ethyl acetate, or washing of pollen grains in organic solvents before pollination, has been reported to be successful in overcoming interspecific incompatibility (Willing and Pryor 1976). In a few crosses, the seed-set following incompatible pollination with this treatment was as good as that following compatible pollination. However, the basis of the effects of these organic solvents is not known. This method has not been successfully applied to any other species. Stettler et al. (1980) reported failure of this method in some crosses of *Populus*.

Use of mentor pollen

Pollination with a mixture of incompatible pollen and mentor pollen (compatible pollen made ineffective to effect fertilization) has been successful in overcoming interspecific incompatibility. Mentor pollen is prepared by exposing compatible pollen to lethal doses of radiation or treating them with methanol or subjecting them to repeated freezing and thawing. Irradiation is more effective compared to the other two methods. Pollen grains withstand a high degree of irradiation (Brewbaker and Emery 1962); irradiation permits pollen germination (although pollen tubes are incapable of effecting fertilization), thus activating some of the post-pollination events of the stigma which are likely to have favourable effects on incompatible pollen.

The mentor pollen technique has been very useful in several crosses of *Populus*: *P. trichocarpa* × *Populus* spp. (Stettler 1968), *P.*

deltoides × *P. alba* (Knox et al. 1972), (*P. alba* × *P. glandulosa*) × *P. maximowiczii* (Koo et al. 1987). This method has also been applied to interspecific crosses of *Nicotiana* (Pandey 1979), *Cucumis* (den Nijs and Oost 1980), *Salix* spp. (Hathaway 1987) and *Brassica* (Sarla 1988). In the cross *Sesamum indicum* × *S. mulayanum* (Sastri and Shivanna 1976a,b), mentor pollen was successful in overcoming the barrier at the level of the stigma but not in the style; mentor pollen permitted germination of incompatible pollen and pollen tube entry into the stigma, but not their further growth in the style. Mentor pollen technique was reported to be ineffective in crosses of *Ipomoea* (Guries 1978) and many other crosses of *Populus* (Stettler and Guries 1976, Gaget et al. 1989). In recent years, however, there have been no successful reports of interspecific hybrids realized through the mentor pollen technique.

Although the basis of the mentor effect is not clearly understood, it is suggested to (a) provide recognition substances to incompatible pollen, (b) provide pollen growth-promoting substances to incompatible pollen, (c) activate the style and/or ovary, and (d) promote fruit retention (Villar and Gaget-Faurobert 1997). These suggestions are based on the well-established fact that pollen grains release a variety of substances onto the stigma, which play a role in post-pollination events (see Chapter 8).

Stump pollination

Pollination of the cut end of the style or the tip of the ovary after removing the stigma and a part or whole of the style is another effective technique for overcoming prefertilization barriers (van Tuyl and De Jeu 1997). Stump pollination markedly reduces the distance through which incompatible pollen tubes have to grow to reach the ovary. This is particularly advantageous in crosses in which the length of the style of the two parents differs markedly (Blakeslee 1945). Stump pollination has been successful in *Solanum* (Swaminathan 1955), *Nicotiana* (Swaminathan and Murty 1957), *Lathyrus*

(Davies 1957, Herrick et al. 1993), *Lilium* (Asano and Myodo 1977, van Tuyl et al. 1991, Janson et al. 1993) and *Fritillaria* (Wietsma et al. 1994). In some crosses, the cut end of the style may not support pollen germination; application of a suitable pollen germination medium to the cut end of the style may overcome this problem (Swaminathan 1955).

Stump pollination has also been a success in crosses between maize and other members of Poaceae (Dhaliwal and King 1978). The style (silk) of maize is about 30 cm long and the whole length is receptive; pollen of other species are often unable to complete this distance. In many such crosses, pollination of the cut surface of the silk just above the ovary is effective (Heslop-Harrison et al. 1985b); pollen tubes are able to reach the ovary before ovules lose receptivity.

Stylar grafting

In stylar grafting, pollination is carried out on a compatible pistil; following pollen germination and initial pollen tube growth, the stigma together with part of the style of the compatibly pollinated pistil is grafted onto an incompatible style/ovary. This grafting technique has been successful in *Lilium* (van Tuyl et al. 1991, van Tuyl and De Jeu 1997) in which grafting has been carried out using a piece of straw filled with the exudate of the compatible stigma. This is a delicate operation and is difficult to perform under in-vivo conditions, especially in species with a delicate style.

Bridge crosses

The bridge-cross technique has been used to circumvent crossability barriers in a number of wide crosses. An amphiploid obtained between the cultivar and one of the wild species has often been used as the bridging species to raise hybrids with another cultivated species. This approach has been very effective in wheat, tobacco, potato, lettuce, sugar-cane and *Brassica* (Table 13.6). Often bridge-cross hybrids are realized through field pollinations without resorting to application of any special technique. The bridge-cross technique is simple and very useful, particularly in transferring cytoplasm from one species to another (Shivanna 2000), al-

Table 13.6 Examples of hybrids realized through bridge-cross method

Cross	Reference
(*Triticum dicoccoides* × *Aegilops umbellulata*) × *T. aestivum*	Sears 1956
(*Triticum aestivum* × *Aegilops turgidum*) × *Ae. ventricosa*	Doussinault et al. 1983
(*Lactuca sativa* × *L. serriola*) × *L. virosa*	Eenink et al. 1982
(*Solanum tuberosum* × *S. acaule*) × *S. bulbocastanum*	Dionne 1963
(*S. tuberosum* × *S. acaule*) × *S. cardiophyllum*	Dionne 1963
(*S. tuberosum* × *S. acaule*) × *S. pinnatisectum*	Dionne 1963
(*S. tuberosum* × *S. lignicaule*) × *S. commersonii*	Chavez et al. 1988a
(*S. tuberosum* × *S. capsicibaccatum*) × *S. commersonii*	Chavez et al. 1988b
(*Solanum acaule* × *S. bulbocastanum*) × *S. phureja*	Hermsen and Ramanna 1973
[(*S. acaule* × *S. bulbocastanum*) × *S. phureja*] × *S. tuberosum*	Hermsen and Rammanna 1973
(*Nicotiana repanda* × *N. sylvestris*) × *N. tabacum*	Burk 1967
(*Gossypium anomalum* × *G. thuberi*) × *G. hirsutum*	Meyer 1974
(*Diplotaxis siettiana* × *Brassica campestris*) × *B. juncea*	Nanda Kumar and Shivanna 1993
(*D. siettiana* × *B. campestris*) × *B. napus*	Nanda Kumar and Shivanna 1993
(*D. tenuisiliqua* × *B. campestris*) × *B. napus*	Vyas 1993
(*D. erucoides* × *B. campestris*) × *B. juncea*	Vyas et al. 1995
(*Erucastrum abyssinicum* × *B. juncea*) × *B. nigra*	Rao et al. 1996
(*E. abyssinicum* × *B. juncea*) × *B. campestris*	Rao et al. 1996
(*E. abyssinicum* × *B. juncea*) × *B. napus*	Rao et al. 1996
(*D. berthauttii* × *B. campestris*) × *B. napus*	Malik et al. 1999

though the transfer of nuclear genes through this method involves a more elaborate breeding programme (Hadley and Openshaw 1980).

Bud pollination

Bud pollination is one of the very successful techniques for overcoming self-incompatibility (see Chapter 12); it has also been used in a few crosses to overcome interspecific incompatibility. Crosses *Nicotiana tabacum* × *N. rustica, N. tabacum* × *N. rependa* and *N. tabacum* × *N. trigonophilla* showed inhibition of pollen tubes in the stigma/style. Bud pollination was used to overcome this inhibition; pollen tubes continued to grow and reached the ovary in all these crosses following bud pollination (Kuboyama et al. 1994). As the technique is simple, it is worthwhile trying this method in other crosses and also combining bud pollination with application of growth substances or mentor pollen.

Intraovarian pollination

Intraovarian pollination involves injection of pollen grains (suspended in a suitable medium) directly into the ovary and achieving fertilization and seed development (Maheshwari and Kanta 1961). Intraovarian pollination has been used to realize hybrids between *Argemone mexicana* and *A. ochroluca* (Kanta and Maheshwari 1963a, b). It has not yet been extended to other crosses.

In-vitro pollination of ovules

In-vitro pollination of ovules aims at eliminating pollen-pistil interaction altogether by bringing pollen grains into direct contact with the ovules. Theoretically, this is the most effective method to overcome pre-fertilization barriers. The techniques of in-vitro pollination of cultured ovules and placental pollination have been described in Chapter 8. The technique of placental pollination has been used to overcome interspecific and intergeneric incompatibility in many species, particularly by Zenkteler and associates (Table 13.7) (Zenkteler 1980, 1990, Zenkteler and

Table:13.7 Examples of interspecific/intergeneric hybrids produced through placental pollination

Cross	Reference
Brassica napus × *B. campestris*	Zenkteler et al. 1992
B. campestris × *B. napus*	Zenkteler et al. 1992
B. oleracea × *B cretica*	Zenktler 1992
B. oleracea × *Sinapis arvensis*	Zenktler 1992
Melandrium album × *M. rubrum*	Zenkteler 1967
M. album × *Viscaria vulgaris*	Zenkteler 1967, 1990
M. album × *Silene schafta*	Zenkteler 1967, 1990
M. rubrum × *M. album*	Zenkteler 1967
Nicotiana alata × *N. debney*	Zenkteler 1990
N. tabacum × *N. debney*	Zenkteler 1990
N. tabacum × *N. knightiana*	Zenkteler 1990

Bagniewska-Zadworna 2001). They obtained hybrid progeny following the cross *Melandrium album* × *M. rubrum* and *M. album* × *Silene schafta.* They also attempted intrafamily hybridization among members of Caryophyllaceae, Brassicaceae, Solanaceae and Campanulaceae. In many of these crosses there was normal fertilization and initiation of embryo and endosperm development, but later the embryo degenerated. Isolation of young embryos and their culture on a suitable nutrient medium would probably enable the production of hybrids in such crosses also. Recently attempts have also been made to realize hybrids between angiosperms and gymnosperms (Zenkteler and Bagniewska-Zadworna 2001). Pollen grains of *Pinus* spp. and *Ephedra* have been shown to germinate on placental cultures of many species of flowering plants. Although pollen tubes were occasionally seen entering the micropyle, no hybrid embryo has been realized to date. Thus the technique of in-vitro pollination of ovules offers immense possibilities for overcoming interspecific incompatibility; the technique can also be used as an effective tool in many fundamental studies concerned with fertilization (see Chapter 8).

In-vitro fertilization

Dramatic progress has been made in recent years in achieving in-vitro fertilization using isolated sperm and egg cells (see Chapter 8). One of the potential applications of this technique is in overcoming crossability barriers. Although success has been confined to date to achieving fertilization between gametes of compatible species and developing embryos from such in-vitro fertilized zygotes for maize, recent studies of Kranz and associates (see Kranz 1997, Scholten and Kranz 2001) have shown that there is no technical difficulty in achieving in vitro fertilization between isolated egg and sperm cells in interspecific and intergeneric combinations. However, further development of zygotes is confined to those belonging to related species (Kranz and Dresselhaus 1996). Zygotes resulting from fusion of the egg of maize with sperms of many other members of Poaceae gave rise to multicellular structures, while those resulting from the egg of maize and sperm of *Brassica* failed to divide. Further studies are needed to extend this technology to other species and to realize interspecific and intergeneric hybrids from in-vitro fused zygotes.

The techniques of in-vitro pollination of cultured ovules and in-vitro fertilization provide powerful technologies to the plant breeder for realizing his aspirations of rearing distant hybrids much more easily than has been possible until now. Besides its potential application in overcoming incompatibility barriers, use of isolated sperm or egg or in-vitro formed zygotes in genetic transformation studies is very promising.

Methods to Overcome Post-fertilization Barriers

Embryo rescue has become the most effective and routinely used technique for overcoming post-fertilization barriers. Details of the embryo rescue technique are beyond the scope of the present volume. The technique has been covered comprehensively by Maheshwari and Rangaswamy (1965), Raghavan (1977, 1999) and Williams et al. (1987). The type of embryo rescue method required depends on the stage of embryo abortion. When embryos abort at an early stage (preglobular/globular), it is difficult

to excise such embryos and to grow them in vitro since the growth requirements of younger embryos are more complex and subtle. In such crosses, culture of ovules or even the entire ovaries has been resorted to with notable success. In a few crosses of legumes, a hybrid embryo has been transplanted into the developing endosperm of compatible seeds (after removing the compatible embryo) and the transplanted embryo-endosperm complex has been successfully cultured (Williams and Lautour 1980). In many interspecific and intergeneric crosses of *Brassica*, a modified technique of embryo rescue termed 'sequential culture' has been more effective than the culture of either ovaries or ovules (Nanda Kumar et al. 1988, Agnihotri 1993, Shivanna 2000). In sequential culture, ovaries are cultured 4–8 days after pollination; cultured ovaries are taken out after 7–10 days of culture, dissected under aseptic conditions and enlarged ovules (young seeds) are then recultured on a fresh medium. In successful crosses, cultured ovules grow further and germinate in vitro. In some crosses it is necessary to dissect the embryo from cultured ovules and reculture it.

MULTIPLICATION OF HYBRIDS

The number of hybrids realized in wide crosses, in spite of using different techniques to overcome crossability barriers, is rather limited. However, a large number of hybrids are needed for morphological and cytological studies, induction of amphiploidy to restore fertility and to raise backcross progeny.

The in-vitro culture technique can be effectively used to multiply hybrids (Nanda Kumar and Shivanna 1991, Agnihotri 1993, Shivanna 2000). This can be achieved through culture of shoot tips or single node segments on a cytokinin medium to induce shoot multiplication. New shoots are isolated and cultured on an auxin medium to induce rooting. The plantlets are then hardened and transferred to the soil. Another method is to induce callus using hybrid embryos or hypocotyl segments and subsequent regeneration of plantlets through the shoot and mct regeneration pathway or through the somatic embryogenesis pathway (Agnihotri 1993).

IDENTIFICATION OF HYBRIDS AT SEEDLING STAGE

Realization of seeds or seedlings following interspecific pollinations is not necessarily an indication of successful hybridization; seeds often develop in incompatible pollinations as a result of contamination with compatible pollen or matromorphy (non-hybrid diploid seeds of entirely maternal origin) (Eenink 1974, Banga 1986). Therefore, the hybrid nature of the resultant plants needs to be confirmed. The conventional method of confirming hybridity is through morphological and cytological studies, for which the plant has to be grown up to flowering. If it turns out to be a non-hybrid, the time, cost and efforts spent on raising such plants would go waste. Any method that can identify hybrids at the seedling stage would be meritorious. This is particularly so in crosses which require the application of embryo rescue technique and in-vitro multiplication.

One of the convenient and effective methods for identifying hybrids has been use of phenotypic markers, which are expressed at the seedling stage. In the absence of such markers, hybrids can be identified through analyses of isozymes or through use of molecular techniques (Fig. 13.4) such as oligonucleotide fingerprinting (Poulsen et al. 1993) RFLP (Rosen et al. 1988), RAPD (Mukhopadhyay et al. 1994) and species-specific repetitive DNA sequences (Saul and Potrykus 1984, Schweizer et al. 1988, Agnihotri et al. 1990a, Rao et al. 1996, Chrungu et al. 1999).

INDUCTION OF AMPHIPLOIDY

Interspecific hybrids, especially of wide crosses are invariably pollen sterile. This is usually due to lack of chromosome pairing during meiosis. Restoration of pollen fertility is essential if such hybrids are to be used in the breeding programme. Induction of amphiploidy through colchicine has become a standard method of restoring pollen fertility. Methods of application and concentration of colchicine used are highly variable and to some extent depend on the prevailing environmental conditions (Eigsti and Dustin 1955, Elliott 1958, Hadley and Openshaw 1980).

Colchicine is usually applied to the axillary buds of hybrid seedlings; amphiploids are realized from new shoots originating from the treated axils. In crosses in which hybrids are multiplied through culture of shoot tips or single node segments, colchicine treatment can be conveniently combined with in-vitro culture. Colchicine is applied to the shoot tips/axils of in-vitro cultured shoot tips or single-node segments. Alternatively, colchicine can also be incorporated into the culture medium on which shoot tips or single node segments are cultured. Often, immersion of only the roots or the entire plantlets raised in vitro, in colchicine solution for a few hours, before they are hardened and transplanted to the soil, is also effective in inducing amphiploidy. Use of DMSO (0.5%) in colchicine solution enhances the penetration of colchicine.

Induction of polyploidy in one of the parents is also helpful in crosses in which the parents differ in number of genomes and/or chromosomes (Hadley and Openshaw 1980). For example, the cross between *Lotus tenuis* (2n = 2X = 12) and *L. corniculatus* (2n = 4X = 24) was unsuccessful. However, the cross between *L. tenuis* (2n = 4x = 24) and *L. corniculatus* (2n = 4X = 24) was successful (Wernsman et al. 1965).

HANDLING OF BACKCROSS GENERATIONS

Following successful production of hybrids and restoration of pollen fertility, the backcross method is the standard approach to transfer one or a few traits from one species to another. The hybrid is repeatedly backcrossed to the cultivated species/variety to eliminate most of the

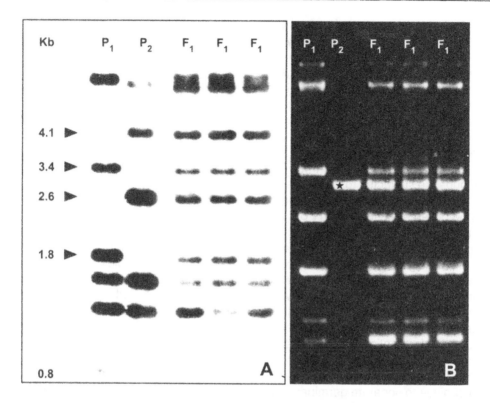

Fig. 13.4 Molecular characterization of wide hybrids. A. Analysis of nuclear DNA sequences of both parents and hybrid of *Eruca sativa* × *Brassica campestris*. Eco RI-digested total DNA from *E. sativa* (P_1 female parent), *B. campestris* (P_2, male parent) and three hybrid plants (F_1) separated on agarose gel and hybridized with wheat rDNA probe pTA—71. The hybrid shows the presence of specific bands from both parents (arrowheads). B. Ethidium bromide stained gels of RAPDs of parents and the F_1 hybrid of *Erucastrum abyssinicum* × *Brassica juncea*. DNA amplification of *E. abyssinicum* (P_1, female parent), *B. juncea* (P_2, male parent), and three hybrid plants (F_1), using oligonucleotide primer C4. Asterisk indicates species-specific band generated by the male parent in all hybrid plants (A—after Agnihotri 1991, B—after Rao 1995).

unwanted genes of the other parent except those imparting specific trait(s). In each backcross (BC) generation, a large number of plants have to be screened to identify those which have desirable traits and show greater resemblance to the cultivar; such plants are used to produce subsequent BC generations (see Fig. 13.1). Screening of plants of BC generations through conventional methods is one of the major constraints in backcross breeding. Because of the limitations of space, time, labour and other facilities, only a limited number of BC plants are used for screening.

Use of pollen to screen plants for the presence of desirable traits, in particular resistance/ tolerance to biotic and abiotic stresses (see page 200), is more convenient and greatly reduces the time and cost of screening BC plants. This would also enable the breeder to use a much larger number of BC plants for screening and thus increase the efficiency of the breeding programme. Use of pollen is possible only when adaptive traits are controlled by genes expressed in both the pollen and the sporophyte (Sari-Gorla and Frova 1997).

Application of Selection Pressure to Pollen

Apart from the use of pollen to screen plants with desirable traits, selection pressure can also be applied to pollen to increase the number of plants with desirable traits in the progeny. Selection pressure on the pollen will result in non-random fertilization; pollen grains carrying genes for adaptive traits produce a larger proportion of seeds, leading to an increase in frequency of desirable alleles in the progeny.

Selection pressure on pollen can be applied at three levels:

1. During pollen development: Anthers or flower buds are subjected to the stress condition during pollen development and the surviving pollen are used to carry out pollinations.

2. After pollen dispersal but before pollination: This can be done by germinating pollen grains in vitro in the presence of the stress condition. Only those pollen grains which have genes resistant to the stress will germinate; pollination with germinated pollen would lead to fertilization by only those pollen grains which have desirable genes. This method requires standardization of a suitable technique to separate germinated pollen from ungerminated pollen. Density gradient centrifugation seems to be a potential technique to achieve such a separation (Mulcahy et al. 1988). Alternatively, pollen grains can be treated with a stress condition (such as pathotoxin) and used directly for pollination.

3. During pollen tube growth in the pistil: Selection pressure is applied to the pistil before or soon after pollination. Only those pollen grains which are tolerant to a particular stress condition will be able to grow and achieve fertilization.

Appropriate stage for application of selection pressure depends on the stage of pollen at which the desired genes are expressed. Treatment during pollen development has been reported to be more effective for tolerance to herbicides (Sari-Gorla et al. 1989) and heavy metals (Searcy and Mulcahy 1985a, b), whereas tolerance to low temperature was more effective when applied during pollen tube growth (Zamir and Vellejos 1983).

As pollen grains from a single heterozygous plant (in which a large number of loci would undergo segregation) would comprise thousands of pollen types, the extent of genetic variability of the pollen populations that can potentially undergo selection is massive compared to sporophytic populations (see Mulcahy 1984, Ottaviano and Mulcahy 1986, Sari-Gorla and Frova 1997). Pollen grains can sustain a high degree of selection pressure; even when a majority of pollen grains are eliminated as a result of selection pressure, a small number of functional pollen is enough to effect fertilization of ovules. Many examples are already available on the efficacy of application of selection pressure (see Sari-Gorla and Frova 1997) of which the following are just a few:

Selection pressure has been applied in tomato to achieve preferential fertilization by pollen genotypes tolerant to cold (Zamir and Gadish 1987, Zamir et al. 1982) and salt (Sacher et al. 1983) stresses. A high altitude ecotype of *Lycopersicon hirsutum* can complete all reproductive stages even when the temperature drops to 6°C, while the cultivated tomato, *L. esculentum*, requires higher temperature. When a mixture of *L. hirsutum* and *L. esculentum* pollen was used to carry out pollination of *L. esculentum* plants maintained at low temperature (12/6°C), the frequency of hybrid seeds more than doubled compared to those formed at higher temperature (24/19°C) (Zamir et al. 1982, Zamir and Gadish 1987). Obviously, low temperature conditions preferentially permitted pollen grains of *L. hirsutum* to reach the ovary and effect fertilization.

Plants of cultivated tomato are not tolerant to salt stress while those of a wild species *Solanum pennellii*, are tolerant to salt stress. With a view to increasing the frequency of genotypes resistant to salt stress, Sacher and colleagues (1983) applied salt stress to vegetatively propagated F$_1$ plants of *Lycopersicon esculentum* ×

S. pinnellii by growing plants in the presence of 100 mM NaCl. Mean dry weight of F_2 plants derived from F_1 plants exposed to salt stress was significantly greater than that of controls. Thus, salt stress preferentially selected pollen with tolerant genes to effect fertilization. It should be possible to apply similar selection pressures to increase the frequency of genotypes showing resistance to many pathogens such as *Helminthosporium* and *Alternaria* (Sari-Gorla and Frova 1997).

In maize (Sari-Gorla et al. 1989, 1994) hybrids were raised between a tolerant line and a susceptible line to the herbicide chlorosulfuron (CS). Selection pressure was applied to the hybrid pollen either during pollen development (treating the excised tassels ca 2 weeks before anthesis in CS solution for 24 h and then culturing them in a medium until pollen shedding) or during pollen germination and pollen tube growth (by spraying the silks with herbicide soon after pollination). The resultant plants exhibited significantly greater tolerance to CS than did controls.

Sari-Gorla and Frova (1997) have given many other examples of gametophytic selection for faster tube growth rate resulting in increased seedling vigour, kernel weight and root growth, and for many biotic and abiotic stresses, resulting in increased sporophytic resistance to these stresses.

It has been possible in some species to grow microspores in vitro up to mature pollen and to successfully use them for pollination (Benito-Moreno et al. 1988). As the bulk of gametophytic gene expression takes place during this phase, in-vitro culture of microspores provides a better method for application of selection pressure; it allows treatment of pollen for a longer period compared to mature pollen. The efficacy of this approach has been elegantly demonstrated in tobacco (Touraev et al. 1995). Pollen from a transgenic plant containing a single hygromycin-resistant gene under the control of dC_3 promoter (active in both pollen and sporophyte) were cultured for 3 days in a hygromycin-containing medium. Only 50% of the cultured pollen could germinate in vitro; pollination of wild-type plants with this pollen yielded 100% transgenic offsprings. When pollen from wild-type plants were grown in hygromycin medium, they failed to germinate, while those grown on hygromycin-free medium showed nearly 100% germination. These results clearly show that 50% of the pollen which did not receive the hygromycin-resistant gene were eliminated during selection pressure; the remaining 50% which contained the hygromycin-resistant gene remained functional and effected fertilization.

STABILIZATION OF RECOMBINANTS

After the desired recombinants are realized, it takes another 4 or 5 generations to achieve homozygosity of the recombinants to make them true breeding. This can be achieved in one step by production of haploids from the recombinants and their diploidization. This aspect is presented in Chapter 14.

USE OF POLLEN FOR GENETIC TRANSFORMATION

At present, cell and protoplast cultures are routinely used to achieve genetic transformation. Plant regeneration from cells and protoplasts is a major limitation of this approach. However, use of pollen for introduction of exogenous DNA would overcome this major limitation since transformed pollen can be used to carry out pollination to realize transformed seeds. Thus, use of pollen for transformation is the most ideal method compared to others since realization of transformed seed through pollination is not species-specific and does not require application of in-vitro culture technology. Initial attempts were made to achieve transformation using mature pollen of a number of species such as tobacco, petunia, wheat and rice (Harwood et al. 1996). The methods employed included soaking pollen in DNA solution, application of DNA-pollen mixtures for pollination, *Agrobacterium*-mediated delivery, elctroporation and particle bombardment.

Mature Pollen Grains

Hess (1980) reported transfer of anthocyanin genes in *Petunia* by using pollen grains incubated in DNA solution before pollination. Pollen grains from a white-flowered line were incubated in DNA from a red-flowered line and used for self-pollination. Some anthocyanin accumulation was found in the progeny. Transformation was also reported through co-cultivation of pollen with *Agrobacterium tumefaciens* (Hess 1987). In maize, pollination with a pollen paste prepared in DNA solution was reportedly successful (Ohta 1986, Ohta and Sudoli 1980). Alternative approaches such as polyethylene glycol-mediated (Kuhlman et al. 1991, Fennell and Hauptmann 1992) and electroporation-mediated (Matthews et al. 1990, Saunders et al. 1990, Jardinaud et al. 1993) pollen transformation have been tried with some success. In *Lilium* pollen, electroporation allowed uptake of 40 kD dextran but not of 80 kD dextran indicating the importance of the size of DNA for electroporation. The rate of germination of electroporated pollen in vitro was low soon after electroporation but increased over time.

Transformation of pollen by particle bombardment has been shown to be more promising in recent years than other methods (Twell et al. 1989, 1991, McCormick et al. 1991, Plegt et al. 1992, Hamilton et al. 1992, Stoger et al. 1992, Nishihara et al. 1993, Morikawa and Nishihara 1997). Anther specific promoters such as *LAT 52*, *LAT 56* and *LAT 59* from tomato, and pollen specific promoters such as *Zm 13* from maize and *PA2* from *Petunia* have been used to drive the expression of the *gus* reporter gene in pollen of many species (Twell et al. 1989, 1991, McCormick et al. 1991, Plegt et al. 1992, Stoger et al. 1992, Nishihara et al. 1993). The efficiency of *gus* expression depended on the type of promoter used. The *Zm 13* promoter from maize has also been shown to work in pollen of

Tradescantia (Hamilton et al. 1992). *Zm 13* promoter was more effective in lily than the *LAT 15* promoter; in pollen of *Nicotiana*, however, *LAT 15* was more effective than *Zm 13* (Nishihara et al. 1993, Morikawa and Nishihara 1997).

Pollen grains of many species show strong endogenous *gus* activity which interferes with detection of expression of the introduced gene (Plegt and Bino 1989, Alwen et al. 1992). However, background *gus* activity of the pollen can be suppressed completely by the addition of 20% methanol and use of neutral pH assay solution (Kosugi et al. 1990, Alwen et al. 1992, Nishihara et al. 1993).

The efficiency of gene delivery and its expression is influenced by many physical (size, shape and amount of metal particles, accelerating pressure and number of shots) and physiological (stage of pollen, composition of the medium and extent of osmotic stress) factors (Klein et al. 1988, Armaleo et al. 1990, Stoger et al. 1992, Nishihara et al. 1993, Vain et al. 1993, Morikawa and Nishihara 1997). Transient expression of the foreign gene has been reported in pollen of several species. Using a pneumatic particle gun, the *gus* gene under the control of the *LAT* promoter and nopaline synthetase polyadenylation terminator was expressed in pollen of *Lilium, Nicotiana* and *Paeonia* ((Twell et al. 1989, Nishihara et al. 1993, see Morikawa and Nishihara 1997). Bombarded pollen grains of many species such as *Nicotiana* (Twell et al. 1989, Stoger et al. 1992), *Tradescantia* (Hamilton et al. 1992) and *Lilium* (Morikawa and Nishihara 1997) have been shown to retain their ability to germinate. In *Lilium* bombarded pollen were able to induce seed-set (Morikawa and Nishihara 1997). However, transgenic seeds have not yet been realized through transformed pollen. Transient expression of the *gus* gene in tobacco pollen has also been reported following electroporation (Matthews et al. 1990, Saunders et al. 1990).

Results using mature pollen for transformation are not unequivocal (Stoger et al. 1992, Matousek and Tupy 1983, Morikawa and

Nishihara 1997). The evidence provided in these studies has been largely phenotypic (Harwood et al. 1996). Alternative explanations have been given to explain evidence presented in some of the above reports (Potrykus 1990,1991).

Microspores and Microspore-derived Embryos

Use of microspores and microspore-derived embryos has been more promising for transformation compared to mature pollen. However, realization of transformed plants through this approach is dependent on induction of in-vitro embryos/plantlets from microspores and thus, this approach can only be used in species in which protocols for the in-vitro androgenic pathway are established. Use of microspores and microspore-derived embryos has been successful in tobacco, *Brassica napus*, maize and barley (Harwood et al. 1996).

In *Brassica napus*, in which conditions have been optimized for a high frequency of microspore embryogenesis, transgenic plants have been recovered from secondary embryos regenerated from microspore embryos injected with DNA (Neuhaus et al.1987) and also following co-cultivation of microspores and microspore-derived embryos with *Agrobacterium tumefaciens* (Pechan 1989, Huang 1992). Biochemical analysis of the progeny exhibited neomycin phosphotransferase II enzymic activity in 7.3% of the plants.

In tobacco, particle bombardment of immature pollen followed by in-vitro culture developed into mature pollen, which expressed the *gus* gene (Stoger et al. 1992). Successful *Agrobacterium*-mediated gene transfer has also been achieved in tobacco using immature pollen-derived embryos (Sangwan et al. 1993). Transgenic haploid plantlets of *Nicotiana rustica* have been realized from the culture of bombarded microspores (Nishihara et al. 1995, Morikawa and Nishihara 1997). The microspores were bombarded with gold particles coated with plasmid DNA containing neomycin phosphotransferase II (NPT II) and the *GUS*

gene under the control of the cauliflower mosaic virus (CaMV) *35S* promoter and nopaline synthetase (NOS) terminator. Pollen embryos derived from bombarded microspores were selected for kanamycin resistance and haploid transgenic plants were realized. Integration of foreign genes in transgenic plants was confirmed by Southern blot analysis and their expression by NPT II and GUS enzyme assays.

Jardinaud et al. (1995) used a maize hybrid, highly responsive to induction of androgenesis in isolated microspores, to optimize the conditions for gene transfer through the biolistic method. *Gus* expression in the microspore was maximum with 1.1 µm diameter tungsten microprojectiles (coated with 4 µg DNA mg^{-1} particles) at 1100 and 1350-psi helium pressure and 6 cm distance. Transformation has also been achieved in maize microspores by using electroporation and PEG-mediated delivery (Fennell and Hauptmann 1992). Another novel approach to transferring DNA in maize (Dupuis and Pace 1993) has been the delivery of gold particles coated with DNA through particle bombardment to tassel primordia. The treated tassels were cultured in vitro and *gus* activity was detected in the anthers that developed from the bombarded tissue. Transformed pollen is likely to be realized through this approach which can then be used for pollination. In barley also, transformation has been successful in microspores and microspore-derived embryos using several delivery methods (Harwood et al 1996), especially particle bombardment (Wan and Lemaux 1994, Jahne et al. 1994).

Rapid homozygosity of an introduced foreign gene can be achieved by diploidization of transformed haploids. As androgenic haploids have been induced in a large number of species, including most crop species, use of microspores may be a more effective approach than use of mature pollen.

It has recently been possible to isolate sperm and egg cells of several species (see Chapter In maize, in-vitro fertilization using isolated sperm and egg cells and growth of in-vitro fertilized zygotes into embryos and plants have been achieved. It has also been possible to culture barley and maize zygotes formed in vivo and to regenerate plants from them (Holm et al. 1994, Mol et al. 1995, Leduc et al. 1996) using a nurse cell culture system. These studies have opened up exciting possibilities of using egg cell or sperm cell or cultured zygotes for genetic transformation. Leduc and colleagues (1996) have already been able to establish conditions for efficient microinjection of DNA into cultured zygotes and have demonstrated transient expression of microinjected genes (*gus* gene and anthocyanin regulatory genes) in transformed zygotes.

14

Induction of Haploids from Pollen Grains

Haploids are sporophytes that have a gameto-phytic chromosome number. Since the discovery of the first natural haploid in *Datura* in 1921, plant breeders have been constantly attempting not only to isolate naturally occurring haploids in different species, but also to induce haploids through several treatments. Natural haploids have now been reported in a large number of species. However, the frequency of natural haploids is too low to be of much use in a plant breeding programme. Many techniques, such as delayed pollination, pollination with aborted pollen or irradiated pollen and distant hybridization have been tried (for details see Kimber and Riley 1963, Maheshwari and Rangaswamy 1965). Although these techniques have been reported to induce haploids occasionally in some taxa, none has become a dependable technique.

Lack of effective techniques to induce haploids was the main constraint for their use in crop improvement (Kimber and Riley 1963). This situation changed dramatically following the report of induction of pollen embryos in cultured anthers of *Datura innoxia* by Guha and Maheshwari (1964, 1966). Haploids were successfully induced thereafter in cultured anthers/microspores in a large number of species. As the techniques for induction of haploids for large numbers are now available for many crop species, haploids have become an effective component of plant breeding programmes.

IMPORTANCE OF HAPLOIDS

Production of Homozygous Diploids

In the conventional method development of homozygous lines through repeated selfing for many generations takes a long time. The most important application of haploids in a plant breeding programme is in development of homozygous diploids in one step by diploidization of haploids. This is particularly useful in plantation crops (such as tea, which are vegetatively propagated and highly heterozygous), self-incompatible species and tree species in which production of pure lines by conventional methods requires extra time and effort. Fixation of suitable recombinants (developed through hybridization) in homozygous condition by the conventional method takes 5–6 generations of selfing. Production of haploids and subsequent chromosome doubling enables development of true breeding recombinants in one generation (Stringam et al. 1995).

Recovery of Novel Recombinants

Apart from fixing recombinants developed through conventional breeding, production of doubled haploids in hybrids and their backcross progeny facilitate not only the recovery of superior recombinants, but also simultaneous fixation of such recombinants. (Foroughi-Wehr and

Friedt 1984, Stringam et al. 1995). This approach, commonly described as hybrid sorting, markedly reduces the time required for developing a new variety.

Combining the technique of induction of haploids from cultured microspores with in-vitro selection provides additional technology to recover desirable recombinants (Wenzel 1980). For example, induction of haploids from the anther/microspore cultures of a hybrid between salt-susceptible cultivar and salt-tolerant species in the presence of a higher level of salt in the medium enabled recovery of salt-tolerant recombinants (Ye et al. 1987). Such an approach would greatly accelerate a breeding programme aimed at transferring genes imparting resistance to biotic and abiotic stresses.

Mutation Research

Another important application of haploids is in mutation breeding. In the absence of alternate allele, all mutations, including recessive ones, induced in haploids are expressed. Further, culture of microspores and cells from haploid plants facilitate application of the techniques of microbial genetics to higher plants.

Model System to Study Embryogenesis

Inaccessibility of zygotic embryos and the difficulty of obtaining them in large numbers at the desired stage have been limitations to studies on the physiology and biochemistry of embryogenesis. Because of the developmental similarities between zygotic and microspore-derived embryos and the feasibility of generating a large number of synchronous microspore embryos in highly responding systems, microspore embryos provide a very convenient system to study a range of physiological and biochemical problems of embryogenesis (Heberle-Bors 1985). Microspore-derived embryos resemble to a large extent the zygotic embryos in biosynthetic pathways of different metabolites. Pollen embryos are therefore presently used for many basic studies such as protein and lipid biosynthesis and their storage, embryo maturation and chilling toler-

ance (see Palmer and Keller 1997). Extensive studies carried out in *Brassica napus* (Pomeroy et al. 1991, Wiberg et al. 1991) have shown that microspore-derived embryos synthesize and accumulate large quantities of storage lipids, and the pattern of fatty acid synthesis and storage, and lipid accumulation of zygotic and microspore embryos are comparable. Also, because of high frequency regeneration of microspore embryos in some species, they can be utilized as a tool for selection and for rapid screening of oil quality in the breeding programme (Pomeroy et al. 1991, Wiberg et al. 1991).

Production of Artificial Seeds and Genetic Transformation

Microspore embryos provide an ideal system for production of artificial seeds (Datta and Potrykus 1989, Senaratna et al. 1991, Brown et al. 1993, Takahata et al. 1993). In *Brassica*, microspore embryos desiccated to about 10% water content through exogenous application of ABA, maintained their ability to produce plants even after 3 months of storage under room conditions (Takahata et al. 1993). Microspore embryos also provide a very effective system for genetic transformation since they exhibit high regenerating potential (see Chapter 13).

PRODUCTION OF HAPLOIDS

Although culture of anthers/microspores has been the most effective technique for production of haploids in a large number of species, some other methods are equally effective at least in a few species (Lacadena 1974, Kasha 1974). Production of haploids through the use of alternative methods is simpler and less expensive than in-vitro techniques. Some of these methods are briefly described here.

Use of Genetic Lines

In some species, genetic lines have been isolated which produce haploids in higher frequency (see Kasha 1974, Sarkar 1974, Clapham 1977, Coe 1959). In maize, haploids occur in nature at a rate of 0.01 to 1.8 per thousand depending on genotype. The genetic strain *stock 6* of maize produces as many as 3.2% gynogenic (derived from unfertilized egg) haploids on selfing. The pollen of *stock 6* stimulates the egg to develop gynogenetically in many other female lines also. By using suitable markers, the haploid kernels can be isolated without even raising the seedlings. For example, use of female parent homozygous for coloured scutellum and aleurone, and *stock 6* homozygous for colour inhibition factors, enables haploid isolation by selecting kernels with coloured scutellum (gynogenic embryos) and colourless aleurone (as a result of triple fusion).

Another genetic line of maize, which induces haploids in high frequency, is the mutant 'indeterminate gametophyte' (*ig*). When the *ig* line is used as female parent, as many as 3% of the resultant embryos are androgenic haploids. By using this mutant it is possible to transfer the cytoplasm of one line to another in a single step without resorting to repeated backcrosses. Effective markers can be used with *ig* mutant also. 'PEM' (purple embryo marker) produces deep purple pigment in the embryo and aleurone layer. When the female parent has *ig* as well as *PEM* genes, kernels with colourless embryos with coloured aleurone can be isolated as haploids.

In Pima cotton, use of the genetic line 'doubled haploid 57-4' resulted in a high frequency of haploids due to semigamy (male and female gametes in the zygote divide separately without fusion to give rise to chimeral sectors of male and female origin in the embryo and the resultant seedling). Markers are used with one or both parents and haploids of male and female parents are recovered by isolating branches of parental genotypes and rooting them. In *Gossypium barbadense*, 'semigamous strain 7' (*Sg-7*) markedly increased the frequency of semigamy (Turcotte and Feaster 1974). The cross between *Sg-7* (female) × non-*Sg* (male) yielded 2.5% chimeral seedlings; the frequency was increased to 62.9% when the male parent also had the *Sg* gene.

Chromosome Elimination

Chromosome elimination is the most important technique of producing haploids in barley (see Kasha 1974). When cultivated barley, *Hordeum vulgare* (2x), was crossed with a wild species, *H. bulbosum* (2x), a high frequency of *H. vulgare* haploids were produced through selective elimination of *H. bulbosum* chromosomes during embryogenesis (Kasha and Kao 1970, Subrahmanyam and Kasha 1973). Recovery of haploids is dependent on application of GA_3 to pollinated florets, followed by culture of embryos (14–16 days after pollination) on a suitable medium. In the absence of embryo rescue, no haploids are realized as the embryos degenerate in vivo. Subsequently success in production of haploids through chromosome elimination has been reported in crosses involving several species of *Hordeum* (Barclay 1975, Subrahmanyam 1978, 1980, 1999).

Production of haploids through chromosome elimination has been reported in wheat (*Triticum* spp. × *H. bulbosum*) (Barclay 1975, Snape et al. 1979, Fedak 1978, 1988), *Aegilops crossa* × *H. bulbosum* (Shigenobu and Sakamoto 1977) and *H. vulgare* × *Secale cereale* (Fedak 1977) Crossability of wheat with *H. bulbosum* is limited to a few genotypes. Wheat haploids have also been realized through wheat × maize crosses (Laurie and Bennett 1986, 1988), which are genotype independent. A comparison between wide crosses and anther culture in the production of wheat polyhaploids indicated that the success rate in the former is higher (Kisana et al. 1993).

Many instances of spontaneous origin of haploids are likely to be the result of chromosome elimination (Bennett et al. 1976). It is possible that the phenomenon of selective elimination of chromosomes is widespread in wide crosses involving other groups of plants also. Lack of recovery of haploids following distant hybridization may probably be due to degeneration of haploid embryos in the absence of embryo rescue.

The mechanism involved in chromosome elimination is not clearly understood. Much evidence indicates that chromosome elimination is the result of malfunctioning of the spindle apparatus (Bennett et al. 1976, Orton and Tai 1977). Detailed cytological studies by Bennett and co-authors (1976) showed that elimination of the *H. bulbosum* chromosome was due to its failure to congress at metaphase and/or to reach the pole at anaphase. The process of chromosome elimination has been shown to be under the control of specific genetic factors located on chromosomes 2 and probably 3 of *H. vulgare*; their effect is neutralized in the presence of sufficient dosage of *H. bulbosum* chromosomes. The dose ratio of these factors to *H. bulbosum* chromosomes is critical for the process.

Another attractive hypothesis to explain chromosome elimination has been the operation of endonucleases in the hybrid embryo, similar to those in the bacterial system, resulting in the fragmentation and subsequent elimination of chromosomes of one of the parents.

Culture of Anthers and Microspores

Induction of haploids from cultured anthers and microspores has been one of the important advances in the area of pollen biotechnology. This has been the most effective method for production of haploids in terms of not only the number of successful species, but also in the number of haploids that can be induced. Haploids have been induced thus far in over 200 species through this method (Ferrie et al. 1995, Palmer and Keller 1997).

Although embryos realized from cultured anthers as well as microspores are generally derived from microspores and not from pollen, they are often referred to as pollen embryos in the literature. The progress on pollen embryogenesis has been reviewed extensively from time to time (Melchers 1972, Clapham 1977, Reinert and Bajaj 1977, Vasil 1980, Wenzel 1980, Maheshwari et al. 1982, Heberle-Bors 1985, Hu and Huang 1987, Keller et al. 1987, Arnison and Keller 1990, Bajaj 1990, Narayanaswamy 1994,

Ferrie et al. 1995, Jain et al. 1996, Palmer and Keller 1997).

Initial success in induction of pollen embryos was through culture of anthers. Anthers were removed from buds of a suitable stage and cultured on the surface of a semisolid medium. In many species a good correlation has been found between stage of microspore development and size of flower bud, so that anthers of the right stage can be selected for culture on the basis of bud size. In some investigations, anthers have been cultured in liquid medium on filter paper bridges. In a few species of Poaceae, inflorescence segments containing one or two spikelets have also been cultured.

Realization of embryos generally involves two phases—induction phase and further development. Anther tissue seems to favour the induction phase but is often inhibitory for further embryo development. This inhibition is thought to be due to release of inhibitory substances from the degenerating tissues of the anther (Sunderland 1974). In many species, the anther tissue tends to callus and competes with embryos/callus derived from the pollen; eventually anther callus overgrows and obscures microspore-derived callus. Also, when microspores in cultured anthers give rise to a callus instead of embryos, the callus is genetically heterogeneous since it originates from the division of many microspores.

Culture of microspores, instead of anthers, eliminates most of these limitations. Microspore culture provides a better technique than anther culture to: (a) regulate and monitor pollen responses, (b) control the factors involved in induction and development of pollen embryos, (c) study the structural and biochemical details associated with pollen embryogenesis, (d) enrich embryogenic microspores through gradient centrifugation or cell sorting which is not feasible with anthers, (e) induce and recover mutations and (f) achieve genetic transformation. Because of these advantages, the trend in recent years has been to culture isolated microspores instead of anthers. Remarkable success has been achieved in culture of microspores, at least in several species such as *Nicotiana* and *Brassica*.

Early attempts in microspore culture combined the techniques of anther and microspore cultures. In *Nicotiana* and *Hyoscyamus* anthers were cultured in a liquid medium for 4–16 days; they were then split open and the microspores gently squeezed into the liquid medium. The microspores continued to grow and eventually developed into green plants. Sunderland and Roberts (1977) devised a 'shed pollen culture' technique in tobacco. Pretreated anthers were floated on shallow liquid medium containing high osmoticum to allow dehiscence of anthers and release of microspores. Embryogenic microspores released into the medium continued development and produced mature embryos even after removal of the original anther (Sunderland and Xu 1982, Kohler and Wenzell 1985). In a few species, centrifugation of anthers before microspore isolation (Nitsch 1977) or of microspores before culture has been shown to be beneficial for induction of embryos.

Microspore isolation and culture has been standardized to a high degree in some species. In *B. napus*, the time required for isolation of microspores has been greatly reduced by macerating the flower buds and recovering the microspores through filtration and centrifugation. In this species, microspore culture has been shown to be about 10 times more efficient than anther culture (Lichter 1989, Kieffer et al. 1993, Arnison and Keller 1990).

One of the major limitations of anther culture in many cereals has been the development of a large number of albino plants due to failure of proplastids to develop normally into chloroplasts (Hu and Huang 1987). This is particularly so when anthers are cultured at the bicellular pollen stage (Huang 1982). Albino plants in wheat have been reported to contain large deletions in chloroplast genome (Sun et al. 1979, Day and Ellis 1985). In many species, it is possible to reduce the proportion of albino plants by culturing microspores at an early stage of development (Huang 1992) and by manipulating cultural conditions.

Optimal conditions for induction of pollen embryos

A large number of investigations have identified the following factors as critical for induction of pollen embryos:

Genotype: As with any other in-vitro response, the responsiveness of cultured anthers/microspores is also greatly influenced by the genotype of the donor plant. In wheat, pollen embryogenesis seems to be controlled by at least 3 dominant genes (Marsolais et al. 1986, Agache et al. 1989). It is possible to transfer the embryogenic responsive trait from more responsive species to the less responsive (Arnison et al. 1990, Zhang and Qifeng 1993). The physiological basis of such genetic control of embryogenesis is not clearly understood. In *Brassica oleracea*, this seems to be mediated through differences in ethylene production in cultured anthers (Biddington and Robinson 1991).

Physiology of donor plant: The age and physiological status of the donor plant, determined by prevailing growth conditions, markedly affect embryogenic response of the cultured anthers/microspores. Although no generalizations can be made on the factor(s) conducive for embryogenesis, growth conditions such as temperature, light intensity and duration, and nutritional status of the plant affect the frequency of embryogenic microspores (see Palmer and Keller 1997). Generally plants grown under lower temperatures respond better (Lo and Pauls 1992) than those grown under higher temperatures. Often, this requirement may be replaced with low temperature pretreatment of flower buds/anthers before culture (Heberle-Bors 1989, Mordhorst and Lorz 1993). Low temperature treatment has been suggested to favour embryogenesis by arresting gametophytic development of the microspores (Foroughi-Wehr and Wenzel 1989) or by inducing dimorphic microspores (Heberle-

Bors 1985, 1989), which is considered to be a prerequisite for the embryogenic pathway (see p. 225).

Composition of medium: Composition of the medium used for induction of haploids is highly variable. Although most of the initial studies used complex nutrient media containing an auxin and a cytokinin, and often other undefined adjuvants such as coconut water (Maheshwari et al. 1982), optimal response was obtained in many species on a simpler defined medium containing low levels of hormones. In a few species such as tobacco, it has been possible to induce pollen embryos on a relatively simple hormone-free medium (Nitsch 1972, Rashid and Street 1973). Addition of higher concentrations of hormones and/or complex undefined additives such as coconut water, potato extract and yeast extract tends to induce callus. Exogenous hormones, specifically auxins, are required in most cereals in which callusing usually precedes embryogenesis.

The carbohydrate component in the medium seems to play an important role (Palmer and Keller 1997). In several species the frequency of embryo induction increases with increasing concentration of carbohydrate source (Huang and Keller 1989). Sucrose has so far been the most effective carbohydrate, although maltose has been reported to be a better carbohydrate source for some species, in particular cereals (Kasha et al. 1990, Hoekstra et al. 1992). Other factors such as pH, nitrogen source and gelling agent also affect the efficacy of induction.

Use of activated charcoal in the medium, which adsorbs toxic metabolites of the medium, has proven beneficial in many species (Buter et al. 1993). In several species, incorporation of an anti-ethylene agent such as silver nitrate in the medium increases the response of cultured anthers (Biddington et al. 1988, Malik et al. 2001). Other cultural conditions which affect embryogenic response are: density of anthers/microspores, temperature, culture aeration, light quality and intensity, and photoperiod. In

several species, the initial culture temperature is critical for pollen embryogenesis; initial culture at 30–35°C for 12–72 h followed by maintenance of cultures at 20–25°C, generally gives optimal response. Initial high temperature seems to facilitate induction of embryos, while subsequent lower temperature promotes further embryo growth (Palmer and Keller 1997).

Stage of microspores at culture: For most species, the stage of microspores at culture is critical for induction of embryogenesis. Generally anthers/microspores cultured at late uninucleate to early binucleate stage are more responsive than earlier or later stages. In a few species, however, the tetrad stage (*Helianthus annuus*—Gurel et al. 1991, *Camellia japonica*—Pedroso and Pais 1993) has been reported to be optimal. It has been suggested that microspores are more amenable to a switchover to the embryogenic pathway than pollen grains since microspores are not yet fully committed for the gametophytic pathway (Sunderland and Huang 1987). Further development of microspores would result in strong commitment to the gametophytic pathway and thus they become less and less amenable to a shift to the sporophytic pathway.

Pretreatment of anthers/flower buds: Pretreatment of anthers/flower buds at low (3–5°C) or high (30–45°C) temperature for different periods (temperature shock) is effective in inducing pollen embryos in several species. Low temperature shock markedly improved embryogenic response in *Nicotiana tabacum*, *Datura metel*, *D. innoxia*, *Brassica napus*, *Petunia hybrida* and *Hyoscyamus niger* (Maheshwari et al. 1982). In *D. innoxia*, cold treatment (3°C for 48 h) of flower buds followed by mild centrifugation (40 g for 5 min) induced embryogenesis in as many as 80% of the anthers. In *P. hybrida*, pretreatment of anthers at 6°C for 48 h doubled the percentage of anthers producing embryos (Malhotra and Maheshwari 1977). Similarly cold treatment of anthers/spikelets/spikes of many cereals for varying periods has been beneficial for embryogenesis (Huang and Sunderland 1982).

Pretreatment with high temperature has been effective in increasing embryogenic response, especially in *Brassica* spp. (Keller and Armstrong 1978, 1979, 1981, Dunwell 1986). In rice, pretreatment at 35°C for 5 min followed by 10°C for 7 days promoted pollen embryogenesis. Much research indicates that temperature shock alters division of the microspore from asymmetric to symmetric, probably through change in microtubule distribution.

Pollen dimorphism

The nature of microspores which develop into embryos, is still controversial. According to one view, embryogenic pollen grains are predetermined following meiosis; embryogenic pollen are thus present along with gametophytic pollen in cultured anthers/microspores (Horner and Street 1978, Herberle-Bors 1985, Rashid et al. 1981,1982). In several species, such pollen types are morphologically distinguishable in anthers (dimorphic pollen). In Burley cultivar of tobacco, for example, one type of pollen grains is smaller with thin exine and shows poor affinity for cytoplasmic stains (embryogenic grains, E-grains). Other type of pollen grains are larger with thick exine and show strong affinity for cytoplasmic stains. Under inductive conditions, embryos develop from E-grains and normal grains degenerate. In vivo, E-grains degenerate and normal grains develop into gametophytic pollen. Similar pollen dimorphism is also found in *Paeonia*. In species which do not show pollen dimorphism in nature, pollen dimorphism is induced under inductive conditions. According to another view, all microspores are potentially embryogenic and embryos can be induced under specific stimuli in culture (Sunderland and Huang 1987, Telmer et al. 1993).

Diploidization

Diploidization of haploids is routinely achieved through colchicine treatment. Colchicine treatment is given to the axillary buds or the whole plantlet. However, the efficiency of such a treatment is rather low in many species. Incorporation of colchicine into the induction medium has been shown to improve diploidization efficiency without negatively affecting embryogenic response of microspores (Mollers et al. 1994); in a few species colchicine in the medium even increases embryogenic response (Iqbal et al. 1994, Hansen and Andersen 1996). In *Brassica napus*, culture of microspores in colchicine-containing inductive medium resulted in diploidization in 80–90% embryos (Mollers et al. 1994). Cold treatment (–40°C) of isolated microspores of *B. napus* also resulted in a high percentage of spontaneous diploids (Chen and Beversdorf 1992). Recently some of the microtubule depolymerizing herbicides such as oryzalin, trifluralin and amiprophosmethyl (APM) have been shown to be equally effective in inducing diploidy in cultured microspores or microspore-derived callus even at 100 times lower concentration than colchicine (Hansen and Andersen 1996). It may be worthwhile to try these chemicals in some of the refractory species.

Developmental pathways

Development of androgenic plantlets follows two main patterns: (a) embryos develop directly from the microspore/pollen without callusing and give rise to plantlets and (b) microspores give rise to callus, and the plantlets are induced subsequently from the callus through organogenesis or embryogenesis. The ontogeny of microspore divisions has been detailed for many species. There is considerable variation in the ontogeny of microspore embryos, not only between species but also within the same species. To some extent these variations are dependent on cultural conditions and the composition of the medium. Broadly speaking, there are two major ontogenic pathways leading to pollen embryogenesis (Fig. 14.1).

1. Microspore undergoes symmetrical division (instead of the characteristic asymmetrical division in the gametophytic pathway, see

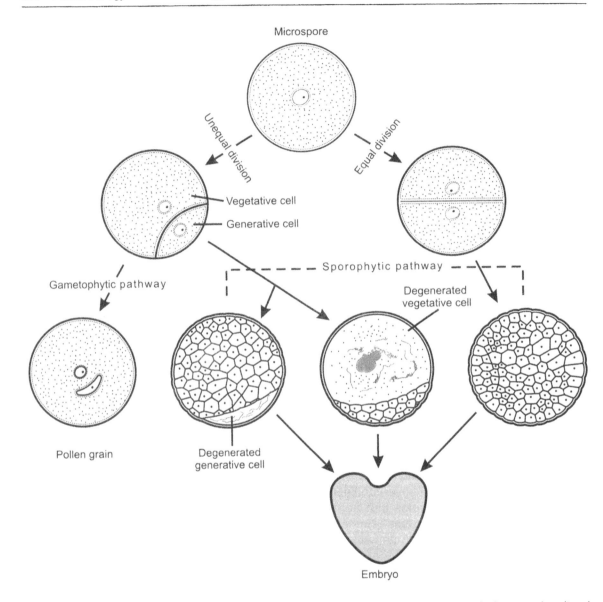

Fig. 14.1 Diagram of cellular details of gametophytic and sporophytic (embryogenic) pathways of microspore in cultured anthers.

Chapter 1) and gives rise to two equal cells. The embryo or callus is derived from the activity of both cells. This pathway is more prevalent and is reported in several species such as *Nicotiana*, *Datura* and *Lycopersicon* (see Nitsch 1977), *Atropa* (Rashid and Street 1973) and many others. Often, initial nuclear divisions are not followed by wall formation; this may lead to fusion of nuclei, resulting in polyploid cells.

2. Initial microspore division is asymmetrical leading to formation of generative and vegetative cells. The resultant embryo/callus may be

derived from the activity of only the vegetative cell or only the generative cell or rarely from both. Development of embryo/callus from the activity of only the vegetative cell is more common than from the other two pathways. In this pathway, the generative cell remains undivided or at most undergoes a few divisions before degeneration. This pathway has been reported in a number of species: *Nicotiana* (Sunderland and Wicks 1971), *Lolium* and *Festuca* (Rose et al. 1987), *Datura* (Iyer and Raina 1972) and *Triticum* (Sun et al. 1974). Embryogenesis resulting from the activity of both vegetative and generative cells was reported in *D. innoxia* (see Sangwan and Sangwan-Norreel 1990). In *Hyoscyamus niger* (Raghavan 1976, 1978), embryos originating from the activity of only the generative cell has been shown to be more prevalent. The other two methods were observed only in a small proportion of microspores.

In general, cell divisions are initiated in a larger number of microspores (Fig. 14.2) than the number of mature embryos realized. Embryo development in many of the microspores is arrested at different stages. The morphological stages in microspore-derived embryos apparently resemble zygotic embryos (Fig. 14.3), although initial divisions in the former are not as precise as in the latter. As the development of embryos in microspores occurs under completely different environmental conditions compared to that inside the ovule, such differences are expected.

Early divisions occur within the original pollen wall (Fig 14.2). Eventually the pollen wall is ruptured, releasing a group of cells which may continue to undergo orderly cell division and differentiation to a give rise to a mature embryo (Fig. 14.3) and eventually the plantlet. Alternatively, the cells released from the pollen may undergo unorganized divisions to give rise to a callus. Shoots and roots or embryos induced from the callus follow basically the same pattern of development as reported for embryos induced in callus derived from somatic tissues. The factors responsible for triggering morphogenesis

towards embryogenesis or callus formation in the microspore are not clear. Presumably the composition of the medium is an important one. Presence of growth substances in the medium tends to shift the morphogenetic pathway towards callusing.

Embryos developing from microspore callus generally show polarity from the beginning as they have a distal pole (which is free) and a proximal pole attached to the non-embryogenic cells. Often, many show a prominent suspensor-like region. However, in embryos arising directly from the pollen, polarity is not apparent during the initial stages of division.

Ploidy of the plants obtained through anther culture is variable; it is rather rare to obtain plants which are exclusively haploids (Nitsch 1969, Narayanaswamy and Chandy 1971). In general, haploids form only a small fraction of the plants realized from microspores, the remainder varying from diploids to pentaploids and aneuploids. Chromosomal instability, leading largely to polyploidy, is well know in callus cultures. Hence variation in ploidy level of plants resulting from the callus is expected. Variations in ploidy level in plants obtained directly from the microspore without callusing is the result of endomitosis and fusion of nuclei in the pollen (Narayanaswamy and Chandy 1971, Engvild 1974). When polyploidy results after reduction division, the plants will continue to be homozygous and hence can be used as pure lines. Diploids originating in this way are more useful as they eliminate the need for diploidization of haploids through colchicine.

Production of Haploids through Gynogenesis

Unlike androgenic haploids which are derived from microspores/pollen grains, gynogenic haploids originate from unfertilized egg or occasionally other cells of the embryo sac from cultured unpollinated ovaries/flowers. Since the latter half of the 1970s (San 1976), gynogenic plants have been reported in vitro in many economically important species, such as barley, wheat, rice, maize and sunflower (see Sita 1997). In-vitro

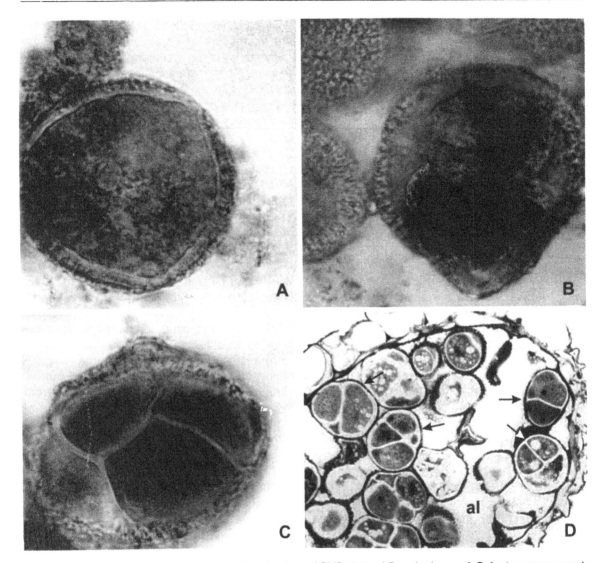

Fig. 14.2 Ontogeny of microspore embryos in cultured anthers of CMS plants of *Brassica juncea.* A-C. Acetocarmine squash preparations of microspores from cultured anthers. A—Activated microspore with dense cytoplasm. B, C - 2—and 4-celled microspores. D. Section of an anther locule (al) of cultured anther to show early divisions in embryogenic microspores (arrows) (after Malik 1998).

gynogenesis has been induced in over 20 species belonging to over 10 families (Yang and Zhou 1990). Details of the various parameters conducive for gynogenesis have not been investigated as extensively as those for androgenesis. One of the advantages of induction of gynogenesis, especially in cereals, has been that most of the gynogenic plants are green, unlike androgenic ones which are mostly albinos. For example, in a cultivar of rice 89.3% of the gynogenic plants were green against only 36.4% from anther cultures (Yang and Zhou 1982). Furthermore, androgenic plants cannot be obtained from male sterile plants and female plants

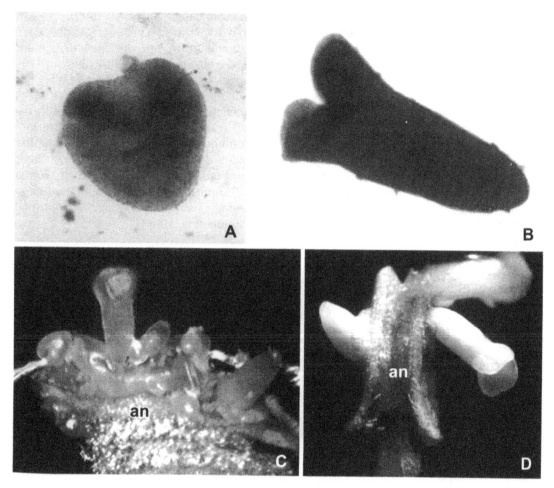

Fig. 14.3 Microspore embryos in cultured anthers of CMS *Brassica juncea*. A, B. Whole-mount preparations of heart-shaped and dicot embryos. C, D. Pollen embryos emerging from cultured anthers (an) along the line of dehiscence (after Malik 1998).

of dioecious species. In such species, induction of gynogenic embryos is the only means of producing haploid plants. For further details see San and Gelebart (1986), Sita (1997).

UTILIZATION OF HAPLOIDS

Although practical utilization of haploids is confined so far to a few crop species, many investigations in recent years have given ample indication that they will form an important component of a crop improvement programme in the coming years. Effective utilization of haploids depends largely upon optimization of conditions for their induction in high frequency. At present, such protocols are available for only a limited number of species. In many crop species such as wheat, rice and tobacco, commercial cultivars have been developed through haploid production (Chaplin et al. 1980, de Buyser et al. 1987, Han and Huang 1987, Deyao and Xian 1990).

Herbicide-tolerant lines have been developed in tobacco and *B. napus* through induction of mutations in microspores or microspore-derived callus and subsequent selection of haploid embryos (Chaleff and Ray 1984, Swanson et al. 1988). This approach has also been used in *B. napus* to isolate mutants with altered storage lipids (Turner and Facciotti 1990, Huang et al. 1991).

Gametoclonal variation among microspore-derived embryos has been used to select variants for disease resistance, alkaloid content and increased protein levels (Sacristan 1982, Foroughi-Wehr and Friedt 1984, Collins et al. 1974, Voorrips and Visser 1990, Witherspoon et al. 1991). Haploids derived from F_1 hybrids such as *Oryza japonica* × *O. indica* and *Triticum aestivum* × *Secale cereale* have been used to obtain fertile lines and chromosome addition and substitution lines (Chu et al. 1982).

In many crop species such as tobacco, rice, wheat and barley, new varieties have been developed with one or more attributes such as disease resistance, high yield, and better quality (Tsuchia 1993, de Buyser et al.1987, Khush and Virmani 1996). A tobacco cultivar F-211 developed in Japan is resistant to bacterial wilt (Nakamura et al. 1975). In China, a number of varieties developed through doubled haploids occupy considerable acreage. In *Brassica napus*, a new blackleg-resistant line has been developed using the doubled haploid method (Stringam et al. 1995).

15

Production of Other Economic Products

Pollen is used in the production of a number of pollen-based economic products. These include health food supplements as well as medicine for humans and domesticated animals. Pollen is also required for diagnosis and treatment of pollen allergy.

POLLEN AS HEALTH FOOD SUPPLEMENT

The most important commercial use of pollen is that of the health food industry. Pollen has long been used as a food supplement (see Linskens and Jorde 1997). Analysis of coprolites recovered in America (1400–200 B.C.) indicates that pollen grains were eaten by American Indians since prehistoric times (Reinhard et al. 1991). Pollen grains are rich in proteins, minerals and vitamins (especially B) and are low in fats, sodium and fat-soluble vitamins (D, K and E). The nutritional composition of pollen surpasses that of any food routinely used (Herbert and Shimamuki 1978, Schmidt and Buchmann 1992). During the last three decades use of pollen as a health food has increased dramatically and developed into a large industry. Pollen is sold in tablet (Fig. 15.1) and liquid form for consumption by humans and domestic animals as a nutritional supplement.

Pollen-based health food is most popular with athletes as it is believed that pollen increases strength and performance (Steben and Boudreaux 1978). Pollen has been shown to increase the mean speed and haemoglobin content of athletes. Also pollen-fed runners did not register a decrease in potassium levels, unlike others who showed net loss in potassium level.

Pollen supplement to the diet of a variety of animals has been reported to show beneficial effects. Use of pollen in racehorse feed seems to be very common (Schmidt and Buchmann 1992). Chickens fed on a balanced diet plus 2.5% pollen showed improved food conversion efficiency (Constantini and Riciardelli d'Albore 1971). Addition of corn pollen to the diet of hens produced egg yolks with deeper colour and higher carotene levels (Tamas et al. 1970). Piglets fed with corn pollen gained weight more quickly and were more efficient in converting food to weight (see Stanley and Linskens 1974). Regular pollen supplement to the diet has also been claimed to prolong the lifespan.

POLLEN AS MEDICINE

Crude pollen, pollen preparations and pollen mixed with honey are used as tonics for general debility and for lack of appetite. There are many other medical claims on the beneficial effects of pollen (see box). Pollen is reported to be effective in the treatment of chronic prostatitis (Ask-Upmark 1967, Linskens and Jorde 1997), in

Fig. 15.1 One of the commercially available pollen products photographed from both sides of the packet.

giving protection against the adverse effects of X-rays (see Schmidt and Buchmann 1992), and in reducing the symptoms of hay fever (Feinberg et al. 1940). Extract of rye pollen is used as an anticongestive. Pollen is also used to treat gastric ulcers (Linskens and Jorde 1997).

Experiments carried out in China have shown the efficacy of pollen in preventing high-altitude sickness. Mice maintained on a pollen-supplemented diet for 3 days prior to high-altitude exposure, showed markedly increased survival rate at an altitude of 12,000 m (Peng-Hong-Fu et al. 1990). Tests on humans also showed that those fed a pollen diet 3–7 days before exposure to high altitude showed no symptoms or fewer symptoms at 5000 m altitude. These re-

Medical benefits of bee pollen (Source: http://www.best-bee-pollen.com—Dr Victor Benilous, Author and Medical Research Director for the website)

Antibiotic qualities of bee pollen
'The antibiotic in honey-bee pollen was extracted and found extremely active. Certain cultures of microbes were killed almost instantly with bee pollen extract, especially those agents responsible in hard-to-control diseases such as salmonella (typhoid types).'

Intestinal functions
'scientists think pollen's capacity for regulating intestinal functions may rest in its ability to destroy harmful intestinal flora.'

Respiratory problems
'Scientists have found bee pollen to be helpful in treating problems such as fatigue, allergies and other respiratory problems (e.g. bronchitis, sinusitis and colds).'

Endocrine system
'Equally impressive, bee pollen helps balance the endocrine system; it is especially beneficial for menstrual and prostate problems. It is also effective in treating constipation and colitis, colibacillosis, anemia, circulatory disorders, neurasthenia and depressive states, skin fragility and hair loss.'

China alone is reported to be 3000–5000 metric tons (Wang 1989). The sales of one company in Sweden, which has been selling pollen products since 1952, totalled $2.5 million during 1984 (see Newman 1984, Schmidt and Buchmann 1992); the company produced 140 million tablets using over 20 metric tons of pollen during that year. The current figures are likely to be much higher.

Pollen is also included in cosmetic preparations such as face creams and pollen-mask facials to remove wrinkles. Pollen is also incorporated in toothpastes (Newman 1984).

One of the constraints in preparing pollen products is the difficulty of collecting large quantities of pollen. Pollen is generally collected by using suitable pollen traps at the entrance of beehives (Waller 1980, Schmidt and Buchmann 1992). The bees have to pass through pollen traps in which pollen corbicular pellets are detached from the legs and fall into a collection tray. The traps should not unduly stress the colony by taking too much pollen. Severe pollen stress may result in reduced brood rearing and decrease in honey production of the colony. Removal of about 60% of the pollen from incoming honey-bees is considered optimal (Levin and Loper 1984). Collected pollen are cleaned to remove contaminants such as insect parts and debris and then maintained under dry conditions (Benson 1984).

sults clearly indicate that consumption of pollen increases adaptability to high altitude (low oxygen tension).

Although many claims on the beneficial effects of pollen are based on clinical trials, most claims are made in non-scientific pamphlets published by manufacturers of pollen products. Notwithstanding this lack of scientific evidence, the demand for pollen products is steadily increasing. The USA, China, Sweden and the USSR are the main producers of pollen products. Although there are no authentic quantitative data available, annual pollen harvest in

POLLEN GRAINS/POLLEN ALLERGENS FOR DIAGNOSIS AND THERAPY

Treatment of Pollen Allergy

Availability of authentic pollen samples and dependable pollen calendars would greatly facilitate diagnosis and treatment of allergy. Avoidance of allergenic pollen is the best way to prevent allergy. Treatment with antiallergic drugs is the predominant management method. Many commercial companies supply authentic pollen samples for diagnosis of pollen allergy. Authentic pollen supply is also required for biochemical characterization of allergens and for understanding their structure.

Immunotherapy (Lichtenstein et al. 1971) is another potential method of treatment. In this therapy, an extract of allergenic pollen is given to patients in very low concentrations; the concentration is increased gradually. The patient develops blocking antibodies against the antigens which compete with IgE antibodies, thus leaving fewer antigens to bind to antibodies located on the surface of mast cells. However, because of the heterogeneity of the extract, immunotherapy induces many side reactions including systemic anaphylaxis.

Presently the main limitations of studies on pollen allergy and effective immunotherapy are: (a) lack of information on the precise structural details of allergens and their relationship to allergy and (b) lack of sufficient amount of pure and standardized preparations of allergens. Pollen extracts presently used for diagnosis and therapy are highly heterogeneous and show a high degree of variation in the amount of active allergen present. The active fraction in the pollen varies according to the plant genotype, prevailing environmental conditions and the age and quality of pollen. Application of crude extract for testing allergic response often leads to the patient developing antibodies to the contaminants. Attempts have been made to biochemically purify allergens in some of the well-known allergenic pollen species. However, similar to the crude extract, biochemically purified extract also causes IgE-mediated side effects (Rogers et al. 1996).

Recent advances in cloning of many major allergenic genes (Table 15.1) and their expression in bacteria and yeast (see Chapter 2) should help in overcoming these major limitations. They have already led to a better understanding of the primary structure of allergens. Isolation of cDNA clones encoding allergens and identification of their deduced amino acid sequences have now become routine (see Smith et al. 1996). Expression of recombinant allergens in bacteria and yeast and purification of recombinant proteins would enable the production of a large amount of recombinant allergens not only for research use, but also for diagnosis and immunotherapy. Further modifications in cDNAs permit the expression of truncated allergens, which can be used to localize allergenic epitopes. Use of chemically modified allergens or peptide fragments containing T-cell epitopes have potential application in diminishing IgE-mediated side effects and thus may open up a new therapeutic approach to allergy desensitization. Attempts are underway to use liposomes as vehicles to deliver allergens. As liposomes have an immunomodulating effect, this approach may prove to be a safer and effective method for immunotherapy. In the coming years pharmaceutical companies are likely to come up with recombinant allergens/modified allergens (in place of pollen being sold at present), which would revolutionize the diagnosis and treatment for pollen allergy.

Pollen Calendars

Pollen calendars are very useful for effective diagnosis and treatment of pollen allergy. Pollen calendars, giving information on diurnal, seasonal and annual variations in the pollen types and their concentration in air, have been prepared for a large number of cities in different geographic locations. A continual check on pollen counts in the air of many cities is kept in several countries in order to forecast pollen incidence. Such studies are useful in finding areas and periods that are comparatively safer for allergic patients.

Various methods are available to sample airborne pollen. In recent years, suction samplers have become most popular. In suction samplers, a vacuum source is used to accelerate air through an opening and pollen grains are impacted on an adhesive surface placed inside the sampler. In the Hirst spore trap (Hirst 1952) an

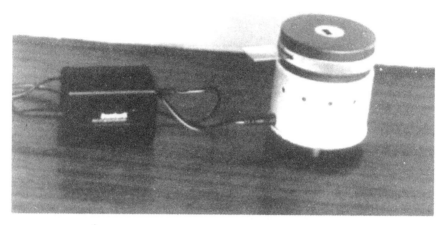

Fig. 15.2 Photograph of Burkard Portable Pollen Sampler with rechargeable battery. A microslide coated with glycerine jelly is inserted through the side slit shown in the Figure, and the slit closed by turning the lid. Air is sucked in through a rectangular orifice located at the top and pollen grains are impacted on the glycerine-coated surface of the slide. (courtesy: Dr A.B. Singh, Centre for Biochemical Technology, Delhi).

Table: 15.1 Some of the major allergenic taxa and their allergens*

Tree species
Alnus glutinosa (alder) Aln g 1
Betula verrucosa (birch) Bet v 1
Corylus avellana (hazel) Cor a 1
Cryptomeria japonica (Japanese red cedar) Cry j 1
Juniperus sabinoides (mountain cedar) Jun s 1
Quercus alba Que a 1
Olea europea (olive) Ole e 1

Grasses
Cynodon dactylon (Bermuda grass) Cyn d 1
Dactylis glomerata Dac g 1
Holcus lanatus Hol l 1
Lolium perenne (rye-grass) Lol p 1, Lol p 5
Phleum pratense (timothy-grass) - Phl p 1, Phl p 5
Poa pratensis (Kentuky bluegrass) Poa p 1
Sorghum halepense Sor h 1
Zea mays (maize) Zea m 1

Weeds
Ambrosia artemisifolia (ragweed) Amb a 1, Amb a 2, Amb a
    5m, Amb a 10, Amb a 12e
A. terifida (giant ragweed) Amb t 5
Artemisia vulgaris (mugwort) Art v 1
Parietaria officinalis Par o 1
Parthenium judaica Par j 1

*Allergens are named according to World Health Organization rules: after the name of the genus (first three letters), space, first letter of the species, space, followed by a number (1 for the most important allergen).

adhesive-coated microslide is placed behind a single impactor slit. The slide moves 2 mm h^{-1}; thus pollen grains are deposited in the form of a band of 48 mm in 24 h. This can be used to obtain daily mean count or continuous 2 h readings by scanning of 4 mm intervals.

The Burkard volumetric spore trap uses a circular drum (instead of a slide), which rotates continuously for 7 days. It sucks air at the rate of 10 L min^{-1}. Pollen is collected on a cellophane tape (adhesive-coated surface) placed in the drum. After exposure, the tape is divided into 7 strips of 24 h each to get daily counts. Alternatively each day's strip can be divided into 12 strips of 2 h each. The Burkard Personal Volumetric Sampler is a handy portable sampler (Fig. 15.2) and can operate on DC as well as AC power. A microslide coated with glycerine jelly is inserted in the sampler and exposed for a definite period.

Pollen grains on the tape/slide of the samplers are identified with the help of reference collections (pollen herbarium) covering pollen grains of all the species of the area and with standard pollen keys. A large number of pollen surveys have been published from different parts of the world.

References

Aas K 1978. What makes an allergen an allergen? Allergy 33: 3–14

Abdalla MMF 1974. Unilateral incompatibility in plant species: analysis and implications. Egypt J Genet Cytol 3: 133–154

Ackerman JD 1997a. Submarine pollination in the marine angiosperm *Zostera marina* (Zosteraceae). I The influence of flower morphology on fluid flow. Am J Bot 84: 1099–1109

Ackerman JD 1997b. Submarine pollination in the marine angiosperm *Zostera marina* (Zosteraceae). II Pollen transport in flow fields and capture by stigmas. Am J Bot 84: 1110–1119

Agache S, de Buyser J, Henry Y, Snape JW 1989. Studies on the genetic relationship between anther culture and somatic tissue culture abilities in wheat. Plant Breed 100: 26–33

Agashe SN (ed) 1994. Recent Trends in Aerobiology and Immunology. Oxford & IBH Publ Co Pvt Ltd, New Delhi

Agnihotri A 1991. Production, evaluation and DNA analysis of three hybrids of oil seed brassicas with their wild allies. Ph.D. Thesis, University of Delhi

Agnihotri A 1993. Hybrid embryo rescue. In: Plant Tissue Culture Manual: Fundamentals and Applications. K Lindsey (ed). Kluwer Academic Publishers, Dordrecht, pp E4: 1–8

Agnihotri A, Gupta M, Lakshmikumaran M, Shivanna KR, Prakash S, Jagannathan V 1990. Production of *Eruca-Brassica* hybrid by embryo rescue. Plant Breed 104: 281–289

Ahloowalia BS 1973. Self- and cross-incompatibility of tetraploid Italian and perennial rye grass *Lolium* spp. Incompatibility Newslett 3: 51–52

Ahokas H 1980. Cytoplasmic male sterility in barley. II. Physiological characterization of the msm 1-Rfm 1a system. Physiol Plant 48: 231–238

Ai Y, Singh A, Coleman CE, Ioerger TR, Kheyrpour A, Kao T-h 1990. Self-incompatibility in *Petunia inflata*: isolation and characterization of cDNAs encoding three S-allele-associated proteins. Sex. Plant Reprod 3: 130–138

Akihama T, Omura M, Kozaki I 1978. Further investigation of freeze-drying for deciduous fruit tree pollen. In: Long-term Preservation of Favourable Germplasm in Arboreal Crops. T Akihama, K Nakajima (eds). Fruit Tree Research Stations, Japan, pp 1–7

Albertini L, Souvre A 1974. Non-histone protein synthesis during microsporogenesis in the *Rhoeo discolor* Hance. In: Fertilization in Higher Plants. HF Linskens (ed). North-Holland Publ, Amsterdam, The Netherlands, pp 49–56

Albertsen MC, Palmer RG 1979. A comparative light- and electron-microscopic study of microsporogenesis in male sterile (ms_1) and male fertile soybeans [*Glycine max* (L.) Merr.]. Am J Bot 66: 253–265

Albertsen MC, Phillips RL 1981. Developmental cytology of 13 genetic male-sterile loci in maize. Can J Genet Cytol 23: 195–208

Alexander MP 1980. A versatile stain for pollen, fungi, yeast and bacteria. Stain Technol 55: 13–18

Allard RW 1960. Principles of Plant Breeding. John Wiley & Sons, New York, NY

Alonso LC, Kimber G 1980. A haploid between *Agropyron junceum* and *Triticum aestivum*. Cereal Res Commun 8: 355–358

Alwen A, Benito MRM, Vicente O, Heberle-Bors E 1992. Plant endogenous β-glucuronidase activity: how to avoid interference with the use of the *E. coli* β-glucuronidase as a reporter gene in transgenic plants. Transgenic Res 1: 63–70

Al Yasiri A, Coyne DF 1964. Effects of growth regulators in delaying pod abscission and embryo abortion in the interspecific cross *Phaseolus vulgaris* × *P. acutifolius*. Crop Sci 4: 433–435

An G, Costa MA, Tepfer D, Gupta HS 1994. Induction of male sterility by pollen-specific expression of *rolB* gene. Abst 4th Intl Cong Plant Mol Biol # 1850, Athens, GA

Ananthakrishnan TN 1993. The role of thrips in pollination. Curr Sci 65: 262–264

Anderson JM, Barrett SCH 1986. Pollen tube growth in tristylous *Pontederia cordata* (Pontederiaceae). Can J Bot 64: 2602–2607

Anderson MA, McFadden GL, Bernatsky R, Atkinson A, Orpin T, Dedman H, Tregear G, Fernley R, Clarke AE 1989. Sequence variability of three alleles of the self-incompatibility gene in *Nicotiana alata*. Plant Cell 1: 483–491

Anderson MA, Cornish EC, Mau SL, Williams EG, Hoggart R, Atkinson A, Bonig I., Grego B, Simpson R, Roch PJ, Haley JD, Penschow JD, Niall HD, Tregear GW, Colghlan JL, Crawford RJ, Clarke AE 1986. Cloning of c-DNA for a stylar glycoprotein associated with expression of self-incompatibility in *Nicotiana alata*. Nature 321: 38–44

Anderson WR 1980. Cryptic self-fertilization in the Malpighiaceae. Science 207: 892–893

Angold RE 1968. The formaion of the generative cell in the pollen grain of *Endymion non-scriptus* (L.). J Cell Sci 3: 573–578

Anonymous 1996. Applied pollination to boost sunflower seed. The Hindu, 4th Oct, p 17

Anthanasiou A, Shore JS 1997. Morph-specific proteins in pollen and styles of distylous *Turnera* (Turneraceae). Genetics 146: 669–679

Arasu NT 1968a. Self-incompatibility in angiosperms: a review. Genetica 39: 1–24

Arasu NT 1968b. Overcoming self-incompatibility by irradiation. Rep E Malling Res Stn (1967): 109–112

Arditti J 1979. Aspects of the physiology of orchids. Adv Bot Res 7: 421–655

Ark PA 1944. Pollen as a source of walnut bacterial blight infection. Phytopathology 34: 330–334

Armaleo D, Ye GN, Klein TM, Shark KB, Sanford JC, Johnston SA 1990. Biolistic nuclear transformation of *Saccharomyces cerevisiae* and other fungi. Curr Genet 17: 97–103

Arnison PG, Keller WA 1990. A survey of the anther culture response of *Brassica oleracea* L. cultivars grown under field conditions. Plant Breed 104: 125–133

Arnison PG, Donaldson P, Jackson A, Semple C, Keller WA 1990. Genotype-specific response of cultured broccoli (*Brassica oleracea* var *italica*) anthers to cytokinins. Plant Cell Tissue & Organ Culture 20: 217–222.

Asano Y, Myodo H 1977. Studies on crosses between distantly related species of lilies. I. The intrastylar pollination technique. J Japan Soc Hort Sci 46: 59–65

Ascher PD 1975. Special stylar property required for compatible pollen tube growth in *Lilium longiflorum* Thunb. Bot Gaz 136: 317–321

Ascher PD 1977. Localization of the self- and inter-specific incompatibility reactions in style sections of *Lilium longiflorum*. Plant Sci Lett 10: 199–203

Ascher PD 1979. Elevated incubation temperature and interspecific pollen tube growth in *Lilium longiflorum* styles. Incompatibility Newslett 11: 50–54

Ascher PD, Peloquin SJ 1966a. Effects of floral ageing on the growth of compatible and incompatible pollen tube in *Lilium longiflorum*. Am J Bot 53: 99–102

Ascher PD, Peloquin SJ 1966b. Influence of temperature on incompatible and compatible pollen tube growth in *Lilium longiflorum*. Can J Genet Cytol 8: 661–664

Ascher PD, Peloquin SJ 1967 Pollen tube growth and incompatibility following intra- and inter-specific pollination in *Lilium longiflorum*. Am J Bot 55: 1230–1234

Ashford AE, Knox RB 1980. Characteristics of pollen diffusates and pollen wall cytochemistry in poplars. J Cell Sci 44: 1–17

Asker SE, Jerling L 1992. Apomixis in Plants. CRC Press, Boca Raton, FL, USA

Ask-Upmark E 1967. Prostatis and its treatment. Acta Med Scand 181: 355–357

Aslam M, Brown MS, Kohel RJ 1964. Evaluation of seven tetrazolium salts as vital pollen stains in cotton (*Gossypium hirsutum* L.). Crop Sci 4: 508–510

Aspinall GD, Rosell KG 1978. The stigmatic exudate from *Lilium longiflorum*. Phytochemistry 17: 919–922

Astrom H, Sorri O, Raudaskoshki M 1995. Role of microtubules in the movement of the vegetative nucleus and generative cell in tobacco pollen tubes. Sex Plant Reprod 8: 61–69

Atkinson AH, Heath RL, Kimpson J, Anderson MA 1993. Proteinase inhibitors in *Nicotiana alata* stigmas are derived from a precursor protein which is processed into five homologous inhibitors. Plant Cell 5: 203–213.

Attia MS 1950. The nature of incompatibility in cabbage. Proc Am Soc Hort Sci 56: 369–371

Avery AG, Satina S, Reitsma J 1959. Blakeslee: the Genus *Datura*. The Ronald Press Co., New York, NY

Azouaou A, Souvre A 1993. Effects of copper deficiency on pollen fertility and nucleic acids in the drum wheat anther. Sex Plant Reprod 6: 199–204

Bagni N, Adamo P, Serafini-Fracassini D, Villanueva VR 1981. RNA, proteins and polyamines during tube growth in germinating apple pollen. Plant Physiol 68: 727–730

Bajaj M, Cresti M, Shivanna KR 1991. Effects of high temperature and humidity stresses on tobacco pollen and their progeny. In: Angiosperm Pollen and Ovule. E Ottaviano, DL Mulcahy, M Sari-Gorla, GB Mulcahy (eds). Springer-Verlag, New York, NY, pp 349–354

Bajaj YPS 1987. Cryopreservation of pollen and pollen embryos, and the establishment of pollen banks. Int Rev Cytol 107:397–420

Bajaj YPS 1990. In vitro production of haploids and their use in cell genetics and plant breeding. In: Biotechnology in Agriculture and Forestry. Vol. 12: Haploids in Crop Improvement. YPS Bajaj (ed), Springer-Verlag, Berlin, pp 1–44

Bajaj YPS, Verma MM, Dhanju MS 1980. Barley x rye hybrids (Hardecale) through embryo culture. Curr Sci 49: 362–363

Baker DA 1978. Proton co-transport of organic solutes by plant cells. New Phytol 81: 485–497

Baker HG 1966. The evolution, functioning and breakdown of heteromorphic incompatibility systems. I. The Plumbaginaceae. Evolution 20: 349–368

Baker HG, Baker I 1979. Starch in angiosperm pollen and its evolutionary significance. Am J Bot 66: 591–600

Baker LR, Chen NC, Park HG 1975. Effect of an immunosuppressant on an interspecific cross of the genus *Vigna*. Hort Sci 10: 313

Balatkova V, Tupy J 1968. Test-tube fertilization in *Nicotiana tabacum* by means of an artificial pollen tube culture. Biol Plant Acad Sci Bohemoslov 10: 266–270

Banda HJ, Paxton RJ 1991. Pollination of greenhouse tomatoes by bees. Acta Hort 288: 194–198

Banga SS 1986. Hybrid pollen-aided matromorphy in *Brassica*. Z Pflanzenzucht 96: 86–89

Banga SS 1992. Heterosis and its utilization. In: Breeding Oilseed Brassicas. KS Labana, SS Banga, SK Banga (eds). Narosa Publ House, New Delhi, pp 21–43

Barclay IR 1975. High frequencies of haploid production in wheat (*Triticum aestivum*) by chromosome elimination. Nature 256: 410–411

Barnabas B 1985. Effect of water loss on germination ability of maize (*Zea mays* L.) pollen. Ann Bot 48: 861–864

Barnabas B 1994. Preservation of maize pollen. In: Biotechnology in Agriculture and Forestry, Vol 25: Maize. YPS Bajaj (ed). Springer-Verlag, Berlin, pp 607–618

Barnabas B, Rajki E 1976. Storage of *Zea mays* L. pollen at −196°C in liquid nitrogen. Euphytica 25: 747–752

Barnabas B, Rajki E 1981. Fertility of deep-frozen maize (*Zea mays* L.) pollen. Ann Bot 48: 861–864

Barnabas B, Fridvalszky L 1984. Adhesion and germination of differently-treated maize pollen grains on the stigma. Acta Bot Hung 30: 329–332

Barnabas B, Kovacs M 1997. Storage of pollen. In: Pollen Biotechnology for Crop Production and Improvement. KR Shivanna, VK Sawhney (eds). Cambridge Univ Press, New York, NY, pp 293–314

Barnes SH, Blackmore S, 1987. Preliminary observation on the formation of the male germ unit in *Catananche caerulea* (Compositae: Lactucea). Protoplasma 138: 187–189

Barone A, del Giudice A, Ng NQ 1992. Barriers to interspecific hybridization between *Vigna unguiculata* and *Vigna vexillata*. Sex Plant Reprod 5: 195–200

Barrett SCH 1990. The evolution and adaptive significance of heterostyly. Trends Ecol Evol 5: 144–148

Barrett SCH 1998. The evolution of mating strategies in flowering plants. Trends Plant Sci 3: 325–340

Barrow JR 1983. Comparisons among pollen viability measurement methods in cotton. Crop Sci 23: 734–736

Bar-Shalom D, Mattsson O 1977 . Mode of hydration, an important factor in the germination of trinucleate pollen grains. Bot Tidsskrift 71: 245–251

Barskaya EI, Balina NV 1971. The role of callose in plant anthers. Soviet Plant Physiol 18: 605–610

Bashe D, Mascarenhas JP 1984. Changes in potassium ion concentrations during pollen dehydration and germination in relation to protein synthesis. Plant Sci Lett 35: 55–60

Bateman AJ 1955. Self-incompatibility systems in angiosperms. III. Cruciferae. Heredity 9: 52–68

Bates LS, Deyoe CW 1973. Wide hybridization and cereal improvement. Econ Bot 27: 401–412

Bates LS, Campos VA, Rodriguez RR, Anderson RG 1979. Progress towards novel cereal grains. Cereal Sci Today 19: 283–286

Batra V, Prakesh S, Shivanna KR 1990. Intergeneric hybridization between *Diplotaxis siifolia*, a wild species, and crop brassicas. Theor Appl Genet 80: 537–541

Batreau L, Gervais A, Quandalle C, Sunderworth S 1991. Le SC-2053, nouvel agent chimique d'hybridation pour le ble. ANPP Annales 3ème Colloque sur les Substances de Croissance et leurs Utilisations dans les Productions Végétales. Ministère de la Recherche et de la Technologie, Paris, pp 75–82

Bawa KS, Webb CJ 1984. Flower, fruit and seed abortion in tropical forest trees: Implications for the evolution of parental and maternal reproductive patterns. Am J Bot 71: 736–751

Bawa KS, Ganeshaiah KN, Uma Shaanker R 1993. Pollination biology of tropics. Curr Sci (special issue) 65: 191–280

Beals TP, Goldberg RB 1997. A novel cell ablation strategy blocks tobacco anther dehiscence. Plant Cell 9: 1527–1545

Beasley CA, Yermanos DM 1976. Effects of storage on in vitro germinability of jojoba pollen. Pollen Spores 18: 471–479

Bednarska E 1989. Localization of calcium on the surface of stigma in *Ruscus aculeatus*. Planta 179: 11–16

Bednarska E 1991. Calcium uptake from the stigma by the germinating pollen in *Primula officinalis* L. and *Ruscus aculeatus* L. Sex Plant Reprod 4: 36–38

Bell PR 1995. Incompatibility in flowering plants: adaptation of an ancient response. Plant Cell 7: 5–16

Bellani LM, Forino LMC, Tagliasacchi AM, Sansavani S 1984. Viability of stored pollen grains of *Malus domestica* Borhk as evaluated by the incorporation of labelled nucleic acid precursors. Caryologia 37: 323–330

Benito-Moreno RM, Macke F, Hauser MT, Alwen A, Heberle-Bors E. 1988. Sporophytes and male gametophytes from in vitro cultured pollen. In: Sexual Reproduction in Higher Plants. M Cresti, P Gori, E Pacini (eds). Springer-Verlag, Berlin, pp 137–142

Bennett MD, Hughes WG 1972. Additional mitosis in wheat pollen induced by ethrel. Nature 240: 566–568

Bennett MD, Smith JB 1976. Nuclear DNA amounts in angiosperms. Phil Trans R Soc Lond B 274: 227–274

Bennett MD, Finch RA, Berclay IR 1976. The time, rate and mechanism of chromosome elimination in *Hordeum* hybrids. Chromosoma 54: 175–200

Bennett MT 1971. The duration of meiosis. Proc R Soc London Ser B 178: 277–299

Benson K 1984. Cleaning and handling pollen. Am Bee J 124: 301–305

Bernhardt P 2000. Convergent evolution and adaptive radiation of beetle-pollinated angiosperms. In: Pollen and Pollination. A Dafni, M Hesse, E Pacini (eds). Springer Wissen, New York, NY, pp 293–320

Bertin RI, Newman CM 1993. Dichogamy in angiosperms. Bot Rev 59: 112–152

Bhadula SK, Sawhney VK 1987. Esterase activity and isozymes during the ontogeny of stamens of male fertile *Lycopersicon esculentum* Mill., and *stamenless-2* mutant and the low temperature reverted mutant. Plant Sci 52: 187–194

Bhadula SK, Sawhney VK 1991. Protein analysis during the ontogeny of normal and male sterile stamenless-2 mutant of tomato (*Lycopersicon esculentum* Mill.). J Exp Bot 40: 789–794

Bhandari NN 1984. The microsporangium. In: Embryology of Angiosperms. BM Johri (ed). Springer-Verlag, Berlin, pp 53–121

Bhaskar MS, Namboodiri AN (1978) Pollen diffusates of *Chlorophytum heyneanum*: promotion of interspecific in vitro pollen germination. Indian J Exp Biol 16: 714–716

Bhatia CR, Mitra R 1998. Biosafety of transgenic crop plants. PINSA B64: 293–318

Biddington NL, Robinson HT 1991. Ethylene production during anther culture of Brussels sprouts (*Brassica oleracea* var. *gemmifera*) and its relationship with factors that affect embryo production. Plant Cell Tissue & Organ Cult 25: 169–177

Biddington NL, Sutherland RA, Robinson TH 1988. Silver nitrate increases embryo production in anther cultures of Brussels sprouts. Ann Bot 62: 181–185

Bino RJ 1985a. Ultrastructural aspects of male sterility in *Petunia hybrida*. Protoplasma 127: 130–140

Bino RJ 1985b. Histological aspects of microsporogenesis in fertile, male sterile and restored fertile *Petunia hybrida*. Theor Appl Genet 69: 423–428

Birnbaum EH, Dugger WM, Beasley BCA 1977. Interaction of boron with components of nucleic acid metabolism in cotton ovules cultured in vitro. Plant Physiol 59: 1034–1038

Blakeslee AF 1945. Removing some of the barriers to crossability in plants. Proc Amer Philos Soc 89: 561–574

Bonhomme S, Budar F, Ferault M, Pelletier G 1991. A 2.5 kb Ncol fragment of Ogura radish mitochondrial DNA is correlated with cytoplasmic male sterility in *Brassica* cybrids. Curr Genet 19: 121–127

Bonner LJ, Dickinson HG 1989. Anther dehiscence in *Lycopersicon esculentum* Mill. I. Structural aspects. New Phytol. 113: 97–115

Bouchard RA, Stern H 1980. DNA synthesized at pachytene in *Lilium*: a non-divergent subclass of moderately repetitive sequences. Chromosoma 81: 349–363

Boutry M, Briquet M 1982. Mitochondrial modifications associated with the cytoplasmic male sterility in faba bean. Eur J Biochem 127: 129–135

Boyer JS 1982. Plant productivity and environment. Science 218: 443–466

Boyes DC, Chen C-H, Tantikanjana T, Esch JJ, Nasrallah JB 1997. Isolation of a second S-locus related cDNA from *Brassica oleracea*: genetic relationship between the S-locus and two related loci. Genetics 127: 221–228

Braun CJ, Siedow JN, Levings CS III 1990. Fungal toxins bind to the URF 13 protein in maize mitochondria and *Escherichia coli*. Plant Cell 2: 153–161

Braun CJ, Siedow JN, Williams ME, Levings CS III 1990. Mutations in the maize mitochondrial *T-urf 13* gene eliminates sensitivity to a fungal pathotoxin. Proc Natl Acad Sci USA 86: 4435–4439

Bredemeijer GMM 1974. Peroxidase activity and peroxidase isozyme composition in self-pollinated, cross-pollinated and unpollinated styles of *Nicotiana alata*. Acta Bot Neerl 23: 149–154

Bredemeijer GMM 1982. Mechanism of peroxidase isozyme induction in pollinated *Nicotiana alata* styles. Theor Appl Genet 62: 305–309

Bredemeijer GMM, Blaas J 1981. S-specific proteins in styles of self-incompatible *Nicotiana alata*. Theor Appl Genet 59: 185–190

Brewbaker JL 1957. Pollen cytology and self-incompatibility systems in plants. J Hered 48: 271–277

Brewbaker JL 1959. Biology of the angiosperm pollen grain. Indian J Genet Plant Breed 19: 121–133

Brewbaker JL, Natarajan AT 1960. Centric fragments and pollen part mutations in incompatibility alleles in *Petunia*. Genetics 45: 699–704

Brewbaker JL, Majumdar SK 1961. Cultural studies of the pollen population effect and the self-incompatibility inhibition. Am J Bot 48: 457–464

Brewbaker JL, Emery GC 1962. Pollen radiobotany. Radiat Bot 1: 101–154

Brewbaker JL, Kwack BH 1963. The essential role of calcium ion in pollen germination and pollen tube growth. Am J Bot 50: 859–865

Brink RA 1924 The physiology of pollen. IV. Chemotropism; effects on growth of grouping grains; formation and function of callose plugs; summary and conclusions. Am J Bot 11: 417–436

Britikov ·EA, Schrauwen J, Linskens HF 1970. Proline as a source of nitrogen in plant metabolism. Acta Bot Neerl 19: 515–520

Brlansky RH, Carroll TW, Zaske SK 1986. Some ultrastructural aspects of the pollen transmission of barley stripe mosaic virus in barley. Can J Bot 64: 853–858

Brock RD 1954. Fertility in *Lilium* hybrids. Heredity 8: 409–420

Brooking IR 1979. Male sterility in *Sorghum bicolor* (L.) Moench induced by low night temperature. II. Genotypic differences in sensitivity. Aust J Plant Physiol 6: 143–147

Brooks J, Shaw G 1971. Recent developments in the chemistry, biochemistry, geochemistry and post-tetrad ontogeny of sporopollenins derived from pollen and spores exines. In: Pollen: Development and Physiology. J Heslop-Harrison (ed). Butterworths, London, pp 99–114

Brooks MH 1962. Comparative analysis of some of the free amino acids in anthers of fertile and genetic-cytoplasmic male sterile sorghum. Genetica 47: 1629–1638

Broothaerts W, Vanvinckenroye P, Decock B, van Damme J, Vendrig JC 1991. *Petunia hybrida* S-proteins: ribonuclease activity and the role of their glycon side chains in self-incompatibility. Sex Plant Reprod 4: 258–266

Brown DCW, Watson EM, Pechan PM 1993. Induction of desiccation tolerance in microspore-derived embryos of *Brassica napus* in vitro. Cell Devel Biol 29: 113–118

Brown GK, Perking RM 1969. Experiments with aircraft methods for pollinating dates. Date Growers Inst Rep 46: 35–40

Brown RC, Lemmon BE 1991. Pollen development in orchids. 3. A novel generative pole microtubule system predicts unequal pollen mitosis. J Cell Sci 99: 273–281

Brugiere N, Cui Y, Rothstein SJ 2000. Molecular mechanisms of self-recognition in *Brassica* self-incompatibility. Trends Plant Sci Rev 5: 432–438

Brummitt RK, Ferguson JK, Poole JW 1980. A unique and extraordinary pollen type in the genus *Crossendra* (Acanthaceae). Pollen Spores 22: 11–16

Bryant VM Jr 1990. Pollen: Nature's fingerprints of plants. Encyclopedia Britannica 93–111

Bucholtz DL 1988. Effect of environment and formulation on the absorption and translocation of fenridazon in wheat *Triticum aestivum* L. Plant Growth Regul 7: 65–74

Burgess, J 1970. Cell shape and meiotic spindle formation in the generative cell of *Endymion non-scriptus*. Planta 95: 72–85

Burk LG 1967. An interspecific bridge-cross *Nicotiana rependa* through *N. sylvestris* to *N. tabacum*. J Hered 58: 215–218

Buss PA Jr, Lersten NR 1975. Survey of tapetal nuclear number as a taxonomic character in Leguminosae. Bot Gaz 136: 388–395

Buter B, Pescitelli SM, Berger K, Schmid JE, Stamp P 1993. Autoclaved and filter-sterilized liquid media in maize anther culture: significance of activated charcoal. Plant Cell Rep 13: 79–82

Cai G, Romagnoli S, Cresti M 2001. Microtubule motor proteins and the organization of the pollen tube cytoplasm. Sex Plant Reprod 14: 27–34

Calder DM, Slater AT 1985. The stigma of *Dendrobium speciosum* Sm (Orchidaceae): a new stigma type comprising detached cells within mucilaginous matrix. Ann Bot 55: 297–307

Campbell RJ, Ascher PD 1972. High temperture removal of self-incompatibility in *Nemesia strumosa*. Incompatibility Newslett 1: 3–5

Capkova-Balatkova V, Hrabetova E, Tupy J 1980. Effect of some mineral ions on pollen tube growth and release of proteins in culture. Biol Plant 22: 294–302

Capkova-Balatkova V, Hrabetova E, Tupy J 1988. Protein synthesis in pollen tubes. Preferential formation of new species independent of transcription. Sex Plant Reprod 1: 150–155

Capkova V, Zbrozek J, Tupy J 1994. Protein synthesis in tobacco pollen tubes: preferential synthesis of cell-wall 69-KDa and 66-KDa glycoproteins. Sex Plant Reprod 7: 57–66

Carlson WR 1969. Factors affecting preferential fertilization in maize. Genetics 62: 543–554.

Carroll TW 1974. Barley stripe mosaic virus in sperm and vegetative cells of barley pollen. Virology 60: 21–28

Carter AL, McNeilly T 1976. Increased atmospheric humidity post pollination: a possible aid to the production of inbred line seed from mature flowers in the Brussels sprouts (*Brassica oleracea* var. *gemmifera*). Euphytica 25: 531–538

Cass DD 1997. Isolation and manipulation of sperm cells. In: Pollen Biotechnology for Crop Production and Improvement. KR Shivanna, VK Sawhney (eds). Cambridge University Press, New York, NY, pp 352–362

Cass DD, Karas I 1975. Development of sperm cells in barley. Can J Bot 53: 1051–1062

Certal AC, Sanchez AM, Kokko H, Broothaerts WJ, Oliveira MM, Feijo JA 1999. S-Rnases in apple are expressed in the pistil along the pollen tube growth path. Sex Plant Reprod 12: 94–98

Chaboud A, Perez R 1992. Generative cells and male gametes: Isolation, physiology and biochemistry. Intl Rev Cytol 140: 205–232

Chaleff RS, Ray TB 1984. Herbicide-resistant mutants from tobacco cell cultures. Science 223: 1148–1151

Chanda S 1994. Pollen grains as aeroallergens. In: Recent Trends in Aerobiology, Allergy and Immunology. SN Agashe (ed). Oxford & IBH Publ Co Pvt Ltd, New Delhi, pp 85–92

Chao CH 1977. Further cytological studies of a periodic acid-Schiff's substance in the ovules of *Paspalum orbiculare*. Am J Bot 64: 920–930

Chaplin JF, Berk LG, Goding AV, Powell NT 1980. Registration of NC744 tobacco germplasm (Reg No. GP18). Crop Sci 20: 677

Chapman GP 1987. The tapetum. Intl Rev Cytol 107: 111–126

Charlesworth D, Charlesworth B 1979. A model for the evolution of distyly. Am Nat 114: 467–498

Charzynska M, Ciampolini F, Cresti M 1988. Generative cell division and sperm cell formation in barley. Sex Plant Reprod 1: 240–247.

Charzynska M, Murgia M, Cresti M 1989. Ultrastructure of the vegetative cell of *Brassica napus* pollen with particular reference to microbodies. Protoplasma 152: 22–28

Charzynska M, Murgia M, Cresti M. 1990. Microspore of *Secale cereale* as a transfer cell type. Protoplasma 158: 26–32

Chase SS 1952. Monoploids in maize. In: Heterosis. JW Gowen (ed). Iowa State College Press, Cedar Falls, IA.

Chaubal R, Reger BJ 1990. Relatively high calcium is localized in synergid cells of wheat ovaries. Sex Plant Reprod 3: 98–102

Chaubal R, Reger BJ 1992. The dynamic of calcium distribution in the synergid cells of wheat after pollination. Sex Plant Reprod 5: 206–213

Chaudhury A, Craig S, Bloemer KC, Farrell L, Dennis ES 1992. Genetic control of male fertility in higher plants. Aust J Plant Physiol 19: 419–426

Chaudhury AM 1993. Nuclear genes controlling male fertility. Plant Cell 5: 1277–1283

Chaudhury R, Shivanna KR 1986. Studies on pollen storage in *Pennisetum typhoides*. Phytomorphology 36: 211–218

Chaudhury R, Shivanna KR 1987. Differential responses of *Pennisetum* and *Secale* pollen. Phytomorphology 37: 181–185

Chauhan SVS, Kinoshita T 1982. Chemically-induced male sterility in angiosperms. Seiken Jiho 30: 54–75

Chavez R, Jackson MT, Schmiediche PE, Franco J 1988a. The importance of wild potato species resistant to potato cyst nematode, *Globodera pallida*, phenotypes Pa-4 and Pa-5 in potato breeding. II. The crossability of resistant species. Euphytica 37: 15–22

Chavez R, Schmiediche PE, Jackson MT, Raman KV 1988b. The breeding potential of wild potato species resistant to the tuber moth *Phthorimaea operculella* (Zeller). Euphytica 39: 123–132.

Chen CH, Nasrallah JB 1990. A new class of S-sequences defined by a pollen recessive self-incompatibility allele of *Brassica oleracea*. Mol Gen Genet 127: 221–228

Chen JL, Beversdorf WD 1992. Production of spontaneous diploid lines from isolated microspores following cryopreservation in spring rapeseed (*Brassica napus* L). Plant Breed 108: 324–327

Chen M, Loewus FA 1977. Myo-inositol metabolism in *Lilium longiflorum* pollen. Uptake and incorporation of myo-inositol-2-³H. Plant Physiol 59: 653–657

Chen M, Loewus MW, Loewus FA 1977. Effect of a myo-inositol antagonist, C-methylene-myo-inositol, on the metabolism of myo-inositol-2-³H and D-glucose-1-¹⁴C in *Lilium longiflorum* pollen. Plant Physiol 59: 658–663

Chen NC, Parrot JF, Jacobs T, Baker LR, Carlson PS 1978. Interspecific hybridization of food grain legumes by unconventional methods of Breeding. In: Intl Mungbean Symp, AVRDC, Taiwan, pp 247–252

Cheung AY 1996. Pollen-pistil interactions during pollen tube growth. Trends Plant Sci Rev 1: 45–51

Cheung AY, Wang H, Wu HM 1995. A floral transmitting tissue-specific glycoprotein attracts pollen tubes and stimulates their growth. Cell 82: 383–393

Cheung AY, May B, Kawata EE, Gu Q, Wu HM 1993. Characterization of cDNAs for stylar transmitting tissue-specific pollen-rich proteins in tobacco. Plant J 3: 151–160

Chiang MS 1974. Cabbage pollen germination and longevity. Euphytica 23: 579–584

Ching TM, Ranzoni MW, Ching KK 1975. ATP content and pollen germinability of some conifers. Plant Sci Lett 4: 331–333

Chopra VL (ed) 1989. Plant Breeding: Theory and Practice. Oxford & IBH Publ Co Pvt Ltd, New Delhi

Chopra VL, Kirti PB, Prakash S 1996. Assessing and exploiting genes of breeding value of distinct relatives of crop brassicas. Genetica 97: 305–312

Christensen JE, Horner HT Jr 1974. Pollen development and its special orientation during microsporogenesis in the grass *Sorghum bicolor*. Am J Bot 61: 604–623

Chrungu B, Mohanty A, Verma N, Shivanna KR 1999. Production and characterization of interspecific hybrids between *Brassica murorum* and crop brassicas. Theor Appl Genet 98: 608–613

Chu C, Liu H-T, Du RH 1982. Microphotometric determination of DNA contents of early developmental pollen grains in tobacco anther culture. Acta Bot Sinica 24: 1–9

Ciampolini F, Moscutelli A, Cresti M 1988. Ultrastructural features of *Aloe ciliaris* pollen. I. Mature grain and its activities in vitro. Sex Plant Reprod 1: 88–96

Ciampolini F, Shivanna KR, Cresti M 1990. The structure and cytochemistry of the pistil of *Sternbergia lutea* (Amaryllidaceae). Ann Bot 66: 703–712

Ciampolini F, Shivanna KR, Cresti M 1991. High humidity and heat stress cause dissociation of endoplasmic reticulum in pollen grains. Bot Acta 104: 110–116

Ciampolini F, Nepi M, Pacini E 1993. Tapetum development in *Cucurbita pepo* (Cucurbitaceae). In: The Tapetum: Cytology, Function, Biochemistry and Evolution.

M. Hesse, E. Pacini, M.T.M. Willemse (eds). Plant Syst Evol Suppl 7: 13–22

Ciha AJ, Ruminski PG 1991. Specificity of pyridine monocarboxylates and benzoic acid analogues as chemical hybridizing agents in wheat. J Agric Food Chem 39: 2072–2076

Clapham DH 1977. Haploid induction in cereals. In: Plant Cell, Tissue and Organ Culture. J Reinert, YPS Bajaj (eds). Springer-Verlag, Berlin, pp 274–298

Clark AG, Kao T-h 1991. Excess nonsynonymous substitution at shared polymorphic sites among self-incompatibility alleles of Solanaceae. Proc Natl Acad Sci USA 88: 9823–9827

Clark KR, Okuley JJ, Collins PD, Sims TL 1990. Sequence variability and developmental expression of S-alleles in self-incompatible and pseudo-self-incompatible Petunia. Plant Cell 2: 815–826.

Clarke AE, Considine JA, Ward R, Knox RB 1977. Mechanism of pollination in Gladiolus: roles of the stigma and pollen tube guide. Ann Bot 41: 15–20

Clarke AE, Gleeson P, Harrison S, Knox RB 1979. Pollen-stigma interactions: identification and characterization of surface components with recognition potential. Proc Natl Acad Sci USA 76: 3358–3362

Clarke AE, Anderson MA, Bacic A, Harris PJ, Mau SL 1985. Molecular basis of cell recognition during fertilization in higher plants. J Cell Sci 2: 261–285

Clauhs RP, Grun P 1977. Changes in plastid and mitochondrion content during maturation of generative cells of Solanum (Solanaceae). Am J Bot 64: 377–383

Clement C, Audran JC 1995. Anther wall layers control pollen sugar nutrition in Lilium. Protoplasma 187: 172–181

Clement C, Pacini E, Audran JC (eds) 1999. Anther and Pollen: From Biology to Biotechnology. Springer-Verlag, Berlin, Heidelberg, New York

Coe EH 1959. A line of maize with high haploid frequency. Am Naturalist 93: 381–382

Coe EH, McCormick SM, Modena SA 1981. White pollen in maize. J Hered 72: 318–320

Colhoun CW, Steer MW 1983. The cytological effect of the gametocides ethrel and RH-531 on microsporogenesis in barley (Hordeum vulgare L.). Plant Cell Environ 6: 21–29

Collins FC, Lertmongkol V, Jones JP 1973. Pollen storage of certain agronomic species in liquid nitrogen. Crop Sci 13: 493–494

Collins GB, Legg PB, Kasperbauer MJ 1974. Use of anther-derived haploids in Nicotiana. I. Isolation of breeding lines differing in total alkaloid content. Crop Sci 14: 77–80

Comstock JC, Martinson CA, Gengenbach BG 1973. Host specificity of a toxin from Phyllosticta maydis for Texas cytoplasmically male sterile maize. Phytopathology 63: 1357–1361

Condeelis, J.S. 1974. The identification of F-actin in the pollen tube and protoplast of Amaryllis belladonna. Exptl. Cell Res 88: 435–439

Conner JA, Tantikanjana T, Stein JC, Kandasamy MK, Nasrallah JB, Nasrallah ME 1997. Transgene-induced silencing of S-locus genes and related genes in Brassica. Plant J 11: 809–823.

Connett MB, Hanson MR, 1990. Differential mitochondrial electron transport through the cyanide-sensitive and cyanide-insensitive pathways in isonuclear lines of cytoplasmic male sterile, male fertile and restored Petunia. Plant Physiol 93: 1634–1640

Cooper DV, Brink RA 1940. Somatoplastic sterility as a cause of seed failure after interspecific hybridization. Genetics 25: 593–617

Corbet SA, Beament J, Eisikowitch D 1982. Are electrostatic forces involved in pollen transfer? Plant Cell Environ 5: 125–129

Cornish EC, Pettitt JM, Bonig I, Clarke AE 1987. Developmentally controlled expression of a gene associated with self-incompatibility in Nicotiana alata. Nature 326: 99–102

Constantini F, Riciardelli d'Albore G 1971. Pollen as an additive to the chicken diet. 23rd Intl Apic Cong, pp 539–542

Cox PA 1988. Hydrophilous pollination. Ann Rev Ecol Syst 19: 261–280

Cox PA 1993. Water-pollinated plants. Sci Amer 269 (4): 68–74

Cox PA, Knox RB 1988. Pollination postulates and two-dimensional pollination in hydrophilous monocotyledons. Ann Missouri Bot Gard 75: 811–818

Cox PA, Humphries CJ 1993. Hydrophilous pollination and breeding system evolution in seagrasses: a phylogenetic approach to the evolutionary ecology of the Cymodoceaceae. Bot J Linn Soc 113: 217–226

Crane MB, Lewis D 1942. Genetical studies in pears. III. Incompatibility and sterility. J Genet 43: 31–49

Crane MB, Marks E 1952. Pear-apple hybrids. Nature 170: 1017

Cresti M, Ciampolini F, Sansavini S 1980. Ultrastructural and histochemical features of pistil of Malus communis: the stylar transmitting tissue. Sci. Hort 12: 327–337

Cresti M, Ciampolini F, Kapil RN 1984. Generative cells of some angiosperms with particular emphasis on their microtubules. J Submicros Cytol 16: 317–326

Cresti M, Lancelle E, Hepler PK 1987. Structure of the generative cell wall complex after freeze-substitution in pollen tubes of Nicotiana and Impatiens. J Cell Sci 88: 373–378

Cresti M, Gori P, Pacini E (eds) 1988. Sexual Reproduction in Higher Plants. Springer-Verlag, Berlin, Heidelberg, New York

Cresti M, Blackmore S, van Went JL (eds) 1992. Atlas of Sexual Reproduction in Flowering Plants. Springer-Verlag, Berlin, Heidelberg, New York

Cresti M, Pacini E, Sarfatti G, Simoncioti C 1975. Ultrastructural features and storage function of *Lycopersicum peruvianum* pollen. In: Gamete Competition in Plants and Animals. DL Mulcahy (ed). North-Holland Publ, Amsterdam, The Netherlands, pp 19–28

Cresti M, Pacini E, Ciampolini F, Sarfatti G 1977. Germination and early tube development in vitro of *Lycopersicum peruvianum* pollen: ultrastructural features. Planta 136: 239–247

Cross JW, Ladyman JAR 1991. Chemical agents that inhibit pollen development: tools for research. Sex Plant Reprod 4: 235–243

Cross JW, Schulz PJ 1997. Chemical induction of male sterility. In: Pollen Biotechnology for Crop Production and Improvement. KR Shivanna, VK Sawhney (eds), Cambridge Univ Press, New York, NY, pp 218–236

Cross JW, Patterson T, Almeida E, Schulz PJ 1989. Effects of chemical hybridizing agents SC-1058 and SC-1271 on the ultrastructure of developing wheat anthers (*Triticum aestivum* L. var.'*Yecora rojo*'). In: Current Topics in Plant Physiology. E Lord, G Bernier (eds). Am Soc Plant Physiol, Rockville, Vol 1, pp 187–188

Cross JW, Herzmark P, Guilford WL, Patterson TG, Labovitz JN 1992. Transport and metabolism of pollen suppressant SC-1271 in wheat. Pest Biochem Physiol 44: 28–39

Crouzillat D, Leroy P, Perrault A, Ledoigt G 1987. Molecular analysis of the mitochondrial genome of *Helianthus annuus* in relation to male sterility and phylogeny. Theor Appl Genet 74: 773–780

Crowe JH, Crowe LM, Chapman D 1984. Preservation of membranes in anhydrobiotic organisms: Role of trehalose. Science 223: 701–703

Crowe JH, Hoekstra FA, Crowe LM 1989. Membrane phase transitions are responsible for imbibitional damage in dry pollen. Proc Nat Acad Sci 86: 520–523

Crowe LK 1954. Incompatibility in *Cosmos bipinnatus*. Heredity 8: 1–11

Crowe LK 1964. The evolution of outbreeding in plants. 1. The Angiosperms. Heredity 19: 293–322

Crowe LK 1971. The polygenic control of outbreeding in *Borago officinalis*. Heredity 27: 111–118

Cruden RW 1977. Pollen-ovule ratios: a conservative indicator of breeding systems in flowering plants. Evolution 31: 32–46

Cui Y, Bi YM, Brugiere N, Arnoldo MA, Rothstein SJ 2000. The S-locus glycoprotein and the S-receptor kinase are sufficient for self-pollen rejection in *Brassica*. Proc Natl Acad Sci USA 97: 3713–3717

Currie RW 1997. Pollination constraints and management of pollinating insects for crop production. In: Pollen Biotechnology for Crop Production and Improvement. KR Shivanna, VK Sawhney (eds). Cambridge University Press, New York, NY, pp 121–151

Currie RW, Winston ML, Slessor KN, Mayer DF 1992a. Effect of synthetic queen mandibular compound sprays on pollination of fruit crops by honeybees (Hymenoptera: Apidae). J Econ Entomol 85: 1293–1299

Currie RW, Winston ML, Slessor KN 1992b. Effect of synthetic queen mandibular pheromone sprays on honeybee (Hymenoptera: Apidae) pollination of berry crops. J Econ Entomol 85: 1300–1306

Dafni A 1992. Pollination Ecology: A Practical Approach. Oxford Univ Press, New York

Dafni A, Maues MM 1998. A rapid and simple procedure to determine stigma receptivity. Sex Plant Reprod 11: 177–180

Dafni A, Firmage D 2000. Pollen viability and longevity: practical, ecological and evolutionary implications. Plant Syst Evol 222: 113–132

Dafni A, Hesse M, Pacini E (eds) 2000. Pollen and Pollination. Springer Wissen, New York, NY

Dahl AO, Rowley JR 1974. A glycocalyx embedded within the pollen exine. J Cell Biol 63: 75a

Darwin C 1876. Cross- and Self-fertilization of Plants. John Murray, London

Darwin C 1877. The Different Forms of Flowers on Plants of the Same Species. John Murray, London

Dashek WV, Rosen WG 1966. Electron microscopical localization of chemical components in the growth zone of lily pollen tubes. Protoplasma 61: 192–204

Dashek WV, Harwood HI 1974. Proline, hydroxyproline and lily pollen tube elongation. Ann Bot 38: 947–959

Dashek WV, Mills RR 1980. Azetidine-2-carboxylic acid and lily pollen tube elongation. Ann Bot 45: 1–12

Datta SK, Potrykus I 1989. Artificial seeds in barley: encapsulation of microspore derived embryos. Theor Appl Genet 77: 820–824

Davies AJS 1957. Successful crossing in the genus *Lathyrus* through stylar amputation. Nature 180: 612

Davis AR 1997. Pollination efficiency of insects. In: Pollen Biotechnology for Crop Production and Improvement. KR Shivanna, VK Sawhney (eds), Cambridge Univ Press, New York, NY, pp. 87–120

Davis GL 1966. Systematic Embryology of Angiosperms. John Wiley, New York, NY

Davis LE, Stephenson AS, Winsor JA 1987. Pollen competition improves performance and reproductive output of the common zucchini squash under field conditions. J Am Soc Hort Sci 112: 712–716

Dawson J, Wilson ZA, Aarts MGM, Braithwaite AF, Briarty LG, Mulligan BJ 1993. Microsporogenesis and pollen development in six male sterile mutants of *Arabidopsis thaliana*. Can J Bot 71: 629–638

Day A, Ellis THN 1985. Deleted forms of plastid DNA in albino plants from cereal anther culture. Curr Genet 9: 671–678

Day C, Miller R, Irish V 1994. Genetic cell ablation to analyze cell interactions during *Arabidopsis* floral development. Abstr 4th Intl Cong Plant Mol Biol # 738, Athens, GA

Dayton DE 1974. Overcoming self-incompatibility in apple with killed compatible pollen. J Am Soc Hort Sci 99: 190–192

de Buyser J, Henry Y, Lonnet P, Hertzog R, Hespel A 1987. Florin: a doubled haploid wheat variety developed by the anther culture method. Plant Breed 98: 53–57.

de Graaf BHJ, Derksen JWM, Mariani C 2001. Pollen-pistil interaction in the progamic phase. Sex Plant Reprod 14: 41–55

De Grandi-Hoffman G 1987. The honeybee pollination component of horticultural crop production system. Hort Sci 9: 237–272.

de Nettancourt D 1977. Incompatibility in Angiosperms. Springer-Verlag, Berlin, Heidelberg, New York

de Nettancourt D 2001. Incompatibility and Incongruity in Wild and Cultivated Plants. Springer-Verlag, Berlin, Heidelberg, New York

de Nettancourt D, Devreux M, Bozzini A, Cresti M, Pacini E, Sarfatti G 1973a. Ultrastructural aspects of the self-incompatibility mechanism in *Lycopersicum peruvianum* Mill. J Cell Sci 12: 403–419

de Nettancourt D, Devreux M, Laneri U, Pacini E, Cresti M, Sarfatti G 1973b. Ultrastructural aspects of unilateral interspecific incompatibility between *Lycopersicum peruvianum* and *L. esculentum.* Caryologia 25: 207–217

de Nettancourt D, Devreux M, Laneri U, Cresti M, Pacini E, Sarfatti G 1974. Genetical and ultrastructural aspects of self- and cross-incompatibility in interspecific hybrids between self-compatible *Lycopersicum esculentum* and self-incompatible *L. peruvianum.* Theor Appl Genet 44: 278–288

de Nettancourt D, Devreux M, Carluccio F, Laneri U, Cresti M, Pacini E, Sarfatti G, van Gastel AJG 1975. Facts and hypothesis on origin of S-mutations and on the function of the S-gene in *Nicotiana alata* and *Lycopersicum peruvianum.* Proc R Soc London Ser B 188: 345–360

De Verna JW, Meyers JR, Collins GB 1987. Bypassing prefertilization barriers to hybridization in *Nicotiana* using in vitro pollination and fertilization. Theor Appl Genet 73: 665–671

Den Nijs APM, Oost EH 1980. Effect of mentor pollen on pollen-pistil incongruities among species of *Cucumis* L. Euphytica 29: 267–272

Denward T 1963. The function of the incompatibility allele in red clover (*Trifolium pratens* L.). Hereditas 49: 189–234

Derksen J 1996. Pollen tubes: a model system for plant cell growth. Bot Acta 109: 341–345

Derksen J, Emons AMC 1990. Microtubles in tip growth systems. In: Tip Growth in Plant and Fungal Cells. JB Heath (ed). Academic Press, San Diego, CA, pp 147–181

Derksen J, Pierson ES, Traas JA 1985. Microtubules in vegetative and generative cells of pollen tubes. Europ J Cell Biol 38: 142–148

Derksen J, Rutten T, Amstel TV, Win AD, Doris F, Steer M 1995. Regulation of pollen tube growth. Acta Bot Neerl 44: 93–119

Deshusses J, Gumber SC, Loewus FA 1981. Sugar uptake in lily pollen. A proton symport. Plant Physiol 67: 793–796

Desikachary TV, Swamy BGL 1976. Isolation and insulation of special cells. Curr Sci 45: 485–486

Deurenberg JJM 1976. In vitro protein synthesis with polysomes from unpollinated, cross- and self-pollinated *Petunia* ovaries. Planta 128: 29–33

Deurenberg JJM 1977. Differentiated protein synthesis with polysomes from *Petunia* ovaries before fertilization. Planta 133: 201–206

Devlin BRJ, Kerr MW 1979. 3,4-methanopyrrolidine derivatives and their use as plant growth regulating agents. Eur Patent Appl 2859

Dewey RE, Levings CS III, Timothy DH 1986. Novel recombinations in the maize mitochondrial genome produce a unique transcriptional unit in the Texas male sterile cytoplasm. Cell 44: 439–449

Dewey RE, Timothy DH, Levings CS III 1987. A mitochondrial protein associated with cytoplasmic male sterility in the T-cytoplasm of maize. Proc Natl Acad Sci USA 84: 5374–5389

Deyao Z, Xian P 1990. Rice (*Oryza sativa* L.): Guan 18—an improved variety through anther culture. In: Biotechnology in Agriculture and Forestry Vol 12: Haploids in Crop Improvement. I. YPS Bajaj (ed). Springer-Verlag, Berlin, pp 204–211

Dhaliwal AS, Malik CP, Singh MB 1981. Overcoming incompatibility in *Brassica campestris* L. by carbon dioxide, and dark fixation of the gas by self- and cross-pollinated pistils. Ann Bot 48: 227–233

Dhaliwal HS 1992. Unilateral incompatibility. In: Distant Hybridization of Crop Plants. G Kalloo, GB Chowdhury (eds), Springer-Verlag, Berlin, pp 32–46

Dhaliwal S, King PJ 1978. Direct pollination of *Zea mays* ovules in vitro with *Z. mays, Z. mexicana* and *Sorghum bicolor* pollen. Theor Appl Genet 53: 43–46

Dickinson DB 1967. Permeability and respiratory properties of germinating pollen. Physiol Plant 20: 118–127

Dickinson DB, Hopper JE, Davies MD 1973. A study of pollen enzymes involved in sugar nucleotide formation. In: Biogenesis of Plant Cell Wall Polysaccharides. F Loewus (ed), Academic Press, New York, NY, pp 29–48

Dickinson HG 1973. The role of plastids in the formation of pollen grain coatings. Cytobios 8: 25–40

Dickinson HG 1994. Simply a social disease? Nature 367: 517–518

Dickinson HG 1995. Dry stigmas, water and self-incompatibility in *Brassica*. Sex Plant Reprod 8: 1–10

Dickinson HG, Heslop-Harrison J 1970. The ribosome cycle, nucleoli, and cytoplasmic nucleoids in the meiocytes of *Lilium*. Protoplasma 69: 181–200

Dickinson HG, Lewis D 1973a. The formaton of the tryphine coating the pollen grains of *Raphanus* and its properties relaing to the self-incompatibility system. Proc R Soc London Ser B 184: 149–165

Dickinson HG, Lewis D 1973b. Cytochemical and ultrastructural differences between intraspecific compatible and incompatible pollinations in *Raphanus*. Proc R Soc London Ser B 183: 21–38

Dickinson HG, Lewis D 1974. Changes in the pollen grain wall of *Linum grandiflorum* following compatible and incompatible intraspecific pollinations. Ann Bot 38: 23–29

Dickinson HG, Lewis D 1975. Interaction between the pollen grain coating and the stigmatic surface during compatible and incompatible intraspecific pollinations in *Raphanus*. In: The Biology of the Male Gamete. JG Duckett, PA Racey (eds). Biol J Linn Soc 7 (Suppl 1): 165–175

Dickinson HG, Lawson J 1975a. Pollen tube growth in the stigma of *Oenothera organensis* following compatible and incompatible intraspecific pollinations. Proc R Soc London Ser B 188: 327–344

Dickinson HG, Potter U 1976. The development of patterning in the alveolar sexine of *Cosmos bipinnatus*. New Phytol 76: 543–550

Dickinson HG, Andrews L 1977. The role of membrane-bound cytoplasmic inclusions during gametogenesis in *Lilium longiflorum* Thunb. Planta 134: 229–240

Dickinson HG, Heslop-Harrison J 1977. Ribosomes, membranes and organelles during meiosis in angiosperms. Phil Trans R Soc Ser B 277: 327–342

Dickinson HG, Moriarty J, Lawson J 1982. Pollen-pistil interaction in *Lilium longiflorum:* the role of the pistil in controlling pollen tube growth following cross- and self-pollinations. Proc R Soc Lond Ser B215:45–62

Dickinson HG, Crabbe MJC, Gaude T 1992. Sporophytic self-incompatibility system: S-gene products. Intl Rev Cytol 140: 525–561

Dieringer G, Cabrera L, Lara M, Loya L, Reyes-Castillot P 1999. Beetle pollination and floral thermogenicity in *Magnolia tamaulina* (Magnoliaceae). Intl J Plant Sci 160: 64–71

Di-Giovanni F, Kevan PG 1991. Factors affecting pollen dynamics and its importance to pollen contamination: a review. Can J For Res 21: 1155–1170

Dionne LA 1958. A survey of the methods for overcoming cross-incompatibility between certain species of the genus *Solanum*. Am Potato J 35: 422–423

Dionne LA 1963. Studies on the use of *Solanum acaule* as a bridge between *Solanum tuberosum* and the species

of the series *Bulbocastana, Cardiophylla* and *Pinnatisecta*. Euphytica 12: 263–269

Dirksen A, Malling H-J, Mosbech H, Soborg M, Biering I 1985. HEP versus PNU standardization of allergen extracts in skin prick testing. Allergy 40: 620–624

Dixon LK, Leaver CJ 1982. Mitochondrial gene expression and CMS in sorghum. Plant Mol Biol 1: 89–102

Dobson HEM 1988. Survey of pollen and pollen kitt lipids—chemical cues to flower visitors? Am J Bot 75: 170–182

Dobson HEM 1994. Floral volatiles in insect biology. In: Insect Plant Interactions, vol V. E Bernays (ed), CRC Press, Boca Raton, FL

Dobson HEM, Groth I, Bergstrom G 1996. Pollen advertisement: Chemical contrasts between whole-flower and pollen odors. Am J Bot 83: 877–885

Dodds PN, Bonig I, Du H, Rodin J, Anderson MA, Newbigin E, Clarke AE 1993. S-RNase gene in *Nicotiana alata* is expressed in developing pollen. Plant Cell 5: 1771–1782

Doughty J, Hedderson F, McCubbin A, Dickinson HG 1993. Interaction between a coating-borne peptide of the *Brassica* pollen grain and stigmatic S (self-incompatibility) locus-specific glycoproteins. Proc Natl Acad Sci USA 90: 467–471

Doughty J, Dixon S, Hiscock SJ, Willis AC, Perkin IAP, Dickinson HG 1998. PCP-A1, a defensin-like *Brassica* pollen coat protein that binds the S-locus glycoprotein, is the product of gametophytic gene expression. Plant Cell 10: 219–236.

Doussinault G, Delibas A, Sanchez-Monge R, Garcia-Olmedo F 1983. Transfer of a dominant gene for resistance to eyespot disease from a wild grass to hexaploid wheat. Nature 303: 698–700

Dreborg S 1989. Position paper on skin tests in Type 1 hypersensitivity. Allergy 44: (Supplement 10)

Driscoll CJ 1986. Nuclear male sterility system in seed production of hybrid varieties. Critical Rev Plant Sci 3: 227–256

Du Toit JT 1992. Winning by a neck. Nat Hist 8: 29–32

Ducker SC, Pettitt J, Knox RB 1978. Biology of Australian seagrasses: pollen development and submarine pollination in *Amphibolis antarctica* and *Thalassodendron ciliata* (Cymodoceaceae). Aust J Bot 26: 265–285

Dugger WM 1973. Functional aspects of boron in plants. In: Trace Elements in the Environment. EL Kothny (ed). Advances in Chemistry Series 123, Amer Chem Soc, Washington DC, pp 112–129

Dulberger R 1964. Flower dimorphism and self-incompatibility in *Narcissus tazetta* L. Evolution 18: 361–363

Dulberger R 1970. Floral dimorphism in *Anchusa hybrida* Ten. Israel J Bot 19: 37–41

Dulberger R 1974. Structural dimorphism of stigmatic papillae in distylous *Linum* species. Am J Bot 61: 238–243

Dulberger R 1992. Floral polymorphisms and their functional significance in heterostylous syndrome. In: Evolution and Function of Heterostyly. SCH Barrett (ed). Springer-Verlag, Berlin, pp. 41–84

Dulieu HL 1966. Pollination of excised ovaries and culture of ovules of *Nicotiana tabacum* L. Phytomorphology 16: 69–75.

Dumas C, Gaude T 1981. Stigma-pollen recognition and pollen hydration. Phytomorphology 31: 191–201

Dumas C, Knox RB 1983. Callose and determination of pistil viability and incompatibility. Theor Appl Genet 67: 1–10

Dumas C, Mogensen HL 1993. Gametes and fertilization: Maize as a model system for experimental embryogenesis in flowering plants. Plant Cell 5: 1337–1348

Dumas C, Kerhoas C, Gay G, Gaude T 1985. Water content, membrane state and pollen physiology. In: Biotechnology and Ecology of Pollen. DL Mulcahy, GB Mulcahy, F Ottaviano (eds). Springer-Verlag, Berlin, pp 333–337

Dumas C, Rougier M, Zandonella P, Ciampolini F, Cresti M, Pacini E 1978. The secretory stigma of *Lycopersicum peruvianum* Mill. ontogenesis and glandular activity. Protoplasma 96: 173–187

Dunbar A 1973. Pollen development in the *Eleocharis palustris* group (Cyperaceae). I. Ultrastructure and ontogeny. Bot Notiser 126: 197–254

Dunwell JM 1986. Pollen, ovule and embryo culture as tools in plant breeding. In: Plant Tissue Culture and Its Agricultural Applications. LA Withers, PG Alderson (eds). Butterworths, London, pp 375–404

Dupuis I 1992. In vitro pollination: A new tool for analyzing environmental stress. Intl Rev Cytol 140: 391–405

Dupuis I, Pace GM 1993. Gene transfer to maize male reproductive structure by particle bombardment of tassel primordia. Plant Cell Rep 12: 607–612

Duvick DN 1965. Cytoplasmic pollen sterility in corn. Advances Genet 13: 1–56

East EM, Mangelsdorf AJ 1925. A new interpretation of the hereditary behaviour of self-sterile plants. Proc Natl Acad Sci USA 11: 166–171

Echlin P 1971. The role of tapetum during microsporogenesis of angiosperms. In: Pollen: Development and Physiology. J. Heslop-Harrison (ed). Butterworths, London, pp 41–61

Echlin P 1972. The ultrastructure and ontogeny of pollen in *Helleborus foetidus* L. IV. Pollen grain maturation. J Cell Sci 11: 111–129

Edwardson JR 1970. Cytoplasmic male sterility. Bot Rev 36: 341–420

Edwardson JR, Warmke HE 1967. Fertility restoration in cytoplasmic male sterile *Petunia*. J Hered 58: 195–196

Edwardson JR, Bond DA, Christie RG 1976. Cytoplasmic sterility factors in *Vicia faba* L. Genetics 82: 443–449

Eenink AH 1974. Matromorphy in *Brassica oleracea* L. I. Terminology, parthenogenesis in Cruciferae and the formation and usability of matromorphic plants. Euphytica 23: 429–433

Eenink AH, Groenwold R, Dielman FL 1982. Resistance of lettuce (*Lactuca*) to the leaf aphid *Nosonovia risbis nigri*. I. Transfer of resistance from *L. virosa* to *L. sativa* by interspecific crosses and selection of resistance breeding lines. Euphytica 31: 291–300

Eigsti OJ, Dustin Jr 1955. Colchicine in Agriculture, Biology and Chemistry. Iowa State College Press, Cedar Falls, IA

El Murabaa AIM 1957. Effect of high temperature on incompatibility in radish. Euphytica 6: 268–270

El-Ghazaly G, Jensen WA 1986. Studies on the development of wheat (*Triticum aestivum*) pollen. 1. Formation of the pollen wall and Ubish bodies. Grana 25: 1–29

El-Ghazaly GA, Jensen WA 1990. Development of wheat (*Triticum aestivum*) pollen wall before and after effect of a gametocide. Can J Bot 68: 2509–2516

Elleman CJ, Dickinson HG 1996. Identification of pollen components regulating pollination-specific responses in the stigmatic papillae of *Brassica oleracea*. New Phytol 133: 197–205

Elleman CJ, Franklin-Tong V, Dickinson HG 1992. Pollination in species with dry stigmas: the nature of the early stigmatic response and the pathway taken by pollen tubes. New Phytol 121: 413–424

Elliott FC 1958. Plant Breeding and Cytogenetics. McGraw-Hill Book Co. Inc., New York, NY

Emerson S 1939. A preliminary survey of the *Oenothera organensis* population. Genetics 24: 524–537

Emsweller SL, Stuart NW 1948. Use of growth regulating substances to overcome incompatibilities in *Lilium*. Proc Am Soc Hort Sci 51: 581–589

Emsweller SL, Uhring J, Stuart NW 1960. The role of naphthalene acetamide and potassium gibberellate in overcoming self-incompatibility in *Lilium longiflorum*. Proc Am Soc Hortic Sci 75: 720–725

Endress SK 1994. Diversity and Evolutionary Biology of Tropical Flowers. Cambridge Univ Press, New York, NY

Engvall E, Pearlman P 1971. Enzyme-linked immunosorbent assay (ELISA): Qualitative assay of immunoglobulin G. Immunochemistry 8: 871–879

Engvild KC 1974. Plantlet ploidy and flower bud size in tobacco anther cultures. Hereiditas 76: 320–322

Erdelska O 1974. Contribution to the study of fertilization in the living embryo sac. In: Fertilization in Higher Plants. HF Linskens (ed). North-Holland Publ, Amsterdam, The Netherlands, pp 191–195

Erdtman G 1956. "Lo-analysis" and "Welcker's rule". Svensk Bot Tidsker 50: 1–7

Erdtman G 1966. Pollen Morphology and Plant Taxonomy, Angiosperms. Hafner, New York, NY

Erdtman G 1969. Handbook of Palynology. Hafner, New York.

Erickson L, Grant I, Beversdorf W 1986. Cytoplasmic male sterility in rapeseed (*Brassica napus* L.) 2. The role of mitochondrial plasmid. Theor Appl Genet 72: 151–157

Evans AM, Denward T 1955. Grafting and hybridization experiments in the genus *Trifolium*. Nature 175: 687–688

Eyster WH 1941. The induction of fertility in genetically self-sterile plants. Science 94: 144–145

Faegri K, van der Pijl L 1979. The Principles of Pollination Ecology. Pergamon Press, Oxford, UK (3rd ed.)

Faegri K, Iversen J 1989. Textbook of Pollen Analysis. K Faegri, PE Kaland, K Krzywinski (eds), John Wiley & Sons, New York, NY (4th ed.)

Fairy DT, Stoskopf NC 1975. Effects of granular ethephon on male sterility in wheat. Crop Sci 15: 29–32

Farkas J, Frank J 1982. Experience gained when using honeybee for pollen dispersion in hybrid sunflower seed production. Acta Agron Acad Scien Hungaricae 31: 267–270

Faure J-E, Mogensen HL, Dumas C, Lorz E, Kranz E 1993. Karyogamy after electrofusion of a single egg and sperm cell protoplasts from maize, cytological evidence and time course. Plant Cell 5: 747–755

Fearon CH, Cornish MA, Hayward MD, Lawrence MJ 1994. Self-incompatibility in ryegrass. X. Number and frequency of alleles in a natural population of *Lolium perenne*. Heredity 73: 262–264

Fedak G 1977. Haploid from barley x rye crosses. Can J Genet Cytol 19: 15–19

Fedak G 1978. Barley-wheat hybrids. In: Interspecific Hybridization in Plant Breeding. E. Sanchez-Monge, F Garcia-Omedo (eds). Proc VIIth Congress Eucarpia, Escuela Technica Superior de Inginieros Agronomos, Madrid, pp 261–267

Fedak G 1988. Haploids in *Triticum ventricasum* via intergeneric hybridization with *Hordeum bulbosum*. Can J Genet Cytol 25: 104–107

Feijo JA, Malho R, Obermeyer G 1995. Ion dynamics and its possible role during in vitro pollen germination and tube growth. Protoplasma 187: 155–167

Feinberg SM, Roran FL, Lichtenstein MR, Padnos E, Rappaport BZ, Sheldon J, Zeller M 1940. Oral pollen therapy in ragweed pollinosis. J Am Med Assoc 115: 23–29

Fennell A, Hauptmann R 1992. Electroporation and PEG delivery of DNA into maize microspores. Plant Cell Rep 11: 567–570

Fernando DD, Cass DD 1994. Plasmodial tapetum and pollen wall development in *Butomus umbellatus* (Butomaceae). Am J Bot 8: 1592–1600

Ferrari TE 1990. 'Enpollination' of honeybees with pre-collected pollen improves pollination of almond flowers. Amer Bee J 130: 801

Ferrari TE, Wallace DH 1975. Germination of *Brassica* pollen and expression of incompatibility in vitro. Euphytica 24: 757–765

Ferrie AMR, Palmer CE, Keller WA 1995. In vitro embryogenesis in plants. In: Plant Embryogenesis. TA Thorpe (ed). Kluwer Acad Publ, Dordrecht, pp 309–344

Fett WF, Paxton JD, Dickinson DB 1976. Studies on self-incompatibility response in *Lilium longiflorum*. Am J Bot 63: 1104–1108

Fitzgerald MA, Knox RB 1995. Initiation of primexine in freeze-substituted microspores of *Brassica campestris*. Sex Plant Reprod 8: 89–99

Fitzgerald MA, Calder DM, Knox RB 1993a. Development and initiation of cohesion between the individual pollen grains of *Acacia* polyads. Ann Bot 71: 51–59

Fitzgerald MA, Calder DM, Knox RB 1993b. Secretory events in the freeze-substiuted tapetum of orchid *Pterostylis concinna*. Plant Syst Evol (Suppl.) 7: 53–62

Fitzgerald MA, Barnes SH, Blackmore S, Calder DM, Knox RB 1994. Exine formation in the pollinium of *Dendrobium*. Protoplasma 179: 121–130

Flavell R 1974. A model for the mechanism of cytoplasmic male sterility in plants with special reference to maize. Plant Sci Lett 3: 259–263

Flavell R 1975. Inhibition of elecron transport in maize mitochondria by *Helminthosporium maydis* race T pathotoxin. Physiol Plant Pathol 6: 107–116

Foote HCC, Ride JP, Franklin-Tong VE, Walder EA, Lawrence MJ, Franklin FCH 1994. Cloning and expression of a distinctive class of self-incompatibility gene(s) from *Papaver rhoeas* L. Proc Natl Acad Sci USA 91: 2265–2269

Ford D, Baldo BA 1986. A reexamination of rye-grass (*Lolium perenne*) pollen allergens. Int Arch Allergy Appl Immunol 81: 193–203

Ford SA, Tovey ER, Baldo BA 1985. Identification of orchard grass (*Dactylis glomerata*) pollen allergens following electrophoretic transfer to nitrocellulose. Int Arch Allergy Appl Immunol 78: 15–21

Forde BG, Oliver RJ, Leaver DJ 1978. Variation in mitochondrial translation products associated with the male sterility cytoplasm in maize. Proc Natl Acad Sci USA 75: 3841–3845

Foroughi-Wehr B, Friedt W 1984. Rapid production of recombinant barley yellow mosaic virus resistant *Hordeum vulgare* lines by anther culture. Theor Appl Genet 67: 377–382

Foroughi-Wehr B, Wenzel G 1989. Androgenetic haploid production. IPTAC Newslett 58: 11–18

Franchi GG, Bellani L, Nepi M, Pacini E 1996. Types of carbohydrate reserves in pollen: Localization, systematic distribution and eco-physiological significance. Flora 191: 143–159

Franke WW, Herth W, van der Woude WJ, Morre DJ 1972. Tubular and filamentous structures in pollen tubes: possible involvement as guide elements in protoplasmic streaming and vectorial migration of secretory vesicles. Planta 105: 317–341

Frankel R 1973. The use of male sterility in hybrid seed production. In: Agricultural Genetics, R. Moav (ed), John Wiley & Sons, New York, NY, pp 85–94

Frankel R, Galun E 1977. Pollination Mechanisms, Reproduction and Breeding. Springer-Verlag, Berlin, Heidelberg, New York.

Franklin FCH, Lawrence MJ, Franklin-Tong VE 1995. Cell and molecular biology of self-incompatibility in flowering plants. Intl Rev Cytol 158: 233–237

Franklin-Tong VE, Franklin FCH 1992. Gametophytic self-incompatibility in *Papaver rhoeas* L. Sex Plant Reprod 5: 1–7

Franklin-Tong VE, Franklin FCH 2000. Self-incompatibility in *Brassica*: The elusive pollen S gene is identified! Plant Cell 12: 305–308

Franklin-Tong VE, Lawrence MJ, Franklin FCH 1988. An in-vitro bioassay for the stigmatic product of the self-incompatibility gene in *Papaver rhoeas* L. New Phytol 110: 109–118

Franklin-Tong VE, Lawrence MJ, Franklin FCH 1990. Self-incompatibility in *Papaver rhoeas* L.: inhibition of incompatible pollen tube growth is dependent on pollen gene expression. New Phytol 116: 319–324

Franklin-Tong VE, Knuth E, Marmey P, Lawrence MJ, Franklin FCH 1989. Characterization of a stigmatic component from *Papaver rhoeas* L. which inhibits the specific activity of self-incompatibility (S-) gene product. New Phytol 112: 307–315

Franklin-Tong VE, Atwal KK, Howell EC, Lawrence MJ, Franklin FCH 1991. Self-incompatibility in *Papaver rhoeas* L.: there is no evidence for the involvement of stigmatic ribonuclease activity. Plant Environ 14: 423–429

Franklin-Tong VE, Ride JP, Read ND, Trewavas AJ, Franklin FCH 1993. The self-incompatibility response in *Papaver rhoeas* is mediated by cytosolic-free calcium. Plant J 4: 163–177

Franssen-Verheijen AAW, Willemse MTM 1993. Micropylar exudate in *Gasteria* (Aloaceae) and its possible function in pollen tube growth. Am J Bot 80: 253–262

Frauen M 1987. Technical and economic aspects of seed production of hybrid varieties of rape. In: Hybrid Seed Production of Selected Cereals, Oil and Vegetable Crops. WP Feistritzer, AF Kelly (eds), FAO, Rome, pp 281–301

Free JB 1993. Insect Pollination of Crops. Academic Press, Orlando, FL (2nd ed.)

Freeman J, Noon L 1911. Further observation on the treatment of hay fever by hypodermic inoculations of pollen vaccine. Lancet ii : 814–817

Friedman WE 1993. The evolutionary history of the seed plant male gametophyte. TREE 8: 15–21

Fritz NK, Hannaman Jr RE 1989. Interspecific incompatibility due to stylar barriers in tuber-bearing and closely related non-tuber-bearing solanums. Sex Plant Reprod 2: 184–192

Fritz SE, Lukaszewski AJ 1989. Pollen longevity in wheat, rye and *Triticale*. Plant Breed 102: 31–34

Fu T, St P, Yang X, Yang G 1992. Overcoming self-incompatibility of *Brassica napus* by salt (NaCl) spray. Plant Breed 109: 255–258

Gabara B 1974. A possible role for the endoplasmic reticulum in exine formation. Grana 14: 16–22

Gadagkar R 1998. Honey: it's just the byproduct. Curr Sci 74: 95

Gaget M, Villar M, Dumas C 1989. The mentor pollen phenomenon in poplars: a new concept. Theor Appl Genet 78: 129–135

Galati BG 1996. Tapetum development in *Segittaria montevidensis* Cham. et Schlech (Alismataceae). Phytomorphology 46: 109–116

Galil J, Eisikowitch D 1968. On the pollination ecology of *Ficus sycomorus* in East Africa. Ecology 49: 259–269.

Ganders FR 1979. The biology of heterostyly. New Zealand J Bot 17: 607–636

Ganeshaiah HN, Shaanker RU 1988. Regulation of seed number and female incitation of male competition by a pH dependent proteinaceous inhibitor of pollen grain germination in *Leucaena leucocephala*. Oecologia 75: 110–113

Ganeshaiah KN, Shaanker RU, Shivashankar G 1986. Stigmatic inhibition of pollen grain germination—its implication for frequency distribution and seed number in pods of *Leucaena leucocephala* (Lam) de Wit. Oecologia 70: 568–572

Gartel W 1974. Micronutrients—their significance in vein nutrition with special regard to boron deficiency and toxicity. Weinberg und Keller 21: 435–508

Gasser CS, Robinson-Beers K 1993. Pistil development. Plant Cell 5: 1231–1239

Geitmann A, Snowman BN, Emons AMC, Franklin-Tong VE 2000. Alterations to the actin cytoskeleton of pollen tubes are induced by the self-incompatibility reaction in *Papaver rhoeas*. Plant Cell 12: 1239–1251

Gengenbach BG, Miller RJ, Koppe DE, Arntzen CJ 1973. The effect of toxin from *Helminthosporium maydis* (race T) on isolated corn mitochondrial swelling. Can J Bot 51: 2119–2125

Ghatnekar SD, Kulkarni AR 1978. Studies on pollen storage of *Capsicum frutescens* and *Ricinus communis*. J Palynol 4: 150–157

Ghosh S 1981. Studies on pollen-pistil interaction in *Linum* and *Zephyranthes*. PhD thesis, Univ Delhi, Delhi

Ghosh S, Shivanna KR 1980. Pollen-pistil interaction in *Linum grandiflorum*: Scanning electron microscopic observations and proteins of the stigma surface. Planta 149: 257–261

Ghosh S, Shivanna KR 1982a. Anatomical and cytochemical studies on the stigma and style in some legumes. Bot Gaz 143: 311–319

Ghosh S, Shivanna KR 1982b. Studies on pollen-pistil interaction in *Linum grandiflorum*. Phytomorphology 32: 385–395

Ghosh S, Shivanna KR 1983. RNA and protein synthesis during pollen germination and tube growth in *Zephyranthes*. Plant Physiol Biochem 10 (Special Vol): 1–10

Ghosh S, Shivanna KR 1984a. Structure and cytochemistry of the stigma and pollen-pistil interaction in *Zephyranthes*. Ann Bot 53: 91–105

Ghosh S, Shivanna KR 1984b. Interspecific incompatibility in *Linum*. Phytomorphology 34: 128–135

Gilissen LJW 1977. The influence of relative humidity on the swelling of pollen grains in vitro. Planta 137: 299–301

Gilissen LJ, Hoekstra FA 1984. Pollination induced corolla wilting in *Petunia hybrida*. Rapid transfer through the style of a wilting-induced substance. Plant Physiol 75: 496–498

Glendinning DR 1960. Selfing of self-incompatible cocoa. Nature 187: 170

Godwin H 1968. The origin of the exine. New Phytol 67: 667–676

Goh C-J, Arditti J 1985. Orchidaceae In: Handbook of Flowering. AH Halevy (ed). CRC Press, Boca Raton, Fl, Vol 1, pp 309–336

Goldberg RB, Beals TP, Sanders PM 1993. Anther development: Basic principles and potential applications. Plant Cell 5: 1217–1229

Goldman MHS, Goldberg RB, Mariani C 1995. Female sterile tobacco plants are produced by stigma-specific cell ablation. EMBO J 13: 2976–2984

Golynskaya EL, Bashkirova NV, Tomchuk NN 1976. Phytohemagglutinins of the pistil in *Primula* as possible proteins of generative incompatibility. Soviet Plant Physiol 23: 69–76

Goodman RM, Hauptli H, Crossway A, Knauf VC 1987. Gene transfer in crop improvement. Science 236: 48–54

Gopinathan MC, Babu CR, Shivanna KR 1986. Interspecific hybridization between rice bean (*Vigna umbellata*) and its wild relative (*Vigna minima*): Fertility-sterility relationships. Euphytica 35: 1017–1022

Gorska-Brylass A 1970. The "callose stage" of the generative cells in pollen grains. Grana 10: 21–30

Gowers S 1980. The production of hybrid oilseed rape using self-incompatibility. Cruciferae Newslett 5: 15–16

Graham RD 1975. Male sterility in wheat plants deficient in copper. Nature 254: 514–515

Graham RD 1986. Induction of male sterility in wheat (*Triticum aestivum* L.) using organic ligands with high specificity for binding copper. Euphytica 35: 621–630

Gray JE, McClure BA, Bonig I, Anderson MA, Clarke AE 1991. Action of the style product of the self-incompatibility gene of *Nicotiana alata* (S-RNase) in in vitro-grown pollen tubes. Plant Cell 3: 271–283

Griffith IJ, Lussier A, Garman R, Koury R, Yeung H, Pollock J 1993. CDNA cloning of Cry j I, the major allergen of *Cryptomeria japonica* (Japanese cedar). J Allergy Clin Immunol 91: 339

Griffith IJ, Smith PM, Pollock J, Theerakulpisut P, Avjioglu A, Davies S, Hough T, Singh MB, Simpson RJ, Ward LD, Knox RB 1991. Cloning and sequencing of *Lol p 1*, the major allergenic protein of rye-grass pollen. FEBS Letters 279: 210–215

Griggs WH, Iwakiri BT 1960. Orchard tests of beehive pollen dispensers for cross-pollination of almonds, sweet cherries and apples. Proc Amer Soc Hort Sci 75: 114–128

Grote M, Dolecek C, van Reen R, Valenta R 1994. Immunogold electron microscope localization of timothy grass (*Phlem pratense*) pollen major allergens *Phl p* I and *Phl p* V after anhydrous fixation in acrolein vapour. J Histochem Cytochem 42: 427–431

Grun P, Aubertin M 1966. The inheritance and expression of unilateral incompatibility in *Solanum*. Heredity 21: 131–138

Grun P, Radlow A 1961. Evolution of barriers to crossing of self-incompatible with self-compatible species of *Solanum*. Heredity 16: 137–143

Gudin S, Arena L 1991. Influence of the pH of the stigmatic exudate on male-female interaction in *Rosa hybrida* L. Sex Plant Reprod 4: 110–112

Guha S, Maheshwari SC 1964. In vitro production of embryos from anthers of *Datura*. Nature 204: 497

Guha S, Maheshwari SC 1966. Cell division and differentiation of embryos in the pollen grains of *Datura* in vitro. Nature 212: 97–98

Guilford WJ, Patterson TG, Vega RD, Fang L, Liang Y, Lewis HA, Labovitz JN 1992. Synthesis and pollen suppressant activity of phenylcinnoline-3-carboxylic acids. J Agric Food Chem 40: 2026–2032

Guilford WJ, Schneider DM, Labovitz J, Opella SJ 1988. High resolution solid state ^{13}C-NMR spectroscopy of sporopollenin from different plant taxa. Plant Physiol 86: 134–136

Guilluy C-M, Trick M, Heizmann P, Dumas C 1991. PCR detection of transcripts homologous to self-incompatibility gene in anthers of *Brassica*. Theoret. Appl. Genet. 82: 466–472

Gundimeda HR, Prakash S, Shivanna KR 1992. Intergeneric hybrids between *Enarthocarpus lyratus*, a wild species, and crop brassicas. Theor Appl Genet 83: 655–662

Gupta P, Shivanna KR, Mohan Ram HY 1996. Apomixis and polyembryony in guggul plant, *Commiphora wightii*. Ann Bot 78: 67–72

Gupta SC, Nanda K 1972. Occurrence and histochemistry of anther tapetal membrane. Grana Palynol 12: 99–104

Gurel A, Kontowski S, Nichterlein K, Friedt W 1991. Embryogenesis in microspore cultures of sunflower (*Helianthus annuus* L.). Helia 14: 123–128

Guries RP 1978. A test of the mentor pollen technique in the genus *Ipomoea*. Euphytica 27: 825–830

Gwyn JJ, Stelly DM 1989. Method to evaluate pollen viability of upland cotton: Tests with chromosome translocations. Crop Sci 29: 1165–1169

Hadley HH, Openshaw SJ 1980. Interspecific and intergeneric hybridization. In: Hybridization in Crop Plants. RW Fehr, HH Hadley (eds). Am Soc Agron, Crop Sci Soc Am, Madison, WI, pp 133–159

Hagemann R, Schroder MB 1989. The cytological basis of plastid inheritance in angiosperms. Protoplasma 152: 57–64

Hakansson G, Glemelius K, Bonnet HT 1990. Respiration in cells and mitochondria of male fertile and male sterile *Nicotiana* sp. Plant Physiol 93: 367–373

Hamdi S, Teller G, Louis JP 1987. Master regulatory genes, auxin level and sexual organogenesis in the dioecious plant *Mercurialis annua*. Plant Physiol 85: 393–399

Hamilton DA, Mascarenhas J 1997. Gene expression during pollen development. In: Pollen Biotechnology for Crop Production and Improvement. KR Shivanna, VK Sawhney (eds). Cambridge Univ Press, Cambridge, pp 40–58

Hamilton DA, Roy M, Rueda J, Sindhu RK, Sanford J, Mascarenhas JP 1992. Dissection of pollen-specific promoter from maize by transient information assays. Plant Mol Biol 18: 211–218

Hamilton RI, Leung E, Nichols C 1977. Surface contamination of pollen by plant viruses. Phytopathology 67: 395–399

Han H, Huang B 1987. Application of pollen-derived plants to crop improvement. Intl Rev Cytol 107: 293–313

Handa H 1993. Molecular genetic studies of mitochondrial genome in rapeseed (*Brassica napus* L.) in relation to cytoplasmic male sterility. Bull Natl Inst Agrobiol Resour 8: 47–105

Handa H, Nakajima K 1992. Different organization and altered transcription of the mitochondrial *atp6* gene in the male-sterile cytoplasm of rapeseed (*Brassica napus* L.). Curr Genet 21: 153–159

Hanna WW 1991. Apomixis in crop plants—cytogenetic basis and role in plant breeding. In: Chromosome Engineering in Plants: Genetics, Breeding, Evolution. Part A. PK Gupta, T Tsuchiya (eds). Elsevier, Amsterdam, The Netherlands, pp 229–242

Hanna WW, Bashaw EC 1987. Apomixis: its identification and use in plant breeding. Crop Sci 27: 1136–1139

Hanna WW, Towill LE 1995. Long-term pollen storage. Plant Breed Rev 13: 179–207

Hanna WW, Wells HD, Burton GW, Monson WG 1983. Long-term pollen storage of pearl millet. Crop Sci 23: 174–175

Hansen NJP, Anderson SB 1996. In vitro chromosome doubling potential of colchicine, oryzalin, trifluralin and APM in *Brassica napus* microspore culture. Euphytica 88: 159–164

Hanson DD, Hamilton DA, Travis JL, Bashe DM, Mascarenhas JP 1989. Characterization of a pollen-specific cDNA clone from *Zea mays* and its expression. Plant Cell 1: 173–179

Hanson MR 1991. Plant mitochondria mutations and male sterility. Ann Rev Genet 25: 461–486

Harder LD, Cruzen MB, Thomson JD 1993. Unilateral incompatibility and the effects of interspecific pollination for *Erythronium americanum* and *Erythronium albidum* (Liliaceae). Can J Bot 71: 353–358

Hardon JJ 1973. Assisted pollination in oil palm: a review. In: Advances in Oil Palm Cultivation. RL Wasti, DA Earp (eds). Proc Intl Oil Palm Conf. Incorporated Soc Planters, Kuala Lumpur, Malaysia, pp 184–195

Hardon JJ, Turner PD 1967. Observations on natural pollination in commercial plantings of oil palm (*Elaeis guineensis* Jasq) in Malaya. Exptl Agric 3: 105–116

Haring V, Gray JE, McClure BA, Anderson MA, Clarke AE 1990. Self-incompatibility: A self-recognition system in plants. Science 250: 937–941

Harlan JR 1976. Genetic resources in wild relatives of crops. Crop Sci 16: 329–333

Harlan JR, de Wet JMJ 1977. Pathways of genetic transfer from *Tripsacum* to *Zea mays*. Proc Natl Acad Sci USA 74: 3494–3497

Harrington JF 1970. Seed and pollen storage for conservation of plant gene resources. In: Genetic Resources in Plants—Their Exploration and Conservation. OH Frankel, E Bennett (eds). IBP Handbook No 11, Blackwell, Oxford, Edinburgh, pp 501–521

Hartley CWS 1988. The Oil Palm. Longman, London, New York (2nd ed.)

Harwood WA, Chen D-F, Creissen GP 1996. Transformation of pollen and microspores: A review. In: In vitro Haploid Production in Higher Plants. SM Jain, SK Sopory, RE Veillers (eds). Kluwer Acad Publ, Dordrecht, The Netherlands, vol 2, pp 53–71

Hathaway RL 1987. Inter-subgeneric hybridization in willows (*Salix*). In: IEA Task II—Biomass Growth and Production Technology. Tech Adv Comm Meet & Activity Workshop, Oulu, Finland and Uppsala, Sweden

Hauser EJP, Morrison JH 1964. Cytochemical reduction of nitroblue-tetrazolium as an index of pollen viability. Am J Bot 51: 748–753

Hawkes JG 1977. The importance of wild germplasm in plant breeding. Euphytica 26: 615–621

Hayman DL 1956. The genetic control of incompatibility in *Phalaris coerulescens*. Aust J Biol Sci 9: 321–331

Heberle-Bors E 1985. In vitro haploid formation from pollen. A critical review. Theor Appl Genet 71: 361–374

Heberle-Bors E 1989. Isolated pollen culture in tobacco: plant reproductive development in a nutshell. Sex Plant Reprod 2: 1–10

Hecht A 1961. Partial reduction of an incompatibility substance in the styles of *Oenothera organensis*. Genetics 46: 869

Herbert EW, Shimamuki M 1978. Chemical composition and nutritive value of bee-collected and bee-stored pollen. Apidiologia 9: 33–40

Hermsen JG 1992. Introductory considerations on distant hybridization. In: Distant Hybridization in Crop Plants. G Kallo, JB Chowdhury (eds). Springer-Verlag, Berlin, pp 1–14

Hermsen JG, Ramanna MS 1973. Double bridge hybrids of *Solanum bulbocastranum* and cultivated *Solanum tuberosum*. Euphytica 22: 457–466

Hernould M, Suharsono S, Litvak S, Arya A, Mouras A 1993. Male-sterility induction in transgenic tobacco plants with an unedited atp9 mitochondrial gene from wheat. Proc Natl Acad Sci USA 90: 2370–2374

Herrero M, Dickinson HG 1979. Pollen-pistil incompatibility in *Petunia hybrida*: changes in the pistil following compatible and incompatible intraspecific crosses. J Cell Sci 36: 1–18

Herrero M, Gascon M 1987. Prolongation of embryo sac viability in pear (*Pyrus communis*) following pollination or treatment with gibberellic acid. Ann Bot 60: 287–293

Herrero M, Arbeloa A 1989. Influence of the pistil on pollen tube kinetics in peach (*Prunus persica*). Am J Bot 76: 1441–1447

Herrero M, Arbeloa A, Gascon M 1988. Pollen-pistil interaction in the ovary in fruit trees. In: Sexual Reproduction in Higher Plants. M Cresti, P Gori, E Pacini (eds). Springer-Verlag, Heidelberg, pp 297–302

Herrick JF, Murray BG, Hammett KRW 1993. Barriers preventing hybridization of *Lathyrus odoratus* and *L. chrysanthus*. New Zeal J Hort Sci 21: 115–121

Herth W, Reiss HD, Hartmann F 1990. Role of calcium ions in tip growth of pollen tubes and moss protonema cells. In: Tip Growth in Plant and Fungal Cells. I.H. Heath (ed), Academic Press, San Diego, CA, pp 91–118

Heslop-Harrison J 1963. An ultrastructural study of pollen wall ontogeny in *Silene pendula*. Grana Palynol 4: 7–24

Heslop-Harrison J 1966. Cytoplasmic connections between angiosperm meiocytes. Ann Bot 30: 221–230

Heslop-Harrison J 1968. Synchronous pollen mitosis and the formation of the generative cell in massulate orchids. J Cell Sci 3: 457–466

Heslop-Harrison J (ed) 1971. Pollen: Development and Physiology. Butterworths, London.

Heslop-Harrison J 1972. Sexuality of angiosperms. In: Plant Physiology: a Treatise. FC Steward (ed), Academic Press, New York, NY, Vol VIC, pp 134–289

Heslop-Harrison J 1975. The Croonian Lecture 1974. The physiology of the pollen grain surface. Proc R Soc London Ser B 190: 275–299

Heslop-Harrison J 1978. Genetics and physiology of angiosperm incompatibility systems. Proc R Soc London Ser B 202: 73–92

Heslop-Harrison J 1979a. An interpretation of the hydrodynamics of pollen. Am J Bot 66: 737–743

Heslop-Harrison J 1979b. Pollen-stigma interaction in grasses: a brief review. New Zealand J Bot 17: 537–546

Heslop-Harrison J 1979c. Aspects of the structure, cytochemistry and germination of the pollen of rye (*Secale cereale* L.). Ann Bot 44 (Suppl): 1–47

Heslop-Harrison J 1979d. Pollen walls as adaptive systems. Ann Mo Bot Gard 66: 813–829

Heslop-Harrison J 1983. Self-incompatibility: phenomenology and physiology. Proc R Soc London Ser B 218: 397–413

Heslop-Harrison J 1988. Pollen germination and pollen tube growth. Intl Rev Cytol 107: 1–78

Heslop-Harrison J, Mackenzie A 1967. Autoradiography of soluble (2-¹⁴C)-thymidine derivatives during meiosis and microsporogenesis in *Lilium* anthers. J Cell Sci 2: 387–400

Heslop-Harrison J, Dickinson HG 1969. Time relationships of sporopollenin synthesis associated with tapetum and microspores in *Lilium*. Planta 84: 199–214

Heslop-Harrison J, Heslop-Harrison Y 1970. Evaluation of pollen viability by enzymatically induced fluorescence; intracellular hydrolysis of fluorescein diacetate. Stain Technol 45: 115–120

Heslop-Harrison J, Heslop-Harrison Y 1975. Enzymic removal of proteinaceous pellicle of the stigma papilla prevents pollen tube entry in the Caryophyllaceae. Ann Bot 39: 163–165

Heslop-Harrison J, Helsop-Harrison Y 1982a. The specialized cuticle of the receptive surfaces of angiosperm stigmas. In: The Plant Cuticle. DF Cutler, KL Alvin, CE Price (eds). Linnean Soc Symp no. 10. Academic Press, London, pp 99–119

Heslop-Harrison J, Heslop-Harrison Y 1982b. Pollen-pistil interaction in the Leguminosae. Constituents of the stylar fluid and stigma secretion of *Trifolium pratense* L. Ann Bot 49: 729–735

Heslop-Harrison J, Heslop-Harrison Y 1985a. Surfaces and secretions in the pollen-stigma interaction: A brief review. J Cell Sci Suppl 2: 287–300

Heslop-Harrison J, Heslop-Harrison Y 1985b. Germination of stress tolerant eucalyptus pollen. J Cell Sci 73: 135–157

Heslop-Harrison J, Heslop-Harrison Y 1986. Pollen tube chemotropism: fact or delusion? In: Biology of

Reproduction and Cell Motility in Plants and Animals. M Cresti, R Dallai (eds), Univ Siena, Siena, Italy, pp 169–174

Heslop-Harrison J, Heslop-Harrison Y 1987. Pollen-stigma interaction in the grasses. In: Grass Systematics and Evolution. TS Sodestrorm, KH Hilu, CS Cambell, ME Barkworth (eds), Smithsonian Inst Press, Wash, pp 133–142

Heslop-Harrison J, Heslop-Harrison Y 1989. Myosin associated with the surface of organelles, vegetative nuclei and generative cells of angiosperm pollen grains and tubes. J Cell Sci 94: 319–325

Heslop-Harrison J, Heslop-Harrison Y 1992. Intracellular motility, the actin cytoskeleton and germinability in the pollen of wheat (*Triticum aestivum* L.). Sex Plant Reprod .5: 247–255

Heslop-Harrison J, Knox RB, Heslop-Harrison Y 1974. Pollen wall proteins: exine-held fractions associated with the incompatibility response in Cruciferae. Theor Appl Genet 44: 133–137

Heslop-Harrison J, Heslop-Harrison Y, Shivanna KR 1984. The evaluation of pollen quality and a further appraisal of the fluorochromatic (FCR) test procedure. Theor Appl Genet 67: 367–375

Heslop-Harrison J, Heslop-Harrison Y, Knox RB, Howlett B 1973. Pollen wall proteins: gametophytic and sporophytic fractions in the pollen walls of the Malvaceae. Ann Bot 37: 403–412

Heslop-Harrison J, Knox RB, Heslop-Harrison Y, Mattsson O 1975. Pollen wall proteins: emission and role in incompatibility responses. In: The Biology of the Male Gamete. JG Duckett, PS Racey (eds). Biol J Linn Soc 7 (Suppl): 189–202

Heslop-Harrison JS, Heslop-Harrison Y, Reger BJ 1987. Anther filament extension in *Lilium*: potassium ion movement and some anatomical features. Ann Bot 59: 505–515

Heslop-Harrison Y 1977. The pollen-stigma interaction. Pollen tube penetration in *Crocus*. Ann Bot 41: 913–922

Heslop-Harrison Y 1981. Stigma characteristics and angiosperm taxonomy. Nordic J Bot 1: 401–420

Heslop-Harrison Y, Shivanna KR 1977. The receptive surface of the angiosperm stigma. Ann Bot 41: 1233–1258

Heslop-Harrison Y, Heslop-Harrison J 1982. Pollen-stigma interaction in the Leguminosae: the secretory system of the style in *Trifolium pratense* L. Ann Bot 50: 635–645

Heslop-Harrison Y, Reger BJ, Heslop-Harrison J 1984. The pollen-stigma interaction in the grasses. 8. The stigma ('silk') of *Zea mays* L. as host to the pollens of *Sorghum bicolor* (L.) Moench and *Pennisetum americanum* (L.) Leike. Acta Bot Neerl 33: 205–227

Heslop-Harrison Y, Heslop-Harrison J, Shivanna KR 1981. Heterostyly in *Primula*. 1. Fine structural and cytochemical features of the stigma and style in *P. vulgaris*. Protoplasma 107: 171–188

Heslop-Harrison Y, Heslop-Harrison J, Reger BJ 1985a. The pollen-stigma interaction in the grasses. 7. Pollen tube guidance and the regulation of tube number in *Zea mays* L. Acta Bot Neerl 34: 193–211

Heslop-Harrison Y, Reger BJ, Heslop-Harrison J 1985b. Wide hybridization: Pollination of *Zea mays* L. by *Sorghum bicolor* (L.) Moench. Theor Appl Genet 70: 252–258

Hess D 1980. Investigations on the intra- and inter-specific transfer of anthocyanin genes using pollen as vectors. J Plant Physiol 98: 321–337

Hess D 1987. Pollen based techniques in genetic manipulation. Int Rev Cytol 107: 367–395

Hess D, Wagner G 1974. Induction of haploid parthenogenesis in *Mimulus luteus* by in vitro pollination with foreign pollen. Z Pflanzenphysiol 72: 466–468

Hesse M 1981. Pollekitt and viscin threads: their role in cementing pollen grains. Grana 20: 145–152

Hesse M 1984. An exine model for viscin threads. Grana 29: 69–75

Higashiyama T, Kuroiwa H, Kawano S, Kuroiwa T 1997. Kinetics of double fertilization in *Torenia fournieri* based on direct observations of the naked embryo sac. Planta 203: 101–110

Higashiyama T, Kuroiwa H, Kawano S, Kuroiwa T 1998. Guidance in vitro of the pollen tube to the naked embryo sac of *Torenia fournieri*. Plant Cell 10: 2019–2031

Higashiyama T, Yabe S, Sasaki N, Nishimura Y, Miagishima S, Kuroiwa H, Kuroiwa T 2001. Pollen tube attraction by the synergid cell. Science 293: 1480–1483.

Hill JP, Lord EM 1987. Dynamics of pollen tube growth in wild raddish, *Raphanus raphanastrum* (Brassicaceae) II Morphology, histochemistry and ultrastructure of the transmitting tissue, and the path of pollen tube growth. Am J Bot 74: 988–997

Hinata K, Nishio T 1981. Con A-peroxidase method: an improved procedure for staining S-glycoproteins in cellulose-acetate electrofocussing in crucifers. Theor Appl Genet 60: 281–283

Hinata K, Watanabe M, Torima K, Isogai A 1993. A review of recent studies on homomorphic self-incompatibility. Int Rev Cytol 143: 257–296

Hirst JM 1952. An automatic volumetric spore trap. Ann Appl Biol 39: 257–265

Hiscock SJ, Doughty J, Dickinson HG 1995. Synthesis and phosphorylation of pollen proteins during the pollen-stigma interaction in self-compatible *Brassica napus* L. and self-incompatible. *Brassica oleracea* L. Sex Plant Reprod 8: 345–353

Hodgkin T 1983. A medium for germinating *Brassica* pollen in vitro. Cruciferae Newslett 8: 62.

Hodgkin T, Lyon GD 1983. Detection of pollen germination inhibitors in *Brassica oleracea* tissue extracts. Ann Bot 52: 781–789

Hodgkin T, Lyon GD, Dickinson HG 1988. Recognition in flowering plants: A comparison of the *Brassica* self-incompatibility system and plant pathogen interactions. New Phytol 110: 557–569

Hoefert LL 1969. Fine structure of sperm cells in pollen grains of *Beta*. Protoplasma 68: 237–240

Hoekstra FA 1979. Mitochondrial development and activity of binucleate and trinucleate pollen during germination in vitro. Planta 145: 25–36

Hoekstra FA 1984. Imbibitional chilling injury in pollen. Plant Physiol 74: 815–821

Hoekstra FA, Bruinsma J 1975a. Viability of Compositae pollen: germination in vitro and influence of climatic conditions during dehiscence. Z Pflanzenphysiol 76: 36–43

Hoekstra FA, Bruinsma J 1975b. Respiration and vitality of binucleate and trinucleate pollen. Physiol Plant 34: 221–225

Hoekstra FA, Bruinsma J 1978. Reduced independence of the male gametophyte in angiosperm evolution. Ann Bot 42: 759–762

Hoekstra FA, Bruinsma J 1979. Protein synthesis of binucleate and trinucleate pollen and its relationship to tube emergence and growth. Planta 146: 559–566

Hoekstra FA, Bruinsma J 1980. Control of respiration of binucleate and trinucleate pollen under humid conditions. Physiol Plant 48: 71–77

Hoekstra FA, Weges R 1986. Lack of control by early pistillate ethylene of the accelerated wilting of *Petunia hybrida* flowers. Plant Physiol 80: 403–408

Hoekstra FA, Crowe LM, Crowe JH 1989. Differential desiccation sensitivity of corn and *Pennisetum* pollen linked to their sucrose contents. Plant Cell Environ 12: 83–91

Hoekstra S, van Zijderveld MH, Louwerse JD, Heidekamp F, van der Mark F 1992. Anther and microspore culture of *Hordeum vulgare* L. cv Igri. Plant Sci 86: 89–96

Hogenboom NG 1972. Breaking breeding barriers in *Lycopersicon*. Breakdown of self-incompatibility in *L. peruvianum* (L.) Mill. Euphytica 21: 228–243

Hogenboom NG 1975. Incompatibility and incongruity. Two different mechanisms for the non-functioning of intimate partner relationships. Proc R Soc London Ser B 188: 361–375

Holden MJ, Sze H 1984. *Helminthosporium maydis* T-toxin increased membrane permiability to Ca^{2+} in susceptible corm mitochondria. Plant Physiol 75: 235–237

Holm PB, Knudsen S, Mouritzen P, Negri D, Olsen FL, Roue C 1994. Regeneration of fertile barley plants from mechanically isolated protoplasts of the fertilized egg cell. Plant Cell 6: 531–543

Holm SN 1966. The utilization and management of bumble bees for red clover and alfalfa seed production. Ann Rev Entomol 11: 155–182

Holman RM, Brubaker F 1926. On the longevity of pollen. Univ Calif Publ Bot 13: 179–204 (cited from Johri BM, Vasil IK 1961)

Hopping ME, Jerram EM 1980a. Supplementary pollination of tree fruits. 1. Development of suspension media. New Zealand J Agric Res 23: 509–515

Hopping ME, Jerram EM 1980b. Supplementary pollination of tree fruits. II. Field trials on Kiwi-fruit and Japanese plums. New Zealand J Agric Res 23: 517–521

Hormaza JI, Herrero M 1992. Pollen selection. Theor Appl Genet 83: 663–672

Horn R, Kohler H, Zetsche K 1991. A mitochondrial 16 kda protein is associated with cytoplasmic male sterility in sunflower. Plant Mol Biol 17: 29–36

Horner HT Jr, Pearson CB 1978. Pollen wall and aperture development in *Helianthus annuus*. Am J Bot 65: 293–309

Horner M, Street FE 1978. Pollen dimorphism—origin and significance in pollen plant formation by anther culture. Ann Bot 42: 763–771

Hotta Y, Stern H 1971. Analysis of DNA synthesis during meiotic prophase in *Lilium*. J Mol Biol 55: 337–355

Hotta Y, Stern H 1974. DNA scission and repair during pachytene in *Lilium*. Chromosoma 46: 279–296

Howell CJ, Slater AJ, Knox RB 1993. Secondary pollen presentation in angiosperms and its biological significance. Aust J Bot 41: 417–438

Howlett BJ, Knox RB, Paxton JH, Heslop-Harrison J 1975. Pollen wall proteins: physicochemical characterization and role in self-incompatibility in *Cosmos bipinnatus*. Proc R Soc London Ser B 188: 167–182

Hrabetova E, Tupy J 1964. The growth effect of some sugars and their metabolism in pollen tubes. In: Pollen Physiology and Fetilization. HF Linskens (ed). North-Holland Publ., Amsterdam, The Netherlands, pp 95–101

Hu H, Huang B 1987. Application of pollen-derived plants to crop improvement. Intl Rev Cytol 107: 397–420

Hu S-Y, Yu H-S 1988. Preliminary observations on the formation of male germ unit in pollen tubes of *Cyphomandra betacea* Sendl. Protoplasma 147: 55–63

Huang B 1986. Ultrastructural aspects of pollen embryogenesis in *Hordeum, Triticum* and *Paeonia*. In: Haploids in Higher Plants. H Hu, H Yang (eds), Springer-Verlag, Berlin, pp 91–117

Huang B 1992. Genetic manipulation of microspores and microspore-derived embryos. In Vitro Cell Devel Biol 28P: 53–58

Huang B, Sunderland N 1982. Temperature-stress pretreatment in barley anther culture. Ann Bot 49: 77–88

Huang B, Keller WA 1989. Microspore culture technology. J Tissue Culture Methods 12: 171–178

Huang B, Swanson EB, Baszcyznski CL, MacRae WD, Bardour E, Armavil V, Wobe L, Arnoldo M, Rozakis S, Westecott M, Keats RF, Kemble R 1991. Application of microspore culture to canola improvement. In: Proc 8th Intl Rapeseed Cong. DJ McGregor (ed). Canola Council, Canada, pp 298–303

Huang B-Q, Russell SD 1992. Female germ unit: organization, reconstruction and isolation. Intl Rev Cytol 140: 233–293

Huang J, Lee SH, Lin C, Medici R, Hack E 1990. Expression in yeast of the T-URF 13 protein from Texas male-sterile mitochondria confers sensitivity to methomyl and to Texas-cytoplasm-specific fungal toxins. EMBO J 9: 339–347

Huang S, Lee HS, Karunanandaa B, Kao T-h 1994. Ribonuclease activity of *Petunia inflata* S proteins is essential for rejection of self-pollen. Plant Cell 6: 1021–1028

Hughes WG, Bodden JJ 1978. Production and yield assessment of F₁ hybrid wheats using ethephon-induced male sterility. Z Pflanzenzucht 81: 17–23

Hulskamp M, Kopozak SD, Horejsi TM, Kihl BK, Pruitt RE 1995a. Identification of genes required for pollen-stigma recognition in *Arabidopsis thaliana*. Plant J 8: 703–714

Hulskamp M, Schneitz K, Pruitt RE 1995b. Genetic evidence for a long range activity that directs pollen tube guidance in *Arabidopsis*. Plant Cell 7: 57–64

Hyde HA 1969. Aeropalynology in Britain—an outline. New Phytol 68: 579–590

Hyde HA 1972. Atmospheric pollen and spores in relation to allergy I. Clin Aller 2: 153–179

Inouye D, Dudash MR, Fenster CB 1994. A model and lexicon for pollen fate. Amer J Bot 81: 1517–1530

Ikeda S, Nasrallah JB, Dixit R, Preiss S, Nasrallah ME 1997. An aquaporin-like gene in the *Brassica* self-incompatibility response. Science 276: 1564–1566.

Ioerger TR, Clark AG, Kao T-h 1990. Polymorphism at the self-incompatibility locus in Solanaceae predates speciation. Proc Natl Acad Sci USA 87: 9732–9735

Ipsen H, Lowenstein H 1983. Isolation and immunochemical characterization of the major allergen of birch pollen (*Betula verrucosa*). J Allergy Clin Immunol 72: 150–159

Iqbal MCM, Mollers C, Robbelen G 1994. Increased embryogenesis after colchicine treatment of microspore culture of *B. napus*. J Plant Physiol 143: 222–226

Islam AS 1964. A rare hybrid combination through application of hormone and embryo culture. Nature 201: 320

Ito M, Stern H 1967. Studies of meiosis in vitro. I. In vitro culture of meiotic cells. Dev Biol 16: 26–53

Ito M, Takegami MH 1982. Commitment of mitotic cells to meiosis during the G2 phase of premeiosis. Plant Cell Physiol 23: 943–952

Ito N 1978. Male sterility caused by cooling treatment at the young microspore stage in rice plants. XVI. Changes in carbohydrates, nitrogenous and phosphorous compounds in rice anthers after cooling treatment. Jpn J Crop Sci 47: 318–323

Iwanami Y 1959. Physiological studies of pollen. J Yokohoma City Uuiv 116: 1–137

Iwanami Y 1972. Retaining the viability of *Camellia japonica* pollen in various organic solvents. Plant Cell Physiol 13: 1139–1141

Iwanami Y 1975. Absolute dormancy of pollen induced by soaking in organic solvents. Protoplasma 84: 267–271

Iwanami Y 1984. The viability of pollen grains of lily (*Lilium auratum*) and the eggs of the brine shrimp (*Artemia salina*) soaked in organic solvents for 10 years. Experientia 40: 568–569

Iwanami Y, Nakamura N 1972. Storage in an organic solvent as a means for preserving viability of pollen grains. Stain Technol 47: 137–139

Iyer RD, Raina SK 1972. The early ontogeny of embryoids and callus from pollen and subsequent organogenesis in anther cultures of *Datura metal* and rice. Planta: 104: 146–156

Izhar S 1975. The timing of temperature effect on microsporogenesis in cytoplasmic male sterile *Petunia*. J Hered 66: 313–314

Izhar S, Frankel R 1971. Mechanism of male sterility in *Petunia*: the relationship between pH, callase activity in the anthers, and the breakdown of the microsporogenesis. Theor Appl Genet 41: 104–108

Izhar S, Frankel R 1973. Mechanism of male sterility in Petunia: II. Free amino acids in male fertile and male sterile anthers during microsporogenesis. Theor Appl Genet 43: 13–17

Jackson JF, Linskens HF 1990. Bioassays for incompatibility. Sex Plant Reprod 3: 207–212

Jaffe LA, Weisenseel MH, Jaffe LF 1975. Calcium accumulations within the growing tips of pollen tubes. J Cell Biol 67: 488–492

Jaggi KS, Gangal SV 1987. Isolation and identification of pollen allergens of *Artemisia scoparia*. J Allergy Clin Immunol 80: 562–572

Jahne A, Becker D, Brettschneider R, Lorz H 1994. Regeneration of transgenic microspore-derived fertile barley. Theor Appl Genet 89: 525–533

Jahnen W, Lush WM, Clarke AE 1989. Inhibition of in vitro pollen tube growth by isolated S-glycoproteins of *Nicotiana alata*. Plant Cell 1: 501–510

Jain A 1989. Storage of *Crotalaria retusa* pollen in organic solvents: Efficacy of the solvents and membrane changes in relation to viability. Ph.D. thesis, Univ Delhi, Delhi

Jain A, Shivanna KR 1987a. Storage of pollen grains in organic solvents: Effects of solvents on pollen viability and membrane integrity. J Plant Physiol 132: 499–502

Jain A, Shivanna KR 1987b. Storage of pollen grains in organic solvents: Effect of organic solvents on leaching of phospholipids and its relationship to pollen viability. Ann Bot 61: 325–330

Jain A, Shivanna KR 1989. Loss of viability during storage is associated with changes in membrane phospholipid. Phytochemistry 28: 999–1002

Jain A, Shivanna KR 1990. Membrane state and pollen viability during storage in organic solvents. In: Proc Intl Cong Plant Physiology. SK Sinha, PV Sane, SC Bhargava, PK Agrawal (eds), Soc Plant Physiol Biochem, New Delhi, pp 1341–1349

Jain SM, Sopory SK, Veilleux RE 1996. In Vitro Haploid Production in Higher Plants. Kluwer Acad. Publ., Dordrecht, The Netherlands, vols 1–5

Janson J, Reinders MC, van Tuyl JM, Keijzer CJ 1993. Pollen tube growth in *Lilium longiflorum* following different pollination techniques and flower manipulations. Acta Bot. Neerl 42: 461–472

Janssen AWB, Hermsen JCT 1980. Estimating pollen viability in *Solanum* species and haploids. Euphytica 25: 577–586

Jardinaud M-F, Souvre A, Alibert G 1993. Transient GUS gene expression in *Brassica napus* electroporated microspores. Plant Sci 93: 177–184

Jardinaud M-F, Souvre A, Beckert M, Alibert G 1995. Optimization of DNA transfer and transient β-glucuronidase expression in electroporated maize (*Zea mays* L.) microspores. Plant Cell Rep 15: 55–58

Jauh GY, Lord EM 1995. Movement of tube cell in the lily style in the absence of the pollen grain and the spent pollen tube. Sex Plant Reprod 8: 168–172

Jay SC 1986. Spatial management of honeybees on crops. Ann Rev Entomol 31: 49–65

Jensen WA 1973. Fertilization in flowering plants. Bioscience 23: 21–27

Jensen WA, Ashton MF 1981. Synergid pollen tube interaction in cotton. XIII Intl Bot Cong, Sydney, Abstr p 61

Jensen WA, Fischer DB 1968. Cotton embryogenesis: the entrance and discharge of the pollen tube in the embryo sac. Planta 78: 158–183

Jensen WA, Ashton M, Heckard LR 1974. Ultrastructural studies of the pollen of subtribe Castilleiinae, family Scrophulariaceae. Bot Gaz 135: 210–218

Jewell AW, Murray BG, Alloway BJ 1988. Light and electron microscopic studies on pollen development in barley (*Hordeum vulgare* L.) grown under copper sufficient and deficient conditions. Plant Cell Environ 11: 273–281

Johns C, Lu M, Lyznik A, Mackenzie S 1992. A mitochondrial DNA sequence is associated with abnormal pollen development in cytoplasmic male sterile bean plant. Plant Cell 4: 435–449

Johnson KD, Lucken KA 1986. Screening hard red spring wheat genotypes for cross-pollinated seed set potential using the CHA, SD84811. In: Agron Abstr, Ann Meetings, Am Soc Agron, Nov 30-Dec 5, New Orelans, LA, p 67

Johri BM (ed) 1984. Embryology of Angiosperms. Springer-Verlag, Berlin, Heidelberg, New York

Johri BM 1992. Haustorial role of pollen tubes. Ann Bot 70: 471–475

Johri BM, Vasil IK 1961. Physiology of pollen. Bot Rev 27: 325–381

Johri BM, Shivanna KR 1977. Physiology of 2- and 3-celled pollen. Phytomorphology 27: 98–106

Jones CE, Little RL (eds) 1983. Handbook of Experimental Pollination Biology. Van Nostrand Reinhold Co, New York, NY

Jones DF 1918. The effect of inbreeding and cross breeding upon development. Agric Exptl Stn Bull 207: 1–100

Jones HA, Davis GN 1944. Inbreeding and heterosis and their relation to the development of new varieties of onion. US Dept Agric Tech Bull 874: 1–28

Kakizaki Y 1930. Studies on the genetics and physiology of self- and cross-incompatibility in the common cabbage. Jpn J Bot 5: 133–208

Kalloo G, 1992. Utilization of wild species. In: Distant Hybridization in Crop Plants. G Kalloo, JB Chowdhury (eds). Springer-Verlag, Berlin, pp 149–167

Kalloo G, Chowdhury JB (eds) 1992. Distant Hybridization in Crop Plants. Springer-Verlag, Berlin, Heidelberg, New York

Kamboj AK, Jackson JF 1986. Self-incompatibility alleles control a low molecular weight basic protein in pistils of *Petunia hybrida*. Theor Appl Genet 71: 815–900

Kandasamy MK, Vivekananda M 1983. Effect of stigmatic exudate and pistil extracts of *Cajanus cajan* (L.) Mills on in vitro pollen tube growth on *Cyamopsis tetragonoloba* Taub. Plant Sci Lett 32: 343–348

Kandasamy MK, Kappler R, Kristen U 1988. Plasma tubules in the pollen tubes of *Nicotiana sylvestris*. Planta 173: 35–41.

Kandasamy MK, Paolillo DJ, Farady CD, Nasrallah JB , Nasrallah ME 1989. The S-locus specific glycoproteins in *Brassica* accumulate in the cell wall of developing stigma papillae. Devel Biol 135: 462–472

Kanno T, Hinata K 1969. An electron microscopic study of the barrier against pollen tube growth in self-incompatible Cruciferae. Plant Cell Physiol 10: 213–216

Kanta K, Maheshwari P 1963a. Intraovarian pollination in some Papaveraceae. Phytomorphology 13: 215–229

Kanta K, Maheshwari P 1963b. Test-tube fertilization in some angiosperms. Phytomorphology 13: 230–237

Kanta K, Rangaswamy NS, Maheshwari P 1962. Test-tube fertilization in a flowering plant. Nature 194: 1214–1217

Kapil RN, Bhatnagar AK 1975. A fresh look at the process of double fertilizaiton in angiosperms. Phytomorphology 25: 334–368

Kappler R, Kristen U 1987. Photometric quantification of in vitro pollen tube growth: a new method suited to determine the cytochemistry of various environmental substances. Environ Exptl Bot 27: 305–307

Kappler R, Kristen U 1988. Photometric quantification of water-insoluble polysaccharides produced by in vitro grown pollen tubes. Environ Exptl Bot 28: 33–36

Karunanandaa B, Huang S, Kao T-h 1994. Carbohydrate moiety of the *Petunia inflata* S_3 protein is not required for self-incompatibility interaction between pollen and pistil. Plant Cell 6: 1933–1940

Kasha KJ (ed) 1974. Haploids in Higher Plants: Advances and Potential. Univ Guelph, Ontario, CAN

Kasha KJ, Kao KN 1970. High frequency haploid production in barley (*Hordeum vulgare* L.). Nature 225: 874–876

Kasha KJ, Ziauddin A, Cho UH 1990. Haploids in cereal improvement: anther and microspore culture. In: Gene Manipulation in Plant Improvement. JP Gustafson (ed). Plenum Press, New York, NY, vol II, pp 213–235

Kathuria P, Ganeshaiah KN, Shaanker U, Vasudeva R 1995. Is there dimorphism for style lengths in monoecious figs? Curr Sci 68: 1047–1049

Kaul MLH 1988. Male Sterility in Higher Plants. Springer-Verlag, Berlin, Heidelberg, New York

Kaul V, Theunis CH, Palser BF, Knox RB, Williams EG 1987. Association of the generative cell and vegetative nucleus in pollen tubes of *Rhododendron*. Ann Bot 59: 227–235

Kearns CA, Inouye DW 1993. Techniques for Pollination Biologists. Univ Press Colorado, Niwot Ridge, CO

Keijzer CJ 1987a. The processes of anther dehiscence and pollen dispersal. I The opening mechanism of longitudinally dehiscing anthers. New Phytol 105: 487–498

Keijzer CJ 1987b. The processes of anther dehiscence and pollen dispersal. II. The formation and transfer mechanism of pollenkitt, cell wall development of the loculus tissues and a function of orbicules in pollen dispersal. New Phytol 105: 499–507

Keijzer CJ 1999. Mechanism of angiosperm pollen dehiscence, a historical review. In: Anther and Ovule: From Biology to Biotechnology. C Clement, E Pacini, J-C Audran (eds). Springer-Verlag, Berlin, pp 55–67

Keijzer CJ, Hoek HIS, Willemse MTM 1987. The process of anther dehiscence and pollen dispersal. 3. The dehydration of the filament tip and the anther in some monocotyledonous species. New Phytol 106: 281–287

Keller WA, Armstrong KC 1978. High frequency production of microspore-derived plants from *Brassica napus* anther cultures. Z Pflanzenzucht 80: 100–108

Keller WA, Armstrong KC 1979. Stimulation of embryogenesis and haploid production in *Brassica campestris* anther cultures by elevated temperature treatments. Theor Appl Genet 55: 65–67

Keller WA, Armstrong KC 1981. Dihaploid plant production by anther culture in autotetraploid narrow stem kale (*Brassica oleracea* L. var *acephala* DC). Can J Genet Cytol 23: 259–265.

Keller WA, Arnison PG, Cardy BK 1987. Haploids from gametophytic cells—recent developments and future prospects. In: Plant Tissue and Cell Culture. CE Green, DA Somers, WP Hackett, DD Biesboer (eds.). Allen R Liss, New York, NY, pp 233–241

Kemble RJ, Gunn RE, Flavell RB 1980. Classification of normal and male sterile cytoplasms in maize. II. Electrophoretic analysis of DNA species in mitochondria. Genetics 95: 451–458

Kendall WA, Taylor NL 1969. Effects of temperature on pseudocompatibility in *Trifolium pratense*. Theor Appl Genet 39: 123–126

Kennell JC, Pring DR 1989. Initiation and processing of atp6, T-urf13, and ORF221 transcripts from mitochondria of T-cytoplasm maize. Mol Gen Genet 216: 16–24

Kenrick J, Knox RB 1981. Post-pollination exudate from stigmas of *Acacia* (Mimosaceae). Ann Bot 48: 103–106

Kenrick J, Knox RB 1989. Pollen-pistil interactions in Leguminosae (Mimosoideae). In: Advances in Legume Biology. CH Stirton, JL Zarucchi (eds). Monogr Syst Bot Missouri Bot Gard 29: 127–156.

Kerhoas D, Gay G, Dumas C 1987. Multidisciplinary approach to the study of the plasma membrane of *Zea mays* pollen during controlled dehydration. Planta 171: 1–10

Kerhoas C, Knox RB, Dumas C 1983. Specificity of the callose response in stigmas of *Brassica*. Ann Bot 52: 597–602

Kevan PG 1997. Pollination biology and plant breeding systems. In: Pollen Biotechnology for Crop Production and Improvement. KR Shivanna, VK Sawhney (eds). Cambridge Univ Press, New York, NY, pp 59–83.

Khan AA (ed) 1982. The Physiology and Biochemistry of Seed Development, Dormancy and Germination. Elsevier Biomedical Press, Amsterdam, The Netherlands

Kheyr-Pour A, Pernes J 1985. A new S-allele specific S-protein associated with two S-allele in *Nicotiana alata*. In: Biotechnology and Ecology of Pollen. DL Mulcahy, D Ottaviano (eds). Springer-Verlag, New York, NY, pp 191–196

Kheyr-Pour A, Bintrim SB, Ioerger TR, Remy R, Hammond SA, Kao T-h 1990. Sequence diversity in pistil S-proteins associated with gametophytic self-incompatibility in *Nicotiana alata*. Sex Plant Reprod 3: 88–97

Khoo U, Stinson HT Jr 1957. Free amino acid differences between cytoplasmic male sterile and normal fertile anthers. Proc Natl Acad Sci USA 43: 603–607

Khosla C, Shivanna KR, Mohan Ram HY 1998. Pollination in the aquatic insectivore *Utricularia inflexa* var. *stellaris*. Phytomorphology 48: 417–425

Khosla C, Shivanna KR, Mohan Ram HY 2000. Reproductive biology of *Polypleurum stylosum* (Podastemonaceae). Aquatic Bot 67: 143–154

Khush GS 1994. Apomixis: Exploiting Hybrid Vigour in Rice. Intl Rice Res Inst, Las Banos, Laguna, Philippines

Khush GS, Brar DS 1992. Overcoming the barriers in hybridization. In: Distant Hybridization of Crop Plants. G Kalloo, GB Chowdhury (eds). Springer-Verlag, Berlin, pp 47–61

Khush GS, Virmani SS 1996. Reflections on doubled haploids in plant breeding. In: In Vitro Haploid Production in Higher Plants Vol 1: Fundamental Aspects. SM Jain, SK Sopory, RE Veilleux (eds). Kluwer Acad. Publ., Dordrecht, The Netherlands, pp 11–33

Kieffer M, Fuller MP, Chauvin TE, Schlesser A 1993. Anther culture of kale (*Brassica oleraceae* L. var *acephala* DC Alef). Plant Cell Tissue Organ Culture 33: 303–313

Kimber G, Riley R 1963. Haploid angiosperms. Bot Rev 29: 480–531

King JR 1960. Peroxidase reaction as an indicator of pollen viability. Stain Technol 35: 225–227

King JR 1961. The freeze drying of pollen. Econ Bot 15: 91–98

King JR 1963. The extension of viability of tree fruit pollens in an uncontrolled atmospheric environment. New Zealand J Sci 6: 163–168

King JR 1965. The storage of pollen particularly by the freeze-drying method. Bull Torrey Bot Club 92: 270–287

Kirby EG, Vasil IK 1979. Effect of pollen-protein diffusates on germination of eluted pollen samples of *Petunia hybrida* in vitro. Ann Bot 44: 361–367

Kirch HH, Uhring H, Lottspeich F, Salamini F, Thompson RD 1989. Characterization of proteins associated with self-incompatibility in *Solanum tuberosum*. Theor Appl Genet 78: 581–585

Kisana NS, Nkongolo KK, Quick JS, Johnson 1993. Production of doubled haploids by anther culture and wheat × maize method in a wheat breeding programme. Plant Breed 110: 96–102

Klein TM, Gradziel T, Fromm ME, Sanford JC 1988. Factors influencing gene delivery into *Zea mays* cells by high-velocity microprojectiles. BioTech 6: 559–563

Knowlton EH 1992. Studies on pollen with special reference to longevity. Cornell Univ Agric Exptl Sta Memoir 52: 747–794

Knox RB 1979. Pollen and Allergy. Studies in Biology, vol 107, Arnold, London

Knox RB 1984a. The pollen grain. In: Embryology of Angiosperms. BM Johri (ed). Springer-Verlag, Berlin, pp 197–272

Knox RB, 1984b. Pollen-pistil interaction. Encyl Plant Physiol 17. 508–608.

Knox RB 1993. Grass pollen, thunderstorms and asthma. Clin Exptl Allergy 23: 354–359

Knox RB, Heslop-Harrison J 1969. Cytochemical localization of enzymes in the wall of the pollen grains. Nature 223: 92–94

Knox RB, Heslop-Harrison J 1970a. Direct demonstration of the low permeability of the angiosperm meiotic tetrad using a fluorogenic ester. Z Pflanzenphysiol 62: 451–459

Knox RB, Heslop-Harrison J 1970b. Pollen-wall proteins: localization and enzymic activity. J Cell Sci 6: 1–27

Knox RB, McConchie CA 1986. Structure and function of compound pollen. In: Pollen Structure and Function. S Blackmore (ed). Acad Press, London, pp 265–282

Knox RB, Singh MB 1987. New perspectives in pollen biology and fertilization. Ann Bot Suppl 4: 15–37

Knox RB, Suphioglu C 1996. Environmental and molecular biology of pollen allergens. Trends Plant Sci Rev 1: 156–164

Knox RB, Dickinson HG, Heslop-Harrison J 1971. Cytoplasmic RNA and enzyme activity during the meiotic prophase in *Cosmos bipinnatus*. In: Pollen: Development and Physiology. J Heslop-Harrison (ed). Butterworths, London, pp 32–35

Knox RB, Willing RR, Ashford AE 1972. Pollen wall proteins: role as recognition substances in interspecific incompatibility in poplars. Nature 237: 381–383

Knox RB, Heslop-Harrison J, Heslop-Harrison Y 1975. Pollen wall proteins: localization and characterization of gametophytic and sporophytic fractions. In: The Biology of the Male Gamete. JG Duckett, PA Racey (eds). Biol J Linn Soc 7 (Suppl 1): 177–187

Knox RB, Williams EG, Dumas C 1986. Pollen, pistil and reproductive function in crop plants. Plant Breed Rev 4: 9–79

Knox RB, Clarke AE, Harrison S, Smith P, Marchalonis JJ 1976. Cell recognition in plants: determinants of the stigma surface and their pollen interactions. Proc Natl Acad Sci USA 73: 2788–2792

Knudsen S, Due IK, Anderson SB 1989. Components of response in barley anther culture. Plant Breed 103: 241–246

Knuth P 1906–1909. Handbook of Flower Pollination. Translated by JR Ainsworth-Davis (3 vols: 1906, 1908, 1909), Oxford Univ Press, Oxford, England

Kobayashi S, Ikeda I, Nakatani M 1978. Long-term storage of *Citrus* pollen. In: Long Term Preservation of Favourable Germplasm in Arboreal Crops. T Akihama, K Nakajima (eds). Fruit Tree Res Stn, Min Agric Fish, Japan, pp 8–12

Koeppe DE, Cox JK, Malone CP 1978. Mitochondrial heredity: a determination in the toxic response of maize to the insecticide methomyl. Science 201: 1227–1229

Kohler F, Wenzel G 1985. Regeneration of isolated barley microspores in conditioned media and trials to characterize the responsible factor. J Plant Physiol 121: 181–191

Kohno T, Shimmen T 1988. Mechanism of Ca^{2+} inhibition of cytoplasmic streaming in lily pollen tubes. J Cell Sci 95: 501–509

Kohno T, Ishikawa R, Nagata T, Kohama K, Shimmen T 1992. Partial purification of myosin from lily pollen tubes by monitoring with in vitro motility assay. Protoplasma 170: 77–85.

Koltunow AM 1993. Apomixis: Embryo sacs and embryos formed without meiosis or fertilization of ovules. Plant Cell 5: 1425–1437

Koltunow AM, Truettner J, Cox K-H, Wallroth M, Goldberg RB 1990. Different temporal and spatial gene expression patterns occur during anther development. Plant Cell 2: 1201–1224

Konar RN, Linskens HF 1966a. The morphology and anatomy of the stigma of *Petunia hybrida*. Planta 71: 356–371

Konar RN, Linskens HF 1966b. Physiology and biochemistry of stigmatic fluid in *Petunia hybrida*. Planta 71: 372–387

Koo YB, Noh ER, Lee SK 1987. Hybridization between incompatible poplar species using mentor pollen. Res Rep Inst Genet Korea 23: 23–29

Kosugi S, Ohashi Y, Nakajima K, Arai Y 1990. An improved assay for β-glucuronidase in transformed cells: Methanol almost completely suppresses a putative endogenous β-glucosidase activity. Plant Sci 70: 133–140

Koul AK, Paliwal CL 1961. Inorganic acid test for pollen viability. Agra Univ J Res Sci 10: 85–90

Koul P, Sharma N, Koul AK 1993. Pollination biology of Apiaceae. Curr Sci 65: 219–222

Kovacs M, Barnabas B, Kranz E 1995. Electro-fused isolated wheat (*Triticum aestivum* L.) gametes develop into multicellular structures. Plant Cell Reports 15: 178–180

Kranz E 1997. In vitro fertilization with single isolated gametes. In: Pollen Biotechnology for Crop Production and Improvement. KR Shivanna, VK Sawhney (eds). Cambridge Univ Press, New York, NY, pp 377–391

Kranz E, Lorz H 1993. In vitro fertilization with isolated single gametes results in zygotic embryogenesis. Plant Cell 5: 739–746

Kranz E, Lorz H 1994. In vitro fertilization of maize by single egg and sperm cell protoplast fusion mediated by high calcium and high pH. Zygote 2: 125–128

Kranz E, Dresselhaus T 1996. In vitro fertilization with isolated higher plant gametes. Trends Plant Sci 1: 82–89

Kranz E, Bautor J, Lorz H 1991. In vitro fertilization of single, isolated gametes of maize mediated by electrofusion. Sex Plant Reprod 4: 12–16

Kremp GOW 1965. Morphologic Encyclopedia of Palynology. Univ Arizona Press, Tuscon, AZ

Kress WJ 1993. Coevolution of plants and animals: Pollination of flowers by primates in Madagascar. Curr Sci 65: 253–257.

Kress WJ, Stone DE 1982. Nature of the sporoderms in monocotyledons with special reference to pollen grains of *Canna* and *Heliconia*. Grana 21: 129–148

Kress WJ, Stone DE, Sellers SC 1978. Ultrastructure of exineless pollen: *Heliconia* (Heliconiaceae). Am J Bot 65: 1064–1076

Kristen U, Kappler R 1990. The pollen test system. INVITTOX IP-55: 1–7

Kristen U, Biedermann M, Liebezeit G, Dawson R 1979. The composition of stigmatic exudate and the ultrastructure of the stigma papillae in *Aptenia cordifolia*. Eur J Cell Biol 19: 281–287

Kroh M, Helsper JPF 1974. Transmitting tissue and pollen tube growth. In: Fertilization in Higher Plants. HF Linskens (ed). North-Holland Publ, Amsterdam, The Netherlands, pp 167–175.

Kruse A 1974. A 2,4-D treatment prior to pollination eliminates the haplontic (gametic) sterility in wide intergeneric crosses with 2-rowed barley, *Hordeum vulgare* subsp *distichum* as maternal species. Hereditas 78: 319

Kuboyama T, Chung CS, Takeda G 1994. The diversity of interspecific pollen-pistil incongruity in *Nicotiana*. Sex Plant Reprod 7: 250–258

Kuhlmann U, Foroughi-Wehr B, Graner A, Wenzel G 1991. Improved culture system for microspores of barley to become a target for DNA uptake. Plant Breed 107: 165–168

Kumar S, Hecht A 1970. Studies on growth and utilization of carbohydrates by pollen tubes and callose development in self-incompatible *Oenothera organensis*. Biol Plant 12: 41–46

Kuwada H, Mabuchi T 1976. Ovule and embryo culture of seeds obtained from the crosses between *Hibiscus asper*, *H. cannabinus* and *H. sabdariffa*. Jpn J Breed 26: 298–306

Kwack BH 1967. Studies on cellular site of calcium action in promoting pollen tube growth. Physiol Plant 20: 825–833

La Fleur GJ, Mascarenhas JP 1978. The dependence of generative cell division in *Tradescantia paludosa* pollen tubes on protein and RNA synthesis. Plant Sci Lett 12: 251–255

Labarca C, Loewus F 1973. The nutritional role of pistil exudate in pollen tube wall formation in *Lilium longiflorum*. II. Production and utilization of exudate from stigma and stylar canal. Plant Physiol 52: 87–92

Labarca C, Kroh M, Loewus F 1970. The composition of the stigmatic exudate from *Lilium longiflorum*. Labelling studies with myo-inositol, D-glucose, and L-proline. Plant Physiol 46: 150–156

Lacadena J-R 1974. Spontaneous and induced parthenogenesis and androgenesis. In: Haploids in Higher Plants: Advances and Potential. KJ Kasha (ed). Univ Guelph, Guelph, CAN pp 43–52

Ladd PG 1994. Pollen presenters in flowering plants—form and function. Bot J Linn Soc 115: 165–195.

Ladyman JAR, Mogensen HL 1987. Mode of action studies on a chemical hybridizing agent. Agron Abstr, Ann Meetings, Amer Soc Agron, Atlanta, GA, p 69

Lalonde BA, Nasrallah ME, Dwyer KG, Chen C-H, Barlow B, Nasrallah JB 1989. A highly conserved *Brassica* gene with homology to the S-locus-specific glycoprotein structural gene. Plant Cell 1: 249–258

Lancelle SA, Hepler PK 1992. Ultrastructure of freeze-substituted pollen tubes of *Lilium longiflorum.* Protoplasma 167: 215–230

Lansac AR, Sullivan CY, Johnson BE, Lee KW 1994. Viability and germination of the pollen of sorghum (*Sorghum bicolor* (L.) Moench). Ann Bot 74: 27–33

Larsen K 1977. Self-incompatibility in *Beta vulgaris* L. I. Four gametophytic, complementary S-loci in sugar beet. Hereditas 85: 227–248

Larsen PB, Asworth EN, Jones ML, Woodson WR 1995. Pollination-induced ethylene in carnation. Plant Physiol 108: 1405–1412

Larter E, Chaubey C 1965. Use of exogenous growth substances in promoting pollen tube growth and fertilization in barley-rye crosses. Can J Genet Cytol 7: 511–518

Laser KD, Lersten NR 1972. Anatomy and cytology of microsporogenesis in cytoplasmic male sterile angiosperms. Bot Rev 38: 425–454

Laughnan JR, Gabay SJ 1973. Reaction of germinating maize pollen to *Helminthosporium maydis* pathotoxins. Crop Sci. 43: 681–684

Laurie DA, Bennett MD 1986. Wheat × maize hybridization. Can J Genet Cytol 28: 313–316

Laurie DA, Bennett MD 1988. Cytological evidence for fertilization in hexaploid wheat × *Sorghum* crosses. Plant Breed 100: 73–82.

Lawrence MJ, Lane MD, O'Donnell S, Franklin-Tong VE 1993. The population genetics of the self-incompatibility polymorphism in *Papaver rhoeas.* V. Cross-classification of the S-alleles of samples from three natural populations. Heredity 71: 581–590

Leduc N, Monnnier M, Douglas DC 1990. Germination of trinucleate pollen: formation of a new medium for *Capsella bursapastoris.* Sex Plant Reprod 3: 228–235

Leduc N, Matthys-Rochon M, Rougier M, Mogensen L, Holm P, Magnard J-L, Dumas C 1996. Isolated maize zygotes mimic in vivo embryonic development and express microinjected genes when cultured in vitro. Develop Biol 177: 190–203

Lee H-S, Huang S, Kao T-h 1994. S-proteins control rejection in incompatible pollen in *Petunia inflata.* Nature 367: 560–563.

Lee S, Aronoff S 1967. Boron in plants. A biochemical role. Science 158: 798–799

Lee SLJ, Warmke HE 1979. Organelle size and number in fertile and T-cytoplasmic male-sterile corn. Amer J Bot 66: 141–148

Lee SLJ, Gracen VE, Earle ED 1979. The cytology of pollen abortion in S-cytoplasmic male sterile corn anthers. Am J Bot 66: 656–667

Lefebvre A, Scalla R, Pfeiffer P 1990. The double-stranded RNA associated with the '447' cytoplasmic male sterility in *Vicia faba* is packaged together with its replicase in cytoplasmic membranous vesicles. Plant Mol Biol 14: 477–490

Legge AP 1976. Hive inserts and pollen dispensers for tree fruits. Bee World 57: 159–167

Leuschner RM 1993. Human biometeorology, Part II. Pollen. Experientia 49: 931–943.

Levin MD, Loper GM 1984. Factors affecting pollen trap efficiency. Am Bee J 124: 721–723

Levings CS III 1993. Thoughts on cytoplasmic male sterility in *cms-T* maize. Plant Cell 5: 1285–1290

Levings CS III, Pring DR 1979. Molecular basis of cytoplasmic male sterility in maize. In: Physiological Genetics. JG Scandalios (ed). Acad. Press, New York, NY, pp 171–183

Lewis D 1942. The physiology of incompatibility in plants. I. The effect of temperature. Proc R Soc Lond B 131: 13–27

Lewis D 1949. Incompatibility in flowering plants. Biol Rev 24: 472–496

Lewis D 1952. Serological reactions of pollen incompatibility substances. Proc R Soc London Ser B 140: 127–135

Lewis D 1954. Comparative incompatibility in angiosperms and fungi. Adv Genet 6: 235–285

Lewis D 1956. Incompatibility and plant breeding. Brookhaven Symp Biol 9: 89–100

Lewis D 1979. Genetic versataility of incompatibility in plants. New Zealand J Bot 17: 637–644

Lewis D, Crowe LK 1953. Theory of reversible mutation. Nature 171: 561

Lewis D, Crowe LK 1958. Unilateral interspecific incompatibility in flowering plants. Heredity 12: 233–256

Lewis D, Verma SC, Zuberi MI 1988. Gametophytic-sporophytic self-incompatibility in the Crucifereae—*Raphanus sativus.* Heridity 61: 355–366

Li Y-Q, Croes AF, Linskens HF 1983. Cell wall proteins in pollen and root of *Lilium longiflorum.* Extraction and partial characterization. Planta 158: 422–427.

Li Y-Q, Bruun L, Pierson ES, Cresti M 1992. Periodic deposition of arabinogalactan epitopes in the cell wall of pollen tubes of *Nicotiana tabacum.* Planta 188: 532–538

Li Y-Q, Chen F, Linskens HF, Cresti M 1994. Distribution of unesterified and esterified pectins in cell walls of pollen tubes. Sex Plant Reprod 7: 145–152

Lichter R 1989. Efficient yield of embryoids by culture of isolated microspores of different Brassicaceae species. Plant Breed 103: 119–123

Lichtenstein LM, Norman PS, Winkenwerder WL 1971. A single year of immunotherapy for ragweed hay fever;

immunologic and clinical studies. Ann Intl Med 75: 663–671

Liedl BE, Mc Cormick S, Mutschler MA 1986. Unilateral incongruity in crosses involving *Lycopersicon pennellii* and *L. esculentum* is distinct from self-incompatibility in expression, timing and location. Sex Plant Reprod 9: 299–308

Lin MS, Tanner E, Lynn J, Friday GA 1993. Non-fatal systemic allergic reactions induced by skin testing and immunotherapy. Ann Allergy 71: 557–562

Linskens HF 1965. Biochemistry of incompatibility. In: SD Geerts (ed). Genetics Today. Proc XI Intl Cong Genet, The Hague, Vol III, pp 629–636

Linskens HF (ed) 1974. Fertilization in Higher Plants. North-Holland Publ, Amsterdam, The Netherlands

Linskens HF 1975a. Incompatibility in *Petunia*. Proc R Soc London Ser B 188: 299–311

Linskens HF 1975b. Uber eine spezifische Anfarbung der Pollenschlauche im Griffel und die Zahl der Kallosepfropfen nach Selbstung und Fremdung. Naturwissenschaften 44: 16

Linskens HF 1995. Pollen deposition in mosses from northwest Greenland. Proc Kon Ned Akad Wetensch 98: 45–53

Linskens HF, Mulleneers JML 1967. Formation of "instant pollen tubes". Acta Bot Neerl 16: 132–142

Linskens HF, Schrauwen JAM 1969. The release of free amino acids from germinating pollen. Acta Bot Neerl 18: 605–614

Linskens HF, Kroh M 1970. Regulation of pollen tube growth. In: Current Topics in Developmental Biology, AA Moscana, A Monroy (eds). Academic Press, London, vol 5, pp 89–113

Linskens HF, Pfahler PL 1973. Biochemical composition of maize (*Zea mays* L.) pollen. III. Effects of allele × storage interactions at the waxy (wx), sugary (su_1) and shrunken (sh_2) loci, on the amino acid content. Theor Appl Genet 43: 49–53

Linskens HF, Jorde W 1997. Pollen as food and medicine—a review. Econ Bot 51: 78–87

Linskens HF, Bargagli R, Focardi S, Cresti M 1991. Antarctic moss turf as pollen trap. Proc Kon Ned Akad Westench 94: 233–241

Liu XC, Jones K, Dickinson HG 1987. DNA synthesis and cytoplasmic differentiation in tapetal cells of normal and cytoplasmically male sterile lines of *Petunia hybrida*. Theor Appl Genet 74: 846–851

Lloyd DG, Webb CJ 1986. The avoidance of interference between the presentation of pollen and stigma in angiosperms. Dichogamy. New Zealand J Bot 24: 135–162.

Lo K-H, Pauls KP 1992. Plant growth environment effects on rapeseed microspore development and culture. Plant Physiol 99: 468–472

Lord EM 1981. Cleistogamy: a tool for the study of floral morphogenesis, function and evolution. Bot Rev 47: 421–449

Lord EM 2001. Adhesion molecules in lily pollination. Sex Plant Reprod 14: 57–62

Lord EM, Heslop-Harrison Y 1984. Pollen-stigma interaction in the Leguminosae: Stigma organization and the breeding system in *Vicia faba* L. Ann Bot 54: 827–836

Louis JP, Augur C, Teller G 1990. Cytokinins and differentiation process in *Mercurialis annua* L. (2n=16). Genetic regulation with auxin, indoleacetic acid oxidases, and sexual expression patterns. Plant Physiol 94: 1535–1541

Lowenstein H, King TP, Goodfriend L, Hussain R, Roebber M, Marsh DG 1981. Antigens of *Ambrosia elatior* (short ragweed) pollen. Immunological identification of known antigens by quantitative immunoelectrophoresis. J Immunol 127: 634–642

Lundqvist A 1956. Self-incompatibility in rye. I. Genetic control in diploid. Hereditas 42: 293–348

Lundqvist A 1965. The genetics of incompatibilty. In: Genetics Today. SD Geerts (ed). Proc XI Intl Cong Genet, The Hague, vol III, pp 637–647

Lundqvist A 1975. Complex self-incompatibility systems in angiosperms. Proc R Soc London Ser B 188: 235–245

Lundqvist A 1990. The complex S-gene system for control of self-incompatibility in the buttercup genus *Ranunculus*. Hereditas 113: 29–46

Lush WM, Grieser F, Wolter-Arts M 1998. Directional guidance of *Nicotiana alata* pollen tubes in vitro and on the stigma. Plant Physiol 118: 733–741

Luu DT, Qin X, Morse D, Cappadocia M 2000. S-RNase uptake by compatible pollen tubes in gametophytic self-incompatibility. Nature 407: 649–651

Luu DT, Marty-Mazars D, Trick M, Dumas C, Heizmann P 1999. Pollen-stigma adhesion in *Brassica* spp involves SLG and SLR glycoproteins. Plant Cell 11: 251–262

Macfarlane RP, Griffin RP, Read PEC 1983. Bumble bee management options to improve "Grasslands Pawera" red clover seed yield. Proc New Zealand Grassland Assoc 44: 47–53

Mackenzie A, Heslop-Harrison J, Dickinson HG 1967. Elimination of ribosome during meiotic prophase. Nature 215: 997–999

Maevskaya AN, Troitskaya EA, Yakovleva NS 1976. Effect of boron starvation on activity of β-glucosidase in plants of the families Leguminosae and Gramineae. Soviet Plant Physiol 23: 1073–1076

Maheshwari P 1950. An Introduction to the Embryology of Angiosperms. McGraw-Hill, New York, NY

Maheshwari P (ed) 1963. Recent Advances in the Embryology of Angiosperms. Intl Soc Plant Morphol, Delhi

Maheshwari P, Kanta K 1961. Intraovarian pollination in *Eschscholtzia californica* Cham., *Argemone mexicana* L., and *Argemone ochroleuca* Sweet. Nature 191: 304

Maheshwari P, Rangaswamy NS 1965. Embryology in relation to physiology and genetics. In: Advances in Botanical Research. Vol II RD Preston (ed). Academic Press, New York, NY, pp 219–312

Maheshwari R, Mahadevan S 1978. Acid induced protrusion from pollen of *Datura innoxia*. Indian J Exptl Biol 16: 884–886

Maheshwari SC, Rashid A, Tyagi AK 1982. Haploids from pollen grains. Retrospect and prospect. Am J Bot 69: 865–879

Maheshwari SC, Maheshwari N, Khurana JP, Sopory SK 1998. Engineering apomixis in crops: A challenge for plant molecular biologists in the next century. Curr Sci 75: 1141–1147

Maiti IB, Loewus FA 1978. Evidence for a functional myoinositol oxidation pathway in *Lilium longiflorum* pollen. Plant Physiol 62: 280–283

Malhotra K, Maheshwari SC 1977. Enhancement by cold treatment of pollen embryoid development in *Petunia hybrida*. Z. Pflanzenphysiol 85: 177–180

Malik M 1998. Two new cytoplasmic male sterile lines in *Brassica juncea*: Development, characterization and induction of androgenesis. Ph.D. Thesis, Delhi Univ, Delhi

Malik M, Rangaswamy NS, Shivanna KR 2001. Induction of microspore embryos in a CMS line of *Brassica juncea* and formation of the androgenic plantlets. Euphytica 120: 195–203

Malik M, Vyas P, Rangaswamy NS, Shivanna KR 1999. Development of two new cytoplasmic male sterile lines in *Brassica juncea* through wide hybridization. Plant Breed 118: 75–78

Malti, Shivanna KR 1984. Structure and cytochemistry of the pistil of *Crotalaria retusa* L. Proc Indian Natl Sci Acad B 50: 92–102

Mandahar CL 1981. Virus transmission through seeds and pollen. In: Plant Diseases and Vectors: Ecology and Epidemiology. K Maramorosch, KF Harris (eds). Acad Press, New York, NY, pp 241–292.

Marcucci MC, Visser T, van Tuyl JM 1982. Pollen and pollination experiments. VI. Heat resistance of pollen. Euphytica 31: 287–290

Margalith R, Lensky Y, Rabinowitch HD 1984. An evaluation of Beeline as a honeybee attractant to cucumbers and its effect on hybrid seed production. J Apicult Res 23: 50–54

Mariani C, De Beuckeleer M, Truettner J, Leemans J, Goldberg RB 1990. Induction of male sterility in plants by a chimaeric ribonuclease gene. Nature 347: 737–741

Mariani C, Gossele V, De Beuckeleer M, De Block M, Goldberg RB, De Greef W, Leemans J 1992. A chimaeric ribonuclease-inhibitor gene restores fertility to male sterile plants. Nature 357: 384–387

Marsolais AA, Wheatley WG, Kasha KJ 1986. Progress in wheat haploid induction using anther culture. Proc DSIR Plant Breed Symp 1986. Agron Soc NZ Special Publ 5: 340–343

Martin FW 1967. The genetic control of unilateral incompatibility between tomato species. Genetics 56: 391–398

Martin FW 1969. Compounds of the stigma of 10 species. Am J Bot 56: 1023–1027

Martin FW 1970. Compounds of the stigmatic surface of *Zea mays* L. Ann Bot 34: 835–842

Martin FW 1972. In vitro measurement of pollen tube growth inhibition. Plant Physiol 49: 924–925

Martin FW, Ruberte R 1972. Inhibition of pollen germination and tube growth by stigmatic susbtances. Phyton 30: 119–126

Marubashi W, Tatsuno I, Onozawa Y 1988. Effect of high temperature on the growth of lethal hybrid seedlings between *Nicotiana suaveolens* Lehm. and *N. tabacum* L. Jap J Breed 38 (Suppl 1): 66–67.

Marumaya L 1968. Electron microscopic observations of plastids and mitochondria during pollen development in *Tradescantia paludosa*. Cytologia 33: 482–497

Marx JL 1978. The mating game: what happens when sperm meets egg. Science 200: 1256–1259

Mascarenhas JP 1975. The biochemistry of angiosperm pollen development. Bot Rev 41: 259–314

Mascarenhas JP 1989. The male gametophyte in flowering plants. Plant Cell 1: 657–664

Mascarenhas JP 1990. Gene activity during pollen development. Ann Rev Plant Physiol Plant Mol Biol 41: 317–338

Mascarenhas JP 1992. Pollen gene expression: molecular evidence. Intl Rev Cytol 140: 3–18

Mascarenhas JP 1993. Molecular mechanism of pollen tube growth and differentiation. Plant Cell 5: 1303–1314

Mascarenhas JP, Machlis L 1962. Chemotropic response of *Antirrhinum majus* to calcium. Nature 196: 292–293

Mascarenhas JP, Machlis L 1964. Chemotropic response of the pollen of *Antirrhinum majus* to calcium. Plant Physiol 39: 70–77

Mascarenhas JP, Bell E 1970. RNA synthesis during development of the male gametophyte of *Tradescantia*. Dev Biol 21: 475–490

Mascarenhas JP, Marmelstein J 1981. Messenger RNAs: Their utilization and degradation during pollen germination and tube growth. Acta Soc Bot Polon 50: 13–20

Mascarenhas JP, Terrena B, Mascarenhas AF, Rueckert L 1974. Protein synthesis during germination and pollen tube growth in *Tradescantia*. In: Fertilization in Higher Plants. HF Linskens (ed). North-Holland Publ, Amsterdam, The Netherlands, pp 137–143

Mathur G, Mohan Ram HY 1978. Significance of petal colour in thrips-pollinated *Lantana camara* L. Ann Bot 42: 1473–1476

Mathur G, Mohan Ram HY 1986. Floral biology and pollination of *Lantana camara.* Phytomorphology 36: 79–100

Matousek J, Tupy J 1983. The release of nucleases from tobacco pollen. Plant Sci Lett 30: 83–89

Matsubara S 1973. Overcoming self-incompatibility by cytokinin treatment on *Lilium longiflorum.* Bot Mag 86: 43–46

Matsubara S 1980. Overcoming self-incompatibility in *Raphanus sativus* L. with high temperature. J Am Soc Hort Sci 105: 842–846

Matthews BF, Abdul-Baki AA, Saunders JA 1990. Expression of a foreign gene in electroporated pollen grains of tobacco. Sex Plant Reprod 3: 147–151

Matthys-Rochon E, Mol R, Faure JE, Digonnet C. Dumas C 1997. Isolation and micromanipulation of embryo sac and egg cell in maize. In: Pollen Biotechnology for Crop Production and Improvement. KR Shivanna, VK Sawhney (eds), Cambridge Univ Press, New York, NY, pp 363–376

Mattsson O, Knox RB, Heslop-Harrison J, Heslop-Harrison Y 1974. Protein pellicle as a probable recognition site in incompatibility reactions. Nature 213: 703–704

Mau SL, Williams EG, Atkinson A, Anderson MA, Cornish EC, Grego B, Simpson RJ, Kheyr-Pour A, Clarke AE 1986. Style proteins of a wild tomato (*Lycopersicum peruvianum*) associated with expression of self-incompatibility. Planta 169: 184–191

Mayfield JA, Preuss D 2000. Rapid initiation of *Arabidopsis* pollination requires the oleosin-domain protein GRP 17. Nature Cell Biol 2: 128–130.

McClure BA, Gray JE, Anderson MA, Clarke AE 1990. Self-incompatibility in *Nicotiana alata* involves degradation of pollen rRNA. Nature 347: 757–760

McClure BA, Haring V, Ebert PR, Anderson MA, Simpson RJ, Sakiyama F, Clarke AE 1989. Style self-incompatibility gene products of *Nicotiana alata* are ribonucleases. 342: 955–957

McConchie CA 1983. The diversity of hydrophilous pollination in monocotyledons. In: Pollination '82. EG Williams, RB Knox, JH Gilbert, P Bernhardt (eds), School of Botany, Univ Melbourne, Melbourne, Australia, pp 148–165

McConchie CA, Jobson S, Knox RB 1985. Computer assisted reconstruction of the male germ unit in pollen of *Brassica campestris.* Protoplasma 127: 57–63

McConchie CA, Russell SD, Dumas C, Tuohy M, Knox RB 1987. Quantitative cytology of the sperm cells of *Brassica campestris* and *B. oleraceae.* Planta 170: 446–452

McCormick, S 1993. Male gametophyte development. Plant Cell 5: 1265–1275

McCormick S, Yamaguchi J, Twell D 1991. Deletion analysis of pollen-expressed promoters. In Vitro Cell Dev Biol 27: 15–20

McCoy K, Knox RB 1988. The plasma membrane and generative cell organization in pollen of mimosoid legume, *Acacia retinoides.* Protoplasma 143: 85–92

McGregor SE 1976. Insect Pollination of Cultivated Plants. USDA Agriculture Handbook No. 496, USDA, Washington DC

McLellan AK 1977. Minerals, carbohydrates and amino acids of pollen from some woody and herbaceous plants. Ann Bot 41: 1225–1232

McRae DH 1985. Advances in chemical hybridization. Plant Breed Rev 3: 503–506

McVetty PBE 1997. Cytoplasmic male sterility. In: Pollen Biotechnology for Crop Production and Improvement. KR Shivanna, VK Sawhney (eds). Cambridge Univ Press, New York, NY, pp 155–182

Meeuse BJD 1975. Thermogenic respiration in aroids. Ann Rev Plant Physiol 26: 117–126

Melchers G 1972. Haploid higher plants for plant breeding. Z Pflanzenzucht 67: 21–32

Mepham RH, Lane GR 1969. Formation and development of the tapetum periplasmodium in *Tradescantia bracteata.* Protoplasma 68: 175–192

Mepham RH, Lane GR 1970. Observations on the fine structure of developing microspores of *Tradescantia bracteata.* Protoplasma 70: 1–20

Meyer VG 1969. Some effects of genes, cytoplasm, and environment on male sterility of cotton (*Gossypium*). Crop Sci 9: 237–242

Meyer VG 1974. Interspecific cotton breeding. Econ Bot 28: 56–60

Mignouna H, Virmani SS, Briquet M 1987. Mitochondrial DNA modifications associated with cytoplasmic male sterility in rice. Theor Appl Genet 74: 666–669

Miki-Hirosige H, Hoek IHS, Nakamura S 1987. Secretion from the pistil of *Lilium longiflorum.* Am J Bot 74: 1709–1715

Mink GI 1993. Pollen and seed transmitted viruses and viroids. Ann Rev Phytopath 31: 375–402

Mishra R, Shivanna KR 1982 Efficacy of organic solvents for storing pollen grains of some leguminous taxa. Euphytica 31: 991–995

Miyake T, Kuroiwa H, Kuroiwa T 1995. Differential mechanism of movement between a generative cell and a vegetative nucleus in pollen tubes of *Nicotiana tabacum* as revealed by additions of colchicine and nananoic acid. Sex Plant Reprod 8: 228–230

Mizelle MB, Sethi R, Ashton ME, Jensen WA 1989. Development of the pollen grain and tapetum of wheat (*Triticum aestivum*) in untreated plants and plants treated with chemical hybridizing agent RH0007. Sex Plant Reprod 2: 231–253

Mo Y, Nagel C, Taylor LP 1992. Biochemical complementation of calchone synthase mutants define a role for flavonols in functional pollen. Proc Natl Acad Sci USA 89: 1713–1717

Moewus F 1950. Zur Physiologie und Biochemie der Selbsterilitat bei *Forsythia*. Biol Zentralblatt 69: 181–197

Mogensen HL 1988. Exclusion of male mitochondria and plastids during syngamy as a basis for maternal inheritance. Proc Natl Acad Sci USA 85: 2594–2597

Mogensen HL 1992. The male germ unit: concept, composition and significance. Intl Rev Cytol 140: 129–147

Mogensen HL, Rusche ML 1985. Quantitative ultrastructural analysis of barley sperm. 1. Occurrence and mechanism of cytoplasm and organelle reduction and the question of sperm dimorphism. Protoplasma 128: 1–13

Mogensen HL, Wagner VT 1987. Associations among components of the male germ unit following in vitro pollination in barley. Protoplasma 138: 161–172

Mogensen HL, Ladyman JAR 1989. A structural study on the mode of action of CHA™ chemical hybridizing agent in wheat. Sex Plant Reprod 2; 173–183

Mohan Ram HY, Rustagi PN 1966. Increase in yield of wheat (*Triticum aestivum* L. variety N.P. 719) treated with Mendok. Agron J 61: 198–207

Mohan Ram HY, Jaiswal VS 1970. Induction of female flowers on male plants of *Cannabis sativa* L. by 2-chloroethanephosphonic acid. Experientia 26: 214–216

Mohan Ram HY, Khosla C 1998. Pollination in aquatic flowering plants. In: Plant Reproduction, Genetics and Biology. RN Gohil (ed). Scientific Publ (India), Jodhpur, pp 75–92

Mohapatra SS, Knox RB (ed) 1996. Pollen Biotechnology: Gene Expression and Allergen Characterization. Chapman & Hall, New York, NY

Mol R, Matthys-Rochon E, Dumas C 1995. Embryogenesis and plant regeneration from maize zygotes by in vitro culture of fertilized embryo sacs. Plant Cell Rep 14: 743–747

Mollers C, Iqbal MCM, Robbelen G 1994. Efficient production of doubled haploid *Brassica napus* plants by colchicine treatment of microspores. Euphytica 75: 95–100

Monteiro AA, Gabelman WH, Williams PH 1988. The use of sodium chloride solution to overcome self-incompatibility in *Brassica campestris*. Cruciferae Newslett 13: 122–123

Moore PD, Webb JA 1978. An Illustrated Guide to Pollen Analysis. Hodder & Stoughton, London

Mordhorst AP, Lorz H 1993. Embryogenesis and development of isolated barley (*Hordeum vulgare* L.) microspores are influenced by the amount and composition of nitrogen sources in culture media. J Plant Physiol 142: 485–492

Morgan JM 1980. Possible role of abscisic acid in reducing seedset in water-stressed wheat plants. Nature 285: 655–657

Morikawa H, Nishihara M 1997. Use of pollen in gene transfer. In: Pollen Biotechnology for Crop Production and Improvement. KR Shivanna, VK Sawhney (eds). Cambridge Univ Press, New York, NY, pp 423–437

Moscatelli A, Del Casino C, Lozzi L, Cai G, Scali M, Tiezzi A, Cresti M 1995. High molecular weight polypeptides related to dynein heavy chains in *Nicotiana tabacum* pollen tubes. J Cell Sci 108: 1117–1125

Moss GI, Heslop-Harrison J 1967. A cytochemical study of DNA, RNA and proteins in the developing maize anther. II. Observations. Ann Bot 31: 555–572

Moss GI, Heslop-Harrison J 1968. Photoperiod and pollen sterility in maize. Ann Bot 32: 833–846

Moutinho A, Hussey PJ, Trewavas A, Malho R 2001. CAMP acts as a second messenger in pollen tube growth and reorientation. Proc Natl Acad Sci USA 98: 10481–10486

Mujeeb-Kazi A 1981. *Triticum timopheevii* × *Secale cereale* crossability. J Hered 72: 227–228

Mujeeb-Kazi A, Rodriguez R 1980. Some intergeneric hybrids in the Triticeae. Cereal Res Commun 8: 469–475

Mukhopadhyay A, Arumugam N, Pradhan AK, Murthy HN, Yadav BS, Sodhi YS, Pental D 1994. Somatic hybrids with substitution type genomic configuration TCBB for the transfer of nuclear and organelle genes from *Brassica tournefortii* TT to allotetraploid oilseed crop *B. carinata* BBCC. Theor Appl Genet 89: 19–25

Mulcahy DL 1979. Rise of the angiosperms: a genecological factor. Science 206: 20–23

Mulcahy DL 1984. Manipulation of gametophytic populations. In: Efficiency in Plant Breeding. W Lange, AC Zeven, C Hogenboom (eds). PUDOC, Wageningen, The Netherlands

Mulcahy DL, Mulcahy GB 1983. Gametophytic self-incompatibility reexamined. Science 220: 1247–1251

Mulcahy DL, Ottaviano E (eds) 1983. Pollen: Biology and Implications for Plant Breeding. Elsevier Biomedical, New York, Amsterdam, Oxford

Mulcahy DL, Mulcahy GB, Ottaviano E (eds) 1986. Biotechnology and Ecology of Pollen. Springer-Verlag, New York, Berlin, Heidelberg

Mulcahy DL, Mulcahy GB, Popp R, Fong N, Pallais N, Kalinowiski A, Marien JN 1988. Pollen selection for stress tolerance or the advantage of selection before pollination. In: Sexual Reproduction in Higher Plants. M Cresti, P Gori, E Pacini (eds). Springer-Verlag, Berlin, pp 43–50

Mulcahy GB, Mulcahy DL 1982. The two phases of growth of *Petunia hybrida* (Hort. Vilm. Andz.) pollen tubes through compatible styles. J Palynol 18: 61–64

Mulcahy GB, Mulcahy DL 1983. Composition of pollen tube growth in bi- and tri-nucleate pollen. In: Pollen Biology and Implications for Plant Breeding. DL Mulcahy, E Ottaviano (eds). Elsevier Biomedical, New York, pp 29–33.

Mulcahy GB, Mulcahy DL 1987. Induced pollen tube directionality. Am J Bot 74: 1458–1459

Murfett J, Atherton T, Mou B, Gasser C, McClure BA 1994. S-RNase expressed in transgenic *Nicotiana* causes S-allele-specific pollen rejection. Nature 367: 563–566

Murgia M, Charzynska M, Rougier M, Cresti M 1991. Secretory tapetum of *Brassica oleracea* L.: polarity and ultrastructural features. Sex Plant Reprod 4: 28–35

Murphy DJ, Ross JHE 1998. Biosynthesis, targeting and processing of oleosin-like proteins, which are major pollen coat components in *Brassica napus*. Plant J 13: 1–16

Murray BG 1986. Floral biology and self-incompatibility in *Lilium*. Bot Gaz 147: 327–333

Muschietti J, Dircks L, Vancanneyt G, McCormick S 1994. LAT52 protein is essential for tomato pollen development: pollen expressing antisense LAT52 RNA hydrates and germinates abnormally and cannot achieve fertilization. Plant J 6: 321–328

Nair PKK 1970. Pollen Morphology of Angiosperms. Scholar Publ House, Lucknow, UP, India

Nakajima M, Yamaguchi I, Kizawa S, Murofushi N, Takahashi N 1991. Semiquantification of GA_1 and GA_4 in male sterile anthers of rice by radioimmunoassay. Plant & Cell Physiol 32: 511–513

Nakamura A, Yamada T, Oka M, Tatemichi Y, Engushi K, Ayabe T, Kobayashi K 1975. Studies by haploid method of breeding by anther culture in tobacco. V. Breeding of wild fluecured variety F211 by haploid method. Bull Iwata Tobacco Exptl Sta 7: 29–39

Nakamura RR 1988. Seed abortion and seed weight variation within fruits of *Phaseolus vulgaris*: Pollen donor and resource limitation effects. Am J Bot 75:1003–1010

Nakanishi T, Hinata K 1975. Self-seed production by CO_2 gas treatment in self-incompatible cabbage. Euphytica 24: 117–120

Namboodiri AN, Bhaskar MS 1983. Pollen wall diffusates of *Chlorophytum* and their role in pollen germination. Phytomorphology 33: 31–36

Nanda Kumar PBA, Shivanna KR 1991. In vitro multiplication of a sterile interspecific hybrid, *Brassica fruiticulosa* × *B campestris*. Plant Cell Tissue Organ Cult 26: 17–22

Nanda Kumar PBA, Shivanna KR 1993. Intergeneric hybridization between *Diplotaxis siettiana* and crop brassicas for the production of alloplasmic lines. Theor Appl Genet 85: 770–776

Nanda Kumar PBA, Prakash S, Shivanna KR 1988. Wide hybridization in *Brassica*: crossability barriers and studies on the hybrids and synthetic amphiploids of *B. fruiticulosa* × *B. campestris*. Sex Plant Reprod 1: 234–239

Narayanaswamy S 1994. Plant Cell and Tissue Culture. Tata McGraw-Hill Publ Co Ltd, New Delhi, pp 323–389

Narayanaswamy S, Chandy LP 1971. In vitro induction of haploid, diploid and triploid androgenic embryoids and plantlets in *Datura metel* L. Ann Bot 35: 535–542

Nasrallah JB 2000. Cell-cell signalling in the self-incompatibility response. Curr Opinion Plant Biol 3: 368–373.

Nasrallah JB, Nasrallah ME 1989. The molecular genetics of self-incompatibility in *Brassica*. Ann Rev Genet 23: 100–107

Nasrallah JB, Nasrallah ME 1993. Pollen-stigma signalling in sporophytic self-incompatibility response. Plant Cell 5: 1325–1335

Nasrallah, JB, Su-may YU, Nasrallah ME 1988. Self-incompatibility genes in *Brassica oleracea*. Expression, isolation and structure. Proc Natl Acad Sci USA 85: 5551–5555

Nasrallah JB, Nishio T, Nasrallah ME 1991. The self-incompatibility genes of *Brassica*: expression and use in genetic ablation of floral tissue. Ann Rev Plant Physiol Plant Mol Biol 42: 393–422

Nasrallah JB, Rundle SJ, Nasrallah ME 1994. Genetic evidence for the requirement of *Brassica* S-locus receptor kinase gene in the self-incompatibility response. Plant J 5: 373–384

Nasrallah JB, Kao T-h, Goldberg ML, Nasrallah ME 1985. A cDNA clone encoding S-locus specific glycoprotein from *Brassica oleracea*. Nature 318: 263–267

Nasrallah JB, Kao T-h, Chen C-H, Goldberg ML, Nasrallah ME 1987. Aminoacid sequence of glycoproteins encoded by three alleles of the S-locus of *Brassica oleracea*. Nature 326: 617–619

Nasrallah ME 1974. Genetic control of quantitative variation in self-incompatibility proteins detected by immunodiffusion. Genetics 76: 49–50

Nasrallah ME, Wallace DH 1967. Immumogenetics of self-incompatibility in *Brassica oleracea*. Heredity 22: 519–527

Nasrallah ME, Barber J, Wallace DH 1970. Self-incompatibility proteins in plants: detection, genetics, and possible mode of action. Heredity 25: 23–27

Nasrallah ME, Wallace DH, Savo RM 1972. Genotype, protein, phenotype relationships in self-incompatibility of *Brassica*. Genet Res 20: 151–160

Neuhaus G, Spangenberg G, Mettelsten Scheid O, Schweiger H.-G 1987. Transgenic rapeseed plants obtained by the microinjection of DNA into microspore-derived embryoids. Theor Appl Genet 75: 30–36

Newbigin E, Anderson MA, Clarke AE 1993. Gametophytic self-incompatibility systems. Plant Cell 5: 1315–1324

Newman C 1984. Pollen: Breath of life and sneezes. National Geography 166: 491–521

Niimi Y 1970. In vitro fertilizaiton in the self-incompatible plant *Petunia hybrida*. J Jpn Soc Hort Sci 39: 345–352

Niklas K 1985. The aerodynamics of wind pollination. Bot Rev 51: 328–386

Nilsson S, Muller J 1978. Recommended palynological terms and definitions. Grana 17: 55–58

Nishihara M, Seki M, Kyo M, Irifune K, Morikawa H 1993. Expression of the β-glucuronidase gene in pollen of lily (*Lilium longiflorum*), tobacco (*Nicotiana tabacum*),

Nicotiana rustica, and peony (*Paeonia lactiflora*) by particle bombardment. Plant Physiol 102: 357–361

Nishihara M, Seki M, Kyo M, Irifune K, Morikawa H 1995. Transgenic haploid plants of *Nicotiana rustica* produced by bombardment-mediated transformation of pollen. Transgenic Res 4: 341–348

Nishio T, Hinata K 1977. Analysis of S-specific proteins in stigma of *Brassica oleracea* L. by iso-electric focussing. Heredity 38: 391–396

Nishio T, Hinata K 1978. Stigma proteins in self-incompatible *Brassica campestris* L. and self-compatible relatives, with special reference to S-allele specificity. Jpn J Genet 53: 27–33

Nishio T, Hinata K 1979. Purification of an S-specific glycoprotein in self-incompatible *Brassica campestris* L. Jpn J Genet 54: 307–311

Nishio T, Hinata K 1980. Rapid detection of S-glycoproteins of self-incompatible crucifers using Con-A reaction. Euphytica 29: 217–221

Nishio T, Hinata K 1982. Comparative studies on S-glycoproteins purified from different S-genotypes in self-incompatible *Brassica* species. I. Purification and chemical properties. Genetics 100: 641–647

Nitsch C 1977. Culture of isolated microspores. In: Plant Cell, Tissue and Organ Culture. J Reinert, YPS Bajaj (eds). Sprnger-Verlag, Berlin, pp 268–278

Nitsch JP 1969. Experimental androgenesis in *Nicotiana*. Phytomorphology 19: 384–404

Nitsch JP 1972. Haploid plants from pollen. Z. Pflanzenzucht 67: 3–18

Nivison HT, Hanson MR 1989. Identification of a mitochondrial protein associated with cytoplasmic male sterility in *Petunia*. Plant Cell 1: 1121–1130

Nogler GA 1984. Gametophytic apomixis. In: Embryology of Angiosperms. BM Johri (ed). Springer-Verlag, Berlin, pp 475–518

Norton JD 1966. Testing plum pollen viability with tetrazolium salts. Proc Am Soc Hortic Sci 89: 132–134

Nygaard P 1977. Utilization of exogenous carbohydrates for tube growth and starch synthesis in pine pollen suspension cultures. Physiol Plant 39: 206–210

Oberle GD, Watson R 1953. The use of 2,3,5-triphenyl tetrazolium chloride in viability tests of fruit pollens. Proc Am Soc Hort Sci 61: 299–303

Obermeyer G, Weisenseel MH 1991. Calcium channel blocker and calmodulin antagonists affect the gradient of free calcium ions in lily pollen tubes. Eur J Cell Biol 56: 319–327

Obermeyer G, Kriechbaumer R, Strasser D, Maschessnig A, Bentrup F-W 1996. Boric acid stimulates the plasma membrane H$^+$-ATPase of ungerminated lily pollen. Physiol Plant 98: 281–290

Ockendon DJ 1974. The value of stored pollen in incompatibility studies in *Brassica*. Incompatibility Newslett 4: 17–19

Ockendon DJ 1975. Dominance relationships between S-alleles in the stigma of Brussels sprouts of *Brassica oleracea* var. *gemmifera*. Euphytica 24: 165–172

Odland ML, Noll CJ 1950. The utilization of cross-incompatibility and self-incompatibility in the produciton of F$_1$-hybrid cabbage. Proc Am Soc Hort Sci 55: 391–398

Ohmasa M, Watanabe Y, Murata N 1976. A biochemical study of cytoplasmic male-sterility of corn. Alteration of cytochrome oxidase and malate dehydrogenase activities during pollen development. Jpn J Breed 26: 40–50

Ohta Y 1986. High efficiency genetic transformation of maize by a mixture of pollen and exogenous DNA. Proc Natl Acad Sci USA 83: 715–719

Ohta Y, Sudoli M 1980. Genetic transformation in *Capsicum annuum* by applying exogenous DNA to pollen grains and to stigma at the time of pollination. Jpn J Breed 30 Suppl 2: 184–185

Olvey JM, Fisher WD, Patterson LL 1981. TD1123: a selective male gametocide. In: Proc Beltwide Cotton Prod Res Conf. JM Brown (ed). Natl Cotton Council of America, Memphis TN, p 84

O'Neill P, Singh MB, Neales TF, Knox BB, Williams FG 1984. Carbon dioxide blocks the stigma callose response following incompatible pollination in *Brassica*. Plant Cell Environ 7: 285–288

O'Neill SD 1997. Pollination regulation of flower development. Ann Rev Plant Physiol Plant Mol Biol 48: 547–574

O'Neill SD, Nadeau JA, Zhang XS, Bui AQ, Halevy AH 1993. Interorgan regulation of ethylene biosynthetic genes by pollination. Plant Cell 5: 419–432

Ong EK, Singh MB, Knox RB 1995a. Seasonal distribution of pollen in the atmosphere of Melbourne. Aerobiologia 11: 51–55

Ong EK, Singh MB, Knox RB 1995b. Aeroallergens of plant origin: molecular basis and aerobiological significance. Aerobiologia 11: 219–229

Ong EK, Knox RB, Singh MB 1996. Molecular characterization and environmental monitoring of grass pollen allergens. In: Pollen Biotechnology: Gene Expression and Allergen Characterization. SS Mohapatra, RB Knox (eds). Chapman & Hall, New York, NY, pp 176–210

Opeke LK, Jacob VJ 1969. Studies on methods of overcoming incompatibility in *Theobroma cacao* L. Proc II Intl Cacao Res Conf, Bahia, pp 356–358

Orton TJ, Tai W 1977. Chromosome elimination in a complex hybrid of the genus *Hordeum*. Can J Bot 55: 3023–3033

Osborne JL, Williams IH, Corbet SA 1991. Bees, pollination and habitat change in the European community. Bee World 72: 99–115

Osterbye U 1975. Self-incompatibility in *Ranunculus acris* L. Genetics, interpretation and evolutionary aspects. Hereditas 80: 91–112

Ottaviano E, Mulcahy DL 1986. Gametophytic selection as a factor of crop plant evolution. In: Origin and Domestication of Cultivated Plants. Elsevier, Amsterdam, The Netherlands, pp 101–120

Ottaviano E, Mulcahy DL 1989. Genetics of angiosperm pollen. Adv Genet 26: 1–64

Ottaviano E, Mulcahy DL, Sari Gorla M, Mulcahy GB 1992. Angiosperm Pollen and Ovule. Springer-Verlag, New York, Berlin, Heidelberg

Overman MA, Warmke HE 1972. Cytoplasmic male sterility in *Sorghum*. II. Tapetal behaviour in fertile and sterile anthers. J Hered 63: 227–234

Owens JN, Molder M 1971. Meiosis in conifers: prolonged pachytene and diffuse diplotene stages. Can J Bot 49: 2061–2064

Owens JN, Takaso T, Runions CJ 1998. Pollination in conifers. Trends Plant Sci Rev 3: 479–485

Owens SJ 1981. Self-incompatibility in the Commelinaceae. Ann Bot 47: 567–581

Owens SJ 1989. Stigma, style and pollen-pistil interaction in Caesalpinioideae. In: Advances in Legume Biology. CS Stirton, JL Zarucchi (eds). Monogr Syst Bot, Missouri Bot Gard 29: 113–126

Owens SJ, Kimmins FM 1981. Stigma morphology in Commelinaceae. Ann Bot 47: 771–783

Owens SJ, Dickinson HG 1983. Pollen wall development in *Gibasis* (Commelinaceae). Ann Bot 51: 1–16

Pacini E 1990. Tapetum and microspore function. In: Microsporogenesis: Ontogeny and Systematics. S Blackmore, RB Knox (eds). Acad Press, London, pp 213–217

Pacini E 1994. Cell biology of anther and pollen development. In: Genetic Control of Self-incompatibility and Reproductive Development in Flowering Plants. EG Williams, AE Clarke, RB Knox (eds). Kluwer Academic Publishers, The Netherlands, pp 289–308

Pacini E 1997. Tapetum character status; analytical keys for tapetum types and activities. Can J Bot 75: 1448–1459

Pacini E 2000. From anther and pollen ripening to pollen presentation. Plant Syst Evol 222: 19–43

Pacini E, Juniper BE 1979. The ultrastructure of pollen grain development in the olive (*Olea europaea*). 1. Proteins in the pore. New Phytol 83: 157–164

Pacini E, Juniper BE 1983. The ultrastructure of the formation and development of the amoeboid tapetum in *Arum italicum* Miller. Protoplasma 117: 405–408

Pacini E, Keijzer CJ 1989. Ontogeny of intruding nonperiplasmodial tapetum in the wild chicory (*Cichorium intybus* L. (Compositae)). Plant Syst Evol 167: 149–164

Pacini E, Franchi GG 1992. Diversification and evolution of the tapetum. In: Pollen Spores: Patterns of Diversification. S Blackmore, S Barnes (eds). Oxford Univ Press, Oxford, pp 301–316

Pacini E, Franci GG 1993. Role of tapetum in pollen and spore dispersal. In: The Tapetum: Cytology, Function, Biochemistry and Evolution. M Hesse, E Pacini, MTM Willemse, AE Clarke, RB Knox (eds). Plant Syst Evol, Suppl No. 7, pp 1–11

Pacini E, Franchi G, Sarfatti G 1981. On the widespread occurrence of poral sporophytic proteins in pollen of dicotyledons. Ann Bot 47: 405–408

Pacini E, Franchi GG, Hesse M 1985. The tapetum: its form, function and possible phylogeny in Embryophyta. Plant Syst Evol 149: 155–185

Pacini E, Taylor PE, Singh MB, Knox RB 1992. Development of plastids, including amyloplasts and starch grains, in pollen and tapetum of rye-grass, *Lolium perenne*. Ann Bot 70: 179–188

Palmer CE, Keller WA 1997. Pollen embryos. In: Pollen Biotechnology for Crop Production and Developnment. KR Shivanna, VK sawhney (eds), Cambridge Univ Press, New York, NY, pp 392–422

Palmer RG, Albersten MC, Horner HT, Skorupska H 1992. Male sterility in soybean and maize: developmental comparisons. The Nucleus 35: 1–18

Pandey KK 1958. Time and site of S-gene action. Nature 181: 1220–1221

Pandey KK 1963. Stigmatic secretion and bud pollinations in self- and cross-incompatible plants. Naturwissenschaften 50: 408–409

Pandey KK 1964. Elements of the S-gene complex. I. The SFI alleles in *Nicotiana*. Genet Res 2: 397–409

Pandey KK 1965. Centric chromosome fragments and pollen part mutation of the incompatibility gene in *Nicotiana alata*. Nature 206: 792–795

Pandey KK 1967. Origin of genetic variability: combination of peroxidase isozymes determine multiple allelism of the S-gene in *Nicotiana alata*. Nature 213: 669

Pandey KK 1968. Compatibility relationships in flowering plants. Role of the S-gene complex. Am Nat 102: 475–489

Pandey KK 1969. Elements of the S-gene complex. V. Interspecific cross-compatibility relationships and theory of the evolution of the S-complex. Genetica 40: 447–474

Pandey KK 1970. Time and site of the S-gene action, breeding system and relationships in incompatibility. Euphytica 19: 364–372

Pandey KK 1973. Heat sensitivity of esterase isozymes in the styles of *Lilium* and *Nicotiana*. New Phytol 72: 839–850

Pandey KK 1977. Evolution of incompatibility systems in plants: complementarity and the mating locus in flowering plants and fungi. Theor Appl Genet 50: 89–101

Pandey KK 1979. Overcoming incompatibility and promoting genetic recombination in flowering plants. New Zealand J Bot 17: 645–664

Pandolfi T, Pacini E, Calder DM 1993. Ontogenesis of monod pollen in *Pterostylis plumosa* (Orchidaceae, Neottioideae). Plant Syst Evol 186: 175–184

Park CH, Walton PD 1990. Intergeneric hybrids and an amphiploid between *Elymus canadensis* and *Psathyrostachys juncea*. Euphytica 45: 217–222

Parkinson B, Pacini E 1995. A comparison of tapetal structure and function in pteridophyta and angiosperms. Plant Syst Evol 198: 55–88

Patel A, Hossaert-Mckey M, Mckey D 1993. *Ficus*-pollinator research in India: Past, present and future. Curr Sci 65: 243–253

Patel CT 1981. Evolution of hybrid-4-cotton. Curr Sci 50: 443–446

Patterson TG, Batreau L, Lewis H, Vega R, Sunderworth S 1991. Factors influencing the purity of hybrid wheat seed produced with a chemical hybridizing agent, SC-2053. Agron Abstr 169

Paul W, Hodge R, Smartt S, Draper J, Scott R 1992. The isolation and characterization of the tapetum-specific *Arabidopsis thaliana A9* gene. Plant Mol Biol 19: 611–622

Pauw A 1998. Pollen transfer on bird's tongue. Nature 394: 731–732

Peach JC, Latche A, Larriguadire C, Reid MS 1987. Control of early ethylene synthesis in pollinated *Petunia* flowers. Plant Physiol Biochem 25: 431–437

Pearson OH 1983. Heterosis in vegetable crops. In: Heterosis: Reappraisal of Theory and Practice. R Frankel (ed). Springer-Verlag, Berlin, pp 138–189

Pechan PM 1989. Successful cocultivation of *Brassica napus* microspores and proembryos with *Agrobacterium*. Plant Cell Rep 8: 387–390

Pedroso MC, Pais S 1993. Regeneration from anthers of adult *Camellia japonica* L. In Vitro Cell Devel Biol 29: 155–159

Peng-Hong-fu, Xue Zheng-sheng, Miao Fang, Pan Dungpi, Liu Zi-ming, Liu Zhong-wen, Liu Shau-rong, Tao Shung-xing 1990. The effect of pollen enhancing tolerance to hypoxis and promoting adaptation to highlands (in Chinese). J Chinese Med 70: 77–81

Pennell RI, Bell PR 1987. Intracellular RNA during meiosis in microsporangia of *Taxus baccata*. Am J Bot 74: 444–450

Perez M, Ishioka GY, Walker LE, Chestnut RW 1990. cDNA cloning and immunological characterization of the ryegrass allergen *Lol p 1*. J Biol Chem 265: 16210–16215

Perry DA 1978. Report of the vigour test committee—1974–77. Seed Sci Technol 6: 159–181

Pettitt JM 1976. Pollen wall and stigma surface in the marine angiosperms *Thalassia* and *Thalassodendron*. Micron 7: 21–32

Pettitt JM 1978. Regression and elimination of organelles during meiosis in *Lycopodium*. Grana 17: 99–105

Pettitt JM 1980. Reproduction in seagrasses: nature of the pollen and receptive surface of the stigma in the Hydrocharitaceae. Ann Bot 45: 257–271

Pettitt JM 1982. Ultrastructural and immunocytochemical demonstration of gametophytic proteins in the pollen tube wall of the primitive gymnosperm *Cycas*. J Cell Sci 57: 189–214

Pfahler PL 1992. Analysis of ecotoxic agents using pollen tubes. In: Modern Methods of Plant Analysis. HF Linskens, JF Jackson (eds). Springer-Verlag, Berlin, vol 13, pp 317–331

Philbrick CT 1984. Pollen tube growth within vegetative tissues of Callitriche (Callitrichaceae). Am J Bot 71: 882–886

Philbrick CT, Anderson GT 1992. Pollination biology in the Callitrichaceae. Syst Bot 17: 282–292

Picton JM, Steer MW 1981. Determination of secretory vesicle production rates by dictyosomes in pollen tubes of *Tradescantia* using cytochalasin D. J Cell Sci 49: 261–272

Pierson ES, Cresti M 1992. Cytoskeleton and cytoplasmic organization of pollen and pollen tubes. Int Rev Cytol 140: 73–125

Pierson ES, Derksen J, Trass JA 1986. Organization of microfilaments and microtubules in pollen tubes grown in vitro and in vivo in various angiosperms. Eur J Cell Biol 41: 14–18

Pierson ES, Miller DD, Callaham DA, Shipley AM, Rivers BA, Cresti M, Hepler PK 1994. Pollen tube growth is coupled to the extracellular calcium ion flux and the intracellular calcium gradient: Effect of BAPTA-type buffers and hypertonic media. Plant Cell 6: 1815–1828

Piffanelli P, Ross JHE, Murphy DJ 1997. Intra- and extracellular lipid composition and associated gene expression patterns during pollen development in *Brassica napus*. Plant J 11: 549–562

Pittarelli GW, Stavely JR 1975. Direct hybridization of *Nicotiana repanda* × *N. tabacum*. J Hered 66: 281–284

Plegt L, Bino RJ 1989. β-glucuronidase activity during development of the male gametophyte from transgenic and non-transgenic plants. Mol Gen Genet 216: 321–327

Plegt NM, Ven BCE, Bino RJ, Slam TPM, Tunen AJ 1992. Introduction and differential use of various promoters in pollen grains of *Nicotiana glutinosa* and *Lilium longiflorum*. Plant Cell Rep 11: 20–24

Poddubnaya-Arnoldi VA 1960. Study of fertilization in the living material of some angiosperms. Phytomorphology 10: 185–198

Polito VS 1983. Calmodulin and calmodulin inhibitors: effects on pollen germination and pollen tube growth. In: Pollen: Biology and Implications for Plant Breeding, DL Mulcahy, E Ottaviano (eds). Elsevier Biomedicals, Amsterdam, The Netherlands, pp 53–60

Pollard AS, Parr A, Loughman BC 1977. Boron in relation to membrane function in higher plants. J Exp Bot 28: 831–941

Pomeroy MK, Kramer JKG, Hunt DJ, Keller WA 1991. Fatty acid changes during development of zygotic and microspore-derived embryos of *Brassica napus*. Physiol Plant 81: 447–454

Pomeroy N 1981. Use of natural sites and field hives by a long-tongued bumble bee *Bombus ruderatus*. New Zealand J Agri Res 24: 409–414

Porter EK, Bird JM, Dickinson HG 1982. Nucleic acid synthesis in microsporocytes of *Lilium* cv cinnabar: events in the nucleus. J Cell Sci 57: 229–246

Porter KB, Foster JP, Peterson GL, Worrall 1985. Evaluation of chemical hybridizing agent SD 84811 in winter wheat. Agron Abstr Annual Meetings, Am Soc Agron, Dec 1–6, Chicago, Ill, p 67

Portnoi L, Horovitz A 1977. Sugars in natural and artificial pollen germination substrates. Ann Bot 41: 21–27

Potrykus I 1990. Gene transfer in plants: assessment and perspectives. Physiol Plant 79: 125–134

Potrykus I 1991. Gene transfer in plants: Assessment of published approaches and results. Ann Rev Plant Physiol Plant Mol Biol 42: 205–225

Potts BM, Potts WC, Cauvin B 1987. Inbreeding and interspecific hybridization in *Eucalyptus gunnii*. Silvae Genet 30. 194–199

Poulsen GB, Kahl G, Weising 1993. Oligonucleotide fingerprinting of resynthesised *Brassica napus*. Euphytica 70: 53–59

Powell GM III 1980. Polyethylene glycol. In: Handbook of Water Soluble Gums and Resins. RE Davidson (ed). McGraw-Hill, New York, NY, pp 18.1–18.31

Prakash L, John P, Nair GM, Pratapasenan G 1988. Effect of spermidine and methylglyoxal *bis* (guanylhydrazone) (MGBG) on in vitro pollen germination and tube growth in *Catheranthus reseus*. Ann Bot 61: 373–375

Prakash S, Chopra VL 1990. Male sterility caused by cytoplasm of *Brassica oxyrrhina* in *B. campestris* and *B. juncea*. Theor Appl Genet 79: 285–287

Prakash S, Kirti PB, Chopra VL 1998. Development of cytoplasmic male sterility-fertility restoration systems of variable origin in mustard, *Brassica juncea*. In: Proc Intl Symp Brassicas. T Gregiore, AA Monteiro (eds), Acta Hort 459: 299–304

Preuss D, Rhee SY, Davies RW 1994. Tetrad analysis possible in *Arabidopsis* with mutation of the QUARTET (QRT) genes. Science 264: 1458–1460

Preuss D, Lemieux B, Yen G, Davis RW 1993. A conditional sterile mutation eliminates surface components from *Arabidopsis* pollen and disrupts cell signalling during fertilization. Genes Devel 7: 974–985

Priestley DA 1986. Seed Ageing: Implications for Seed Storage and Persistence in the Soil. Comstock, Ithaca, London

Pring DR, Levings CS III, Hu WWL, Timothy DH 1977. Unique DNA associated with mitochondria in the "S"-type cytoplasm of male sterile maize. Proc Natl Acad Sci USA 74: 889–892

Quiros CF 1975. Exine pattern of an hybrid between *Lycopersicon esculentum* and *Solanum pennellii*. J Hered 66: 45–47

Raff JW, Knox RB, Clarke AE 1981. Style antigens of *Prunus avium* L. Planta 153: 125–129

Raghavan V 1976. Role of generative cell in androgenesis in henbane. Science 191: 388–389

Raghavan V 1977. Applied aspects of embryo culture. In: Applied and Fundamental Aspects of Plant Cell, Tissue and Organ Culture. J Reinert, YPS Bajaj (eds). Springer-Verlag, Berlin, pp. 357–397

Raghavan V 1978. Origin and development of pollen embryoids and pollen calluses in cultured anther segments of *Hyoscyamus niger* (Henbane). Am J Bot 65: 984–1002

Raghavan V 1986. Variability through wide crosses and embryo rescue. In: Cell Culture and Somatic Cell Genetics of Plants. IK Vasil (ed), Academic Press, New York, NY, pp 613–663

Raghavan V 1999. Molecular Embryology of Flowering Plants. Cambridge Univ Press, New York, NY

Raghavan V, Jacobs WP 1961. Studies on the floral histogenesis and physiology of *Perilla*. II Floral induction in cultured apical buds of *P. frutescens*. Am J Bot 48: 751–760

Rajam MV 1989. Restriction of pollen germination and tube growth in lily pollen by inhibitors of polyamine metabolism. Plant Sci 59: 53–56

Rangaswamy NS 1963. Control of fertilization and embryo development. In: Recent Advances in the Embryology of Angiosperms. P. Maheshwari (ed). Intl Soc Plant Morphologists, Delhi, pp 327–353

Rangaswamy NS 1977. Applicatons of in vitro pollination and in vitro fertilization. In: Applied and Fundamental Aspects of Plant Cell, Tissue and Organ Culture. J Reinert, YPS Bajaj (eds), Springer-Verlag, Berlin, pp 412–425

Rangaswamy NS, Shivanna KR 1967. Induction of gamete compatibility and seed formation in axenic cultures of a diploid self-incompatible species of *Petunia*. Nature 216: 937–939

Rangaswamy NS, Shivanna KR 1969. Test-tube fertilization in *Dicranostigma franchetianum* (Prain) Fedde. Curr Sci 38: 257–259

Rangaswamy NS, Shivanna KR 1971a. Overcoming self-incompatibility in *Petunia axillaris* (Lam.) BSP II. Placental pollination in vitro. J Indian Bot Soc 50A: 286–296

Rangaswamy NS, Shivanna KR 1971b. Overcoming self-incompatibility in *Petunia axillaris* III. Two-site pollinations in vitro. Phytomorphology 21: 284–289

Rao GU 1995. Development and characterization of new alloplasmics in drop brassicas. Ph.D. thesis, Univ Delhi, Delhi

Rao GU, Jain A, Shivanna KR 1992. Effects of high temperature stress on *Brassica* pollen: Viability, germination and ability to set fruits and seeds. Ann Bot 69: 193–198

Rao GU, Sawhney VK, Shivanna KR 1995. Thermotolerance of pollen of *Petunia* and *Nicotiana*. Curr Sci 69: 809–810

Rao GU, Lakshmikumaran M, Shivanna KR 1996. Production of intergeneric hybrids, amphiploids and backcross progeny between a cold tolerant wild species, *Erucastrum abyssinicum* and crop brassicas. Theor Appl Genet 92: 786–790

Rao GU, Batra-Sarup V, Prakash S, Shivanna KR 1993. Development of a new cytoplasmic male sterile system in *Brassica juncea* through wide hybridization. Plant Breed 112: 171–174

Rao MK, Uma Devi K, Arundhati A 1990. Applications of genic male sterility in plant breeding. Plant Breeding 105: 1–25

Rao PS 1965. In vitro fertilization and seed formation in *Nicotiana rustica* L. Phyton 22: 165–167

Rashid A, Street HE 1973. The development of haploid embryoids from anther cultures of *Atropa belladonna* L. Planta 113: 263–270

Rashid A, Siddiqui AW, Reinert J 1981. Ultrastructure of embryogenic pollen of *Nicotiana tabacum* var. "Badischer Burley". Protoplasma 107: 375–385

Rashid A, Siddiqui AW, Reinert J 1982. Subcellular aspects of origin and structure of pollen embryos of *Nicotiana*. Protoplasma 113: 202–208

Raskin I, Ehmann A, Melander WR, Meeuse BJD 1987. Salicilic acid: a natural inducer of heat production in Arum lilies. Science 237: 1601–1602.

Read SM, Clarke AE, Bacic A 1993. Stimulation of growth of cultured *Nicotiana tabacum* W 38 pollen tubes by polyethylene glycol and Cu$_{(II)}$salts. Protoplasma 177: 1–14

Read SM, Newbegin E, Clarke AE, McClure BA, Kao T-h 1995. Disputed ancestry: comments on a model for the origin of self-incompatibility in flowering plants. Plant Cell 7: 661–665

Real L (ed) 1983. Pollination Biology. Academic Press, Orlando, FL.

Reed SM, Collins GB 1978. Interspecific hybrids in *Nicotiana* through in vitro culture of fertilized ovules. J Hered 69: 311–315

Reger BJ, Chaubal R, Pressey R 1992. Chemotropic responses by pearl millet pollen tubes. Sex Plant Reprod 5: 47–56

Reijnen WH, van Herpen MMA, de Groot PFM, Olmedilla A, Schrauwen JAM, Weterings KAP, Wullems GJ 1991. Cellular localization of pollen-specific mRNA by in situ hybridization and confocal scanning microscopy. Sex Plant Reprod 4: 254–257

Reimann-Phillipp R 1965. The application of incompatibility in plant breeding. In: Genetics Today. SJ Geerts (ed), Proc XI Intl Cong Genet, The Hague, vol III, pp 656–663

Reimann-Phillipp R 1983. Heterosis in ornamentals. In: Heterosis: Reappraisal of Theory and Practice. R Frankel (ed). Springer-Verlag, Berlin, pp 234–257

Reinert J, Bajaj YPS 1977. Anther culture, haploid production and its significance. In: Applied and Fundamental Aspects of Plant Cell Tissue and Organ Culture. J Reinert, YPS Bajaj (eds). Springer-Verlag, Berlin, pp 91–122

Reinhard KJ, Hamilton DL, Hevly RH 1991. Use of pollen concentration in paleopharmacology, coprolite evidence in medical plants. J Ethnobiol 11: 117–132

Reiser L, Fischer RL 1993. The ovule and embryo sac. Plant Cell 5: 1291–1301

Reiss HD, Herth W 1979. Calcium ionophore A 23187 affects localized wall secretion in the tip region of pollen tubes of *Lilium longiflorum*. Planta 145: 225–232

Reynaerts A, van de Wiele H, De Sutter G, Jansens J 1993. Engineered genes for fertility control and their application in hybrid seed production. Sci Hort 55: 125–139

Reynolds TV, Raghavan V 1982. An autoradiographic study of RNA synthesis during maturation and germination of pollen grains of *Hyoscyamus niger*. Protoplasma 111: 177–188

Reznickova SA, Bogdanov YF 1972. Meiosis in excised anthers of *Lilium candidum*. Biol Zentralblatt 91: 409–423

Richards AJ 1986. Plant Breeding Systems. Allen & Unwin Publ Ltd, London

Richards AJ, Ibrahim HB 1982. The breeding system in *Primula veris* L. II. Pollen tube growth and seed-set, New Phytol 90: 305–314

Rick CM 1948. Genetics and development of nine male sterile tomato mutants. Hilgardia 18: 599–633

Rick CM, Boynton JE 1967. A temperature sensitive male-sterile mutant of the tomato. Am J Bot 54: 601–611

Riggs CD, Hasenkampf CA 1991. Antibodies directed against a meiosis-specific, chromatin associated protein identify conserved meiotic epitopes. Chromosoma 101: 92–98

Roberts AM, Beven LJ, Flora PS, Jepson I, Walker MR 1993. Nucleotide sequence of cDNA encoding the group II allergen of cock's foot orchard grass (*Dactylis glomerata*), Dac g II. Allergy 48: 615–623

Roberts AM, van Ree R, Emly J, Cardy SM, Rottier MMA, Walker MR 1992. N-terminal amino acid sequence homologies in group V grass pollen allergens. Int Arch Allergy Appl Immunol 98: 178–180

Roberts EH 1972. Viability of Seeds. Chapman & Hall, London

Roberts IN, Stead AD, Ockendon DJ, Dickinson HG 1979. A glycoprotein associated with the acquisition of the self-incompatibility system in maturing stigmas of *Brassica oleracea*. Planta 146: 179–183

Roberts IN, Stead AD, Ockendon DJ, Dickinson HG 1980. Pollen stigma interactions in *Brassica oleracea*. Theor Appl Genet 58: 241–246

Roberts IN, Gaude TC, Harrod G, Dickinson HG 1983. Pollen stigma interactions in *Brassica oleracea*; a new pollen germination medium and its use in elucidating the mechanism of self-incompatibility. Theor Appl Genet 65: 231–238

Robinson WS 1979. Effect of apple cultivar on foraging behaviour and pollen transfer by honey bees. J Amer Soc Hort Sci 104: 596–598

Rocha OJ, Stephenson AG 1990. The effects of ovule position within the ovary on the probability of seed maturation, seed weight and progeny performance in *Phaseolus coccineus* (Leguminosae). Am J Bot 77: 1320–1329

Rocha OJ, Stephenson AG 1991. Order of fertilization within the ovary in *Phaseolus coccineus* L. (Leguminosae). Sex Plant Reprod 4: 124–131

Rogers BL, Bond JF, Morgenstern JP, Counsell CM, Griffith IW 1996. Immunological characterization of the major ragweed allergens *Amb a* I and *Amb a* II. In: Pollen Biotechnology: Gene Expression and Allergen Characterization. SS Mohapatra, RB Knox (eds). Chapman & Hall, New York, NY, pp 211–134

Rogers CM, Harris BD 1969. Pollen exine deposition: a clue to its control. Am J Bot 56: 1209–1211

Roggen HPJR, Linskens HG 1967. Pollen tube growth and respiration in incompatible intergeneric crosses. Naturwissenschaften 54: 542–543

Roggen HPJR, van Dijk AJ 1972. Breaking incompatibility in *Brassica oleracea* L. by steel brush pollination. Euphytica 21: 424–425

Roggen HPJR, van Dijk AJ, Dorsman C 1972. Electric aided pollination: a method of breaking incompatibility in *Brassica oleracea*. Euphytica 21: 181–184

Rose JB, Dunwell JM, Sunderland N 1987. Anther culture of *Lolium temelentum*, *Festuca pratensis* and *Lolium* × *Festuca* hybrids. II. Anther and pollen development in vivo and in vitro. Ann Bot 60: 203–214

Rosen B, Hallden C, Heneen WK 1988. Diploid *Brassica napus* somatic hybrids: characterization of nuclear and organelle DNA. Theor Appl Genet 76: 197–203

Rosen WG 1971. Pollen tube growth and fine structure. In: Pollen: Development and Physiology, J Heslop-Harrison (ed). Butterworths, London, pp 177–185

Rosen WG, Thomas HR 1970. Secretory cells of lily pistils. I. Fine structure and function. Am J Bot 57: 1108–1114

Rosenfield C-L, Fann C, Loewus FA 1978. Metabolic studies on intermediates in the myoinositol pathway in *Lilium*

longiflorum pollen. I. Conversion to hexoses. Plant Physiol 61: 89–95

Roubik DW (ed) 1995. Pollination of Cultivated Plants in the Tropics. FAO Agric Serv Bulletin 118

Rowley JR 1964. Formation of the pore in pollen of *Poa annua*. In: Pollen Physiology and Fertilization. Linskens HF (ed). North-Holland Publ, Amsterdam, The Netherlands, pp 59–69

Rowley JR 1973. Formation of pollen exine bacules and micro-channels on a glycocalyx. Grana 13: 129–138

Rowley JR 1975. Germinal aperture formation in pollen. Taxon 24: 17–25

Royo R, Kunz C, Kowyama Y, Anderson M, Clarke AE, Newbigin E 1994. Loss of histidine residue at the active site of S-locus ribonuclease is associated with self-incompatibility in *Lycopersicon peruvianum*. Proc Natl Acad Sci, USA 91: 6511–6514

Rubinstein AL, Broadwater AH, Lowrey KB, Bedinger PA 1995. *Pex 1*, a pollen-specific gene with an extensin-like domain. Proc Natl Acad Sci USA 92: 3086–3090

Rudd JJ, Franklin FCH, Lord JM, Franklin-Tong VE 1997. Ca^{2+}-independent phosphorylation of a 60 kD protein is stimulated by the self-incompatibility response. Plant J 12: 507–514

Rusche ML, Mogensen HL 1988. The male germ unit of *Zea mays*: quantitative ultrastructure and three-dimensional analysis. In: Sexual Reproduction in Higher Plants. M Cresti, P Gori, E Pacini (eds). Springer-Verlag, Berlin, pp 221–226

Russell SD 1982. Fertilization in *Plumbago zeylanica*: entry and discharge of the pollen tube in the embryo sac. Can J Bot 60: 2219–2230

Russell SD 1984. Ultrastructure of the sperm of *Plumbago zeylanica*: 2. Quantitative cytology and three-dimensional reconstruction. Planta 162: 385–391

Russell SD 1985. Preferential fertilization in *Plumbago*: Ultrastructural evidence for gamete-level recognition in an angiosperm. Proc Natl Acad Sci USA 82: 6129–6132

Russell SD 1991. Isolation and characterization of sperm cells in flowering plants. Ann Rev Plant Physiol Plant Mol Biol 42: 189–204

Russell SD 1992. Double fertilization. Int Rev Cytol 140: 357–388

Russell SD 1993. The egg cell: development and role in fertilization and early embryogenesis. Plant Cell 5: 1349–1359

Russell SD, Cass DD 1981. Ultrastructure of the sperms of *Plumbago zeylanica*: I. Cytology and association with the vegetative nucleus. Protoplasma 107: 85–108

Russell SD, Dumas C (ed) 1992. Sexual Reproduction in Higher Plants. Intl Rev Cytol, vol 140

Rustagi PN, Mohan Ram HY (1971) Evaluation of Mendok and Dalapon as male sterility compounds in linseed. New Phytol 70: 117–130

Sacher RF, Mulcahy DL, Staples RC 1983. Developmental selection for salt tolerance during self-pollination of *Lycopersicon* × *Solanum* F_1 for salt tolerance of F_2. In: Pollen: Biology and Implications for Plant Breeding. DL Mulcahy, E Ottaviano (eds). Elsevier Biomedical, Amsterdam, The Netherlands, pp 329–334

Sacristan MD 1982. Resistance response of *Phoma lingam* plants regeneratd from selected cells and mutagenized embryogenic culture of haploid *Brassica napus*. Theor Appl Genet 61: 193–200

Saini HS, Aspinall D 1981. Effect of water deficit on sporogenesis in wheat (*Triticum aestivum* L.). Ann Bot 48: 623–633

Sampson DR 1962. Intergeneric pollen-stigma incompatibility in Cruciferae. Can J Genet Cytol 4: 38–49

Sampson FB 1981. Synchronous versus asynchronous mitosis within permanent pollen tetrads of the Winteraceae. Grana 20: 19–23

San LH 1976. Haploides d'*Hordeum vulgare* L. par culture in vitro d'ovaries non fecondes. Ann Amelior Plant 26: 751–754

San LH, Gelebart P 1986. Production of gynogenic haploids. In: Cell Culture and Somatic Cell Genetics. IK Vasil (ed). Academic Press, Orlando, FL, vol 3, pp 305–322

Sanders LC, Lord EM 1989. Directed movement of latex particles in the gynoecia of three species of flowering plants. Science 243: 1606–1608

Sanders LC, Lord EM 1992. A dynamic role for the stylar matrix in pollen tube extension. Intl Rev Cytol 140: 297–318

Sanders LC Wang CS, Lord EM 1991. A homolog of the substance adhesion molecule vitronectin occurs in four species of flowering plants. Plant Cell 3: 629–635

Sanger R, Jackson WT 1971a. Fine structure study of pollen development in *Haemanthus katherinae* Baker II. Microtubules and elongation of generative cells. J Cell Sci 8: 303–305

Sanger R, Jackson WT 1971b. Fine structure study of pollen development of *Haemanthus katherinae* Baker. III. Changes in organelles during development of the vegetative cell. J Cell Sci 8: 317–329

Sangwan RS, Sangwan-Norreel BS 1990. Anther and pollen culture. In: Plant Tissue Culture: Applications and Limitations. SS Bhojwani (ed). Elsevier, Amsterdam, The Netherlands, pp 220–241

Sangwan RS, Mathivet V, Vasseur G 1989. Ultrastructural localization of acid phosphatase during male meiosis and sporogenesis in *Datura*: evidence for digestion of cytoplasmic structures in the vacuoles. Protoplasma 149: 38–46.

Sangwan RS, Ducrocq C, Sangwan-Noreel B 1993. *Agrobacterium*-mediated transformation of pollen embryos in *Datura innoxia* and *Nicotiana tabacum*: production of transgenic haploid and fertile homozygous diploid plants. Plant Sci 95: 99–115

Sareen PK, Chowdhury JB, Chowdhury VK 1992. Amphiploids/synthetic crop species. In: Distant Hybridization of Crop Plants. G Kalloo, JB Chowdhury (eds). Springer-Verlag, Berlin, pp 62–81

Sari-Gorla M, Frova C 1997. Pollen tube growth and pollen selection. In: Pollen Biotechnology for Crop Production and Improvement. KR Shivanna, VK Sawhney (eds). Cambridge University Press, New York, NY, pp 333–351

Sari-Gorla M, Ottaviano E, Frascaroli E, Landi P 1989. Herbicide-tolerant corn by pollen selection. Sex Plant Reprod 2: 65–69

Sari-Gorla M, Ferrario S, Frascaroli E, Frova C, Landi P, Villa M 1994. Sporophytic response to pollen selection for Alachlor tolerance in maize. Theor Appl Genet 88: 812–817

Sarkar KR 1974. Genetic selection techniques for production of haploids in higher plants. In: Haploids in Higher Plants: Advances and Potential. KJ Kasha (ed.) Univ Guelph, Guelph, pp 33–41

Sarker RH, Elleman CJ, Dickinson HG 1988. Control of pollen hydration in *Brassica* requires continued protein synthesis and glycosylation is necessary for intraspecific incompatibility. Proc Natl Acad Sci, USA 85: 4340–4344

Sarla N 1988. Overcoming interspecific incompatibility in the cross *Brassica campestris* ssp. *japonica* × *Brassica oleracea* var. *botrytis* using irradiated mentor pollen. Biol Plant 30: 384–386

Sassa H, Hirano H, Ikehashi H 1992. Self-incompatibility-related RNases in styles of Japanese pear (*Pyrus serotina* Rehd.). Plant Cell Physiol 33: 811–814

Sastri DC 1979. Studies on pollen-pistil interaction in some angiosperms. Ph.D. Thesis, Univ Delhi, Delhi

Sastri DC 1985. Incompatibility in angiosperms: significance in crop improvement. Adv Appl Biol 10: 71–111

Sastri DC, Shivanna KR 1976a. Recognition pollen alters incompatibility in *Petunia*. Incompatibility Newslett 7: 22–24

Sastri DC, Shivanna KR 1976b. Attempts to overcome inter-specific incompatibility in *Sesamum* by using recognition pollen. Ann Bot 40: 891–893

Sastri DC, Shivanna KR 1978. Seedset following compatible pollination in incompatibly pollinated flowers of *Petunia*. Incompatibility Newslett 9: 91–93

Sastri DC, Shivanna KR 1979. Role of pollen-wall proteins in intraspecific incompatibility in *Saccharum bengalense*. Phytomorphology 29: 324–330

Sastri DC, Shivanna KR 1980. Efficacy of mentor pollen in overcoming intraspecific incompatibility in *Petunia*, *Raphanus* and *Brassica*. J Cytol Genet 15: 107–112

Sastri DC, Moss JP 1982. Effects of growth regulators on incompatible crosses in the genus *Arachis* L. J Exp Bot 33: 1293–1301

Sastri DC, Moss JP, Nalini MS 1983. The use of in vitro methods in groundnut improvement. In: Proc Intl Symp on Plant Cell Culture in Crop Improvement. SK Sen,

KL Giles (eds). Plenum Press, New York, NY, pp 365–370

Saul MW, Potrykus I 1984. Species-specific repetitive DNA used to identify interspecific somatic hybrids. Plant Cell Rep 3: 65–67

Saunders JA, Matthews BF, Van Wert SL 1990. Pollen electrotransformation for gene transfer in plants. In: Guide to Electroporation and Electrofusion. DC Chang (ed). Academic Press, New York, NY, pp 227–247

Sauter JJ 1971. Physiology and biochemistry of meiosis in anther. In: Pollen: Development and Physiology. J Heslop-Harrison (ed). Butterworths, London, pp 3–15

Savithri KS, Ganapathy PS, Sinha SK 1980. Sensitivity to low temperature in pollen germination and fruit set in *Cicer arietinum* L. J Exptl Biol 31: 475–481

Savoor RR 1998. Pollination management: an ecofriendly green revolution eludes India. Curr Sci 74: 121–125.

Sawhney VK 1983. Temperature control of male sterility in a tomato mutant. J Hered 74: 51–54

Sawhney VK 1984. Hormonal and temperature control of male-sterility in a tomato mutant. Proc VIII Intl Symp Sexual Reproduction in Seed Plants, Ferns and Mosses. PUDOC Publ, Wageningen, pp 36–38

Sawhney VK 1994. Genic male sterility in tomato and its manipulation in breeding. In: Genetic Control of Self-incompatibility and Reproductive Development in Flowering Plants. EG Williams, AE Clarke, RB Knox (eds). Kluwer Academic Publ, Dordrecht, The Netherlands, pp 443–458

Sawhney VK 1997. Genic male sterility. In: Pollen Biotechnology for Crop Production and Improvement. KR Shivanna, VK Sawhney (eds). Cambridge Univ Press, New York, NY, pp 183–198

Sawhney VK, Shukla A 1994. Male sterility in flowering plants: Are plant growth substances involved? Am J Bot 81: 1640–1647

Scalla R, Duc G, Rigaud J, Lefebre A, Meignoz R 1981. RNA containing intracellular particles in cytoplasmic male sterile faba bean (*Vicia faba* L.). Plant Sci Lett 22: 269–277

Schlichting CD, Stephenson AB, Davis LE, Winsor JA 1987. Pollen competition and offspring variance. Evol Trends Plants 1: 35–39

Schmidt JO 1982. Pollen foraging preferences of honey bees. South-Western Entomol 7: 255–259

Schmidt JO, Buchmann SL 1992. Other products of the hive. In: The Hive and the Honey Bee. JM Graham (ed), Dadant & Sons, Hamilton, IL, pp 927–982

Schmidt R, Alpert PH 1977. A test of Burck's hypothesis relating anther dehiscence to nectar secretion. New Phytol 78: 487–498

Schmucker TH 1933. Zur Blutenbiologie tropischer *Nymphaea* Arten. H. Bor ale entscheidences Factor. Planta 18: 641–650

Scholten S, Kranz E 2001. In vitro fertilization and expression of transgenes in gametes and zygotes. Sex Plant Reprod 14: 35–40

Schopfer CR, Nasrallah ME, Nasrallah JB 1999. The male determinant of self-incompatibility in *Brassica*. Science 286: 1697–1700

Schulz PJ, Cross JW, Almeida E 1993. Chemical agents that inhibit pollen development: effects of the phenyl-cinnoline carboxylates SC-1058 and SC-1271 on the ultrastructure of developing wheat anthers (*Triticum aestivum* L. var. *Yecora rojo*). Sex Plant Reprod 6: 108–212

Schweizer G, Ganal M, Ninnemann H, Hemelben V 1988. Species-specific DNA sequences for identification of somatic hybrids between *Lycopersicon esculentum* and *Solanum acaule*. Theor Appl Genet 75: 679–684

Scott R, Dagless E, Hodge R, Paul W, Soufleri I, Draper J 1991. Patterns of gene expression in developing anthers of *Brassica napus*. Plant Mol Biol 17: 195–207

Scribailo RW, Barrett CH 1991. Pollen-pistil interaction in tristylous *Pontederia sagittata* (Pontederiaceae). I. Floral heteromorphism and structural features of the pollen tube pathway. Am J Bot 78: 1643–1661

Scutt CP, Gates PJ, Gatehouse JA, Boulter D, Croy RRD 1990. A cDNA-encoding an S-locus specific glycoprotein from *Brassica oleracea* plants containing the S_5 self-incompatiblity allele. Mol Gen Genet 220: 409–413

Searcy KB, Mulcahy DL 1985a. The parallel expression of metal tolerance in pollen and sporophytes of *Silene dioica* (l.) Clairv., *Silene alba* (Mill.) Krause and *Mimulus guttatus* DC. Theor Appl Genet 69: 597–602

Searcy KB, Mulcahy DL 1985b. Pollen selection and the gametophytic expression of metal tolerance in *Silene dioica* (Caryophyllaceae) and *Mimulus guttatus* (Scrophulariaceae). Am J Bot 72: 1700–1706

Searle RJG, Day JA 1980. Azabicyclohexane derivatives for pollen suppressing conditions. Eur Patent Application 12,471

Sears ER 1956. The transfer of leaf rust resistance from *Aegilops umbellulata* to wheat. Brookhaven Symp Biol 9:1–22

Seavy SR, Bawa KS 1986. Late-acting self-incompatibility in angiosperms. Bot Rev 52: 195–219

Sedgley M 1975. Flavonoids in pollen and stigma of *Brassica oleracea* and their effect on pollen germination in vitro. Ann Bot 39: 1091–1095

Sedgley M 1976. Control by the embryo sac over pollen tube growth in the avocado (*Persea americana* Mill.). New Phytol 77: 149–152

Sedgley M, Scholefield PB 1980. Stigma secretion in the watermelon before and after pollination. Bot Gaz 141: 428434

Sedgley M, Harbard J 1993. Pollen storage and breeding system in relation to controlled pollination of four species of *Acacia* (Leguminosae: Mimosiodae). Aust J Bot 41: 601–609.

Senaratna T, Kott L, Beversdorf WD, McKersie BC 1991. Desiccation of microspore-derived embryos of oilseed rape (*Brassica napus* L.). Plant Cell Rep 10: 342–344

Seymour RS, Schultze-Motel P 1996. Thermoregulating *Lotus* flowers. Nature 383: 305

Seymour RS, Schultz-Motel P 1997. Heat-producing flowers. Endeavour 21: 125–129

Shaanker RU, Ganeshaiah KN 1989. Stylar plugging by fertilized ovules in *Kleinhovia hospita* (Sterculiaceae)—a case of vaginal sealing in plants. Evol Trends Plants 3: 59–64

Shapiro BM, Eddy EM 1980. When sperm meets egg: biochemical mechanisms of gamete interaction. Intl Rev Cytol 66: 257–302

Sharma N 1986. Self-incompatibility recognition and inhibition in *Petunia* and *Nicotiana*. Ph.D. thesis, Univ Delhi, Delhi

Sharma N, Shivanna KR 1982. Effect of pistil extract on in vitro responses of compatible and incompatible pollen in *Petunia hybrida* Vilm. Indian J Exptl Biol 20: 255–256

Sharma N, Shivanna KR 1983a. Lectin-like components of pollen and complementary saccharide moiety of the pistil are involved in self-incompatibility recognition. Curr Sci 52: 913–916

Sharma N, Shivanna KR 1983b. Pollen diffusates of *Crotalaria retusa* and their role in pH regulation. Ann Bot 52: 165–170

Sharma N, Bajaj M, Shivanna KR 1985. Overcoming self-incompatibility through the use of lectins and sugars in *Petunia* and *Eruca*. Ann Bot 55: 139–141

Shayk M, Kolattukudy PE, Davis R 1977. Production of a novel extracellular cutinase by the pollen and the chemical composition and ultrastructure of the stigma cuticle of nasturtium (*Tropaeolum majus*). Plant Physiol 60: 907–915

Sheldon JM, Dickinson HG 1983. Determination of patterning in the pollen wall of *Lilium henryi*. J Cell Sci 63: 191–208

Sheridan WF, Stern H 1967. Histones of meiosis. Exp Cell Res 45: 323–346

Shi L, Mogensen HL, Zhu T, Smith SE 1991. Dynamics of nuclear pore density and distribution within developing pollen: implications for a functional relationship between the vegetative nucleus and the generative cell. J Cell Sci 99: 115–120

Shi MS 1986. The discovery, determination and utilization of the Hubei photosensitive genic male-sterile rice (*Oryza sativa* subsp. *japonica*). Acta Genet Sinica 13: 107–112

Shigenobu T, Sakamoto S 1977. Production of a polyhaploid plant of *Aegilops crassa* (6x) pollinated by *Hordeum bulbosum*. Jpn J Genet 52: 397–401.

Shivanna KR 1965. In vitro fertilization and seed formation in *Petunia violacea* Lindl. Phytomorphology 15: 183–185

Shivanna KR 1969. Elimination of self-incompatibility in *Petunia axillaris* (Lam.) B.S.P. Ph.D. Thesis, Univ Delhi, Delhi

Shivanna KR 1971. Overcoming self-incompatibility in *Petunia*: Differential treatment in vitro of the two placentae. Experientia 27: 864–865

Shivanna KR 1977. Pollen-stigma interaction—recognition, acceptance and rejection. In: Proc Symp Basic Sciences and Agriculture. Indian Natl Sci Acad, New Delhi, pp 53–61

Shivanna KR 1978. Elution of wall-bound proteins does not affect pollen germination and pollen tube growth in vitro in *Crotalaria retusa*. Incompatibility Newslett 10: 40–42

Shivanna KR 1979. Recognition and rejection phenomena during pollen-pistil interaction. Proc Indian Acad Sci 88 B: 115–141

Shivanna KR 1982. Pollen-pistil interaction and control of fertilization. In: Experimental Embryology of Vascular Plants. Johri BM (ed), Springer-Verlag, Berlin, Heidelberg, New York, pp 131–174

Shivanna KR 1995. Incompatibility and wide hybridization. In: Oilseed and Vegetable Brassicas. VL Chopra, S Prakash (eds), Oxford & IBH Publ Co Pvt Ltd, New Delhi, pp 77–102

Shivanna KR 1997. Barriers to hybridization. In: Pollen Biotechnology for Crop Production and Improvement. KR Shivanna, VK Sawhney (eds), Cambridge Univ Press, New York, NY, pp 261–272

Shivanna KR 2000. Wide hybridization in Brassicas. In: The Changing Scenario in Plant Sciences. VS Jaiswal, AK Rai, Uma Jaiswal, JS Singh (eds), Allied Publ Ltd, New Delhi, pp 197–211

Shivanna KR, Rangaswamy NS 1969. Overcoming self-incompatibility in *Petunia axillaris* (Lam.) B.S.P. I. Delayed pollination, pollination with stored pollen and bud pollination. Phytomorphology 19: 372–380

Shivanna KR, Heslop-Harrison J 1981. Membrane state and pollen viability. Ann Bot 47: 759–770

Shivanna KR, Sastri DC 1981. Stigma-surface proteins and stigma receptivity in some taxa characterized by wet stigma. Ann Bot 47: 53–64

Shivanna KR, Sharma N 1984. Self-incompatibility recognition in *Petunia hybrida*. Micron and Microscopica Acta 16: 233–245

Shivanna KR, Johri BM 1985. The Angiosperm Pollen: Structure and Function: Wiley Eastern, New Delhi

Shivanna KR, Cresti M 1989. Effects of high humidity and temperature stress on pollen membrane and pollen vigour. Sex Plant Reprod 2: 137–141.

Shivanna KR, Owens SJ 1989. Pollen-stigma interactions (Papilionoidae). In: Advances in Legume Biology. CH Stirton, JL Zarucchi (eds). Monogr Syst Bot Missouri Bot Gard 29: 157–182

Shivanna KR, Rangaswamy NS 1992. Pollen Biology: A Laboratory Manual. Springer-Verlag, Berlin, Heidelberg, New York

Shivanna KR, Sawhney VK 1993. Pollen selection for *Alternaria* resistance in oilseed brassicas: Responses of pollen grains and leaves to a toxin of *Alternaria brassicae*. Theor Appl Genet 86: 339–344.

Shivanna KR, Sawhney VK 1995. Polyethylene glycol improves the in vitro growth of *Brassica* pollen tubes without loss of germination. J Exptl Bot 46: 1771–1774

Shivanna KR, Rangaswamy NS 1997. Fertilization in flowering plants—What is new? Curr Sci 72: 300–301

Shivanna KR, Sawhney VK (eds) 1997. Pollen Biotechnology for Crop Production and Improvement. Cambridge University Press, New York, NY

Shivanna KR, Rangaswamy NS 1999. Pollen tube growth. What guides it to the embryo sac? J Plant Biol 26: 111–118

Shivanna KR, Jaiswal VS, Mohan Ram HY 1974a. Inhibition of gamete formation by cycloheximide in pollen tubes of *Impatiens balsamina*. Planta 117: 173–177

Shivanna KR, Jaiswal VS, Mohan Ram HY 1974b. Effect of cycloheximide in cultured pollen grains of *Trigonella foenumgraecum*. Plant Science Lett 3: 335–339

Shivanna KR, Heslop-Harrison Y, Heslop-Harrison J 1978a. Pollen stigma interaction: bud pollination in the Cruciferae. Acta Bot Neerl 27: 107–119

Shivanna KR, Heslop-Harrison Y, Heslop-Harrison J 1978b. Inhibition of the pollen tube in the self-incompatibility response of grasses. Incompatibility Newslett 10: 5–7

Shivanna KR, Heslop-Harrison J, Heslop-Harrison Y 1981. Heterostyly in *Primula*. 2. Sites of pollen inhibition and effects of pistil constituents on compatible and incompatible pollen tube growth. Protoplasma 107: 319–338

Shivanna KR, Heslop-Harrison Y, Heslop-Harrison J 1982. Pollen-pistil interaction in the grasses. 3. Features of self-incompatibility response. Acta Bot Neerl 31: 307–319

Shivanna KR, Heslop-Harrison J, Heslop-Harrison Y 1983. Heterostyly in *Primula*. 3. Pollen water economy: a factor in the intramorph-incompatibility response. Protoplasma 117: 175–184

Shivanna KR, Linskens HF, Cresti M 1991a. Responses of tobacco pollen to high humidity and heat stress: germination in vitro and in vivo. Sex Plant Reprod 4: 104–109

Shivanna KR, Linskens HF, Cresti M 1991b. Pollen viability and pollen vigour. Theor Appl Genet 81: 38–42

Shivanna KR, Saxena NP, Seetharama N 1998. An improved medium for in vitro pollen germination and pollen tube growth of chickpea (*Cicer aeritinum* L.) Chickpea and Pigeonpea Newslett 4: 28–29

Shivanna KR, Xu H, Taylor P, Knox RB 1987. Isolation of sperms from pollen tubes of flowering plants during fertilization. Plant Physiol 87: 647–650

Shivanna KR, Cresti M, Ciampolini F 1997. Pollen development and pollen-pistil interaction. In: Pollen Biotechnology for Crop Production and Improvement. KR Shivanna and VK Sawhney (eds). Cambridge Univ Press, New York, NY, pp 15–39.

Shoup JR, Overton J, Ruddat M 1980. Ultrastructure and development of the sexine in the pollen wall of *Silene alba* (Caryophyllaceae). Bot Gaz 141: 379–388

Shukla A, Sawhney VK 1993. Metabolism of dihydrozeatin in floral buds of wild-type and genic male sterile line of rapeseed (*Brassica napus*). J Exptl Bot 44: 1497–1505

Shukla A, Sawhney VK 1994. Abscisic acid: one of the factors affecting male sterility in *Brassica napus*. Physiol Plant 91: 522–528

Sicuella L, Palmer JD 1988. Physical and gene organization of mitochondrial DNA in fertile and male sterile sunflower. CMS-associated alterations in structure and transcription of the *atpA* gene. Nucleic Acids Res 16: 3787–3799

Silva NF, Stone SL, Christie LN, Sulaman W, Nazarian KAP, Burnett LA, Arnoldo M, Rothstein SJ, Goring DR 2001. Expression of the S receptor kinase in self-compatible *Brassica napus* cv Westar leads to the allele-specific rejection of incompatible *Brassica napus* pollen. Mol Genet Genomics 265: 522–559

Simon EW 1974. Phospholipids and plant membrane permeability. New Phytol 73: 337–420

Sims TL 1993. Genetic regulation of self-incompatibility. Crit Rev Plant Sci 12: 129–167

Singh A, Paolillo DJ Jr 1989 Towards an in vitro bioassay for the self-incompatibility response in *Brassica oleracea*. Sex Plant Reprod 2: 277–280

Singh A, Ai Y, Kao T-h 1991. Characterization of ribonuclease activity of three S-allele-associated proteins of *Petunia inflata*. Plant Physiol 96: 61–68

Singh AB and Malik P 1992. Pollen aerobiology and allergy: An integrated approach. Indian J Aerobiol 5: 1–12

Singh A, Evensen KB, Kao T 1992. Ethylene synthesis and floral senescence following compatible and incompatible pollinations in *Petunia inflata*. Plant Physiol 49: 38–45

Singh AB, Singh A 1994. Pollen allergy: A global scenario. In: Recent Trends in Aerobiology and Immunology. SN Agashe (ed), Oxford & IBH Publ Co Pvt Ltd, New Delhi pp 143–170

Singh AB, Rawat A 2000. Aeroallergens in India. In: Advances in Allergy and Asthma. R. Prasad (ed), Shivam Publications, Lucknow, UP, India, pp 23–46

Singh AB, Malik P, Gangal SV, Babu CR 1992. Intraspecific variations in pollen extracts of castor bean *Ricinus communis* prepared from different source materials. Grana 31: 229–235

Singh AK, Sastri DC, Moss JP 1980. Utilization of wild *Arachis* species at ICRISAT. Proc Intl Groundnut Workshop.

RW Gibbson (ed). ICRISAT, Patancheru, India, pp 82–90

Singh F, Thimmappaiah 1980. Nature and breakdown of self-incompatibility in *Dendrobium aggregatum* (Orchidaceae) with pollen irradiation. Incompatibility Newslett 12: 30–35

Singh M, Brown G 1991. Suppression of cytoplasmic male sterility by nuclear genes alters expression of novel mitochondrial gene region. Plant Cell 3: 1349–1362

Singh MB, Hough T, Theerakulpisut P, Avjioglu A, Davis S, Smith PM, Taylor P, Simpson RJ, Ward LD, McCluskey J, Puy R, Knox RB 1991. Isolation of cDNA encoding a newly identified major allergenic protein of rye-grass pollen: intracellular targtting to the amyloplast. Proc Natl Acad Sci USA 88: 1384–1388

Sita GL 1997. Gynogenic haploids in vitro. In: In Vitro Haploid Production in Higher Plants. SM Jain, SK Sopory, RE Veilleus (eds). Kluwer Academic Publ, Dordrecht, The Netherlands, vol V, pp 175–193

Slater AT, Calder DM 1990. Fine structure of the wet, detached cell stigma of the orchid *Dendrobium speciosum* Sm. Sex Plant Reprod 3: 61–69

Smith BN, Meeuse BJD 1966. Production of volatile amines and skatole at anthesis in some *Arum* lily species. Plant Physiol 41: 343–347

Smith GA 1986. Sporophytic screening and gametophytic verification of phytotoxin tolerance in sugarbeet (*Beta vulgaris* L.). In: Biotechnology and Ecology of Pollen. D Mulcahy, GB Mulcahy, E Ottaviano (eds). Springer-Verlag, New York, NY, pp 83–88

Smith PM, Knox RB, Singh MB 1996. Molecular characterization of group I allergens of grass pollen. In: Pollen Biotechnology: Gene Expression and Allergen Characterization. SS Mohapatra, RB Knox (eds), Chapman & Hall, New York, NY, pp 125–143

Snape JW, Chapman V, Moss J, Blanchard CE, Miller TE 1979. The crossability of wheat varieties with *Hordeum bulbosum*. Heredity 42: 291–298.

Snow AA, Roubik DW 1987. Pollen deposition and removal by bees visiting two species in Panama. Biotropica 19: 57–63

Southworth D 1971. Incorporation of radioactive precursors into developing pollen walls. In: Pollen: Development and Physiology. J Heslop-Harrison (ed). Butterworths, London, pp 115–120

Southworth D 1983. PH change during pollen germination in *Lilium longiflorum*. In: Biotechnology and Ecology of Pollen. DL Mulcahy, GB Mulcahy, E Ottaviano (eds). Springer-Verlag, Berlin, pp 61–65

Spangers AW 1978. Voltage variation in *Lilium longiflorum* pistils induced by pollination. Experientia 34: 36–37

Spena A, Estruch JJ, Prinsen E, Nacken W, van Onckelen H, Sommer H 1994. Anther-specific expression of the *rolB* gene of *Agrobacterium rhizogenes* increases IAA content in anthers and alters anther development and whole flower growth. Theor Appl Genet 84: 520–527

Speranza A, Calzoni GL 1980. Compounds released from incompatible apple pollen during in vitro germination. Z Pflanzenphysiol 97: 95–102

Speranza A, Calzoni GL, Bagni N 1983. Effect of exogenous polyamines on in vitro germination of apple pollen. In: Pollen: Biology and Implications for Plant Breeding. DL Mulcahy, E Ottaviano (eds). Elsevier Biomedical, Amsterdam, The Netherlands, pp 21–27

Speranza A, Calzoni GL, Pacini E 1997. Occurrence of mono- or di-saccharide and polysaccharide reserves in mature pollen grains. Sex Plant Reprod 10: 110–115

Spillane C, Steimer A, Grossniklaus U 2001. Apomixis in agriculture: the quest for clonal seeds. Sex Plant Reprod 14: 179–187

Stace HM 1995. Protogyny, self-incompatibility and pollination in *Anthocercis gracilis* (Solanaceae). Aust J Bot 43: 451–459

Stahl RJ, Arnoldo MA, Glavin TL, Goring DR, Rothstein SJ 1998. The self-incompatibility phenotype in *Brassica* is altered by transformation of a mutant S-locus receptor kinase. Plant Cell 10: 209–218.

Stalker HT 1980. Utilization of wild species for crop improvement. Adv Agron 33: 111–147

Stanley RG 1971. Pollen chemistry and tube growth. In: Pollen: Development and Physiology. Heslop-Harrison (ed). Butterworths, London, pp 131–155

Stanley RG, Linskens HF 1965. Protein diffusion from germination pollen. Physiol Plant 18: 47–53

Stanley RG, Linskens HF 1974. Pollen: Biology, Biochemistry and Management. Springer-Verlag, Berlin, Heidelberg, New York

Stead AD, Roberts IN, Dickinson HG 1979. Pollen-pistil interaction in *Brassica oleracea*. Events prior to pollen germination. Planta 146: 211–216

Stebbins GL 1958. The inviability, weakness and sterility of interspecific hybrids. Adv Genet 9: 147–215

Steben RE, Boudreaux P 1978. The effects of pollen and protein extracts on selected blood factors and performance of athletes. J Sports Med Phys Fitness 18: 221–226

Steer MW 1977. Differentiation of the tapetum in *Avena*. 1. The cell surface. J Cell Sci 25: 125–138

Steer MW, Steer JM 1989. Pollen tube tip growth. New Phytol 111: 323–358

Stein JC, Howlett B, Boycs DC, Nasrallah ME, Nasrallah JB 1991. Molecular cloning of a putative receptor protein kinase encoded at the self-incompatibility locus of *Brassica oleracea*. Proc Natl Acad Sci USA 88: 8816–8820

Stern H 1966. Meiosis: some considerations. J Cell Sci. Suppl 4: 29–43.

Stern H, Hotta Y 1974. Biochemical controls of meiosis. Ann Rev Genet 7: 37–66

Stettler RF 1968. Irradiated mentor pollen: its use in remote hybridization in black cotton wood. Nature 219: 746–747

Stettler RF, Guries RP 1976. The mentor pollen phenomenon in black cotton wood. Can J Bot 54: 820–830

Stettler RF, Koster R, Steenackens V 1980. Interspecific crossability studies in poplars. Theor Appl Genet 58: 273–282

Stevens JP, Kay QON 1989. The number, dominance relationships and frequencies of self-incompatibility alleles in a natural population of *Sinapis arvensis* L. in South Wales. Heredity 62: 199–205

Stieglitz H, Stern H 1973. Regulation of ß-1,3-glucanase activity in developing anthers of *Lilium*. Dev Biol 34: 169–173

Stiles FG 1976. Taste preferences, color preferences and flower choice in humming birds. Condor 78: 10–28

Stoger E, Moreno RMB, Ylstra B, Vicente O, Heberle-Bors E 1992. Comparison of different techniques for gene transfer into mature and immature tobacco pollen. Transgenic Res 1: 71–78

Stone JL, Thomson JD, Dent-Acosta S 1995. Assessment of pollen viability in hand-pollination experiments: A review. Am J Bot 82: 1186–1197

Stone SL, Goring DR 2001. The molecular biology of incompatibility systems in flowering plants. Plant Cell Tissue and Organ Culture 67: 93–114

Stone SL, Arnoldo MA, Goring DR 1999. A breakdown of *Brassica* self-incompatibility in ARC1 antisense transgenic plants. Science 286: 1729–1731.

Stout AB, Chandler C 1941. Change from self-incompatibility accompanying the change from diploid to tetraploid. Science 94: 114.

Straver WA, Plowright 1991. Pollination of greenhouse tomatoes by bumblebees. Greenhouse Canada 11: 10–12.

Stringham GR, Bansal VK, Thiagarajah MR, Degenhardt DF, Tewari JP 1995. Development of an agronomically superior blackleg resistant canola cultivar in *Brassica napus* L. using doubled haploidy. Can J Plant Sci 75: 437–439

Subbaiah CC 1984. A polyethylene glycol based medium for in vitro germination of cashew pollen. Can J Bot 62: 2473–2475

Subba Rao PV 1984. Clinical and experimental studies on the alien allergenic weed *Parthenium hysterophorus* Asp. Allergy Appl Immunol 17: 183–203

Subramanya S, Radhamani TR 1993. Pollination by birds and bats. Curr Sci 65: 201–209

Subrahmanyam NC 1978. Haploids and hybrids following interspecific crosses. Barley Genet Newslett 8: 97–99

Subrahmanyam NC 1980. Haploidy from *Hordeum* interspecific crosses. Theor Appl Genet 56: 257–263

Subrahmanyam NC 1999. Haploids from wide hybridizations in grass genera: progress and perspectives. In: Plant Tissue Culture and Biotechnology: Emerging Trends. PB Ravi Kishore (ed). Universities Press, Hyderabad, India, pp 11–21

Subrahmanyam NC, Kasha KJ 1973. Selective chromosomal elimination during haploid formation in barley following interspecific hybridization. Chromosoma 42: 111–125

Suda M, Marubashi W 1990. Effects of culture temperatures and IAA on the growth of hybrid seedlings of *Nicotiana rependa* × *N. tabacum* L. Jpn J Breed 40 (Suppl 1): 58–49

Sulaman W, Arnoldo MA, Yu KF, Tulsieram L, Rothstein SJ, Goring DR 1997. Loss of callose in the stigma does not affect the *Brassica* self-incompatibility phenotype. Planta 203: 327–331

Sun CS, Wang CC, Chu CC 1974. Cell division and differentiation of pollen grains of *Triticale* anther cultured in vitro. Sci Sinica 17: 47–54

Sun CS, Wu SC, Wang CC, Chu CC 1979. The deficiency of soluble proteins and plastid ribosomal RNA in the albino pollen plantlets of rice. Theor Appl Genet 55: 193–197

Sunderland N 1974. Anther culture as a means of haploid induction. In: Haploids in Higher Plants. Advances and Potential. KJ Kasha (ed). Univ Guelph, Ontario, CAN, pp 91–122

Sunderland N, Wicks FM 1971. Embryoid formation in pollen grains of *Nicotiana tabacum*. J Exptl Bot 22: 213–226

Sunderland N, Roberts M 1977. New approaches to pollen culture. Nature 270: 236–238

Sunderland N, Xu Z-H 1982. Shedding pollen culture in *Hordeum vulgare*. J Exptl Bot 33: 1086–1095

Sunderland N, Huang B 1987. Ultrastructural aspects of pollen dimorphism. Int Rev Cytol 107: 175–220

Sunnichan VG 1998. Reproductive biology of two forest trees of India: *Sterculia urens* (gum karaya) and *Boswellia serrata* (salai guggul). Ph.D. Thesis, Univ Delhi, New Delhi

Suphioglu C, Smith PM, Singh MB, Knox RB 1996. Expression of recombinant grass pollen allergens in the yeast *Pichia pastoris*. In: Plant Reproduction '96, 14th Intl Cong Sexual Plant Reprod, Lorne, Australia, Abst p 52

Suphioglu C, Singh MB, Taylor P, Bellomo R, Holmes P, Puy R, Knox RB 1992. Mechanism of grass-pollen induced asthma. Lancet 339: 569–572

Suss J, Tupy J 1978. tRNA synthesis in germinating pollen. Biol Plant 20: 70–72

Suss J, Tupy J 1979. Poly(A)⁺ RNA synthesis in germinating pollen of *Nicotiana tabacum* L. Biol Plant 21: 365–371

Swaminathan MS 1955. Overcoming cross-incompatibility among Mexican diploid species of *Solanum*. Nature 176: 887–888

Swaminathan MS, Murty BR 1957. One-way incompatibility in some species crosses in the genus *Nicotiana*. Indian J Genet Plant Breed 17: 23–26

Swanson EB, Coumans MP, Brown GL, Patel JD, Beversdorf WD 1988. The characterization of herbicide tolerant plants in *Brassica napus* L. after in vitro selection of microspores and protoplasts. Plant Cell Rep 7: 83–87

Swaboda I, Hoffmann-Sommergruber K, O'Riordain G, Scheiner O, Heberle-Bors E, Vicente O 1996. Molecular analysis of birch pollen allergens. In: Plant Reproduction '96. 14th Intl Cong Sexual Plant Reproduction, Lorne, Australia, Abst p 52

Syed RA 1979. Studies on oil palm pollination by insects. Bull Entomol. Res 69: 213–224

Syed RA, Low IH, Corley RHV 1982. Insect pollination of oilpalm: Introduction, establishment and pollinating efficiency of *Elaeidobius kamerunicus* Faust. Malaysia Planter, Kualalumpur 58: 547–561

Takahashi M 1989. Pattern determination of the exine in *Caesalpinia japonika* (Leguminosae: Caesalpinoidae). Am J Bot 76: 1615–1626

Takahashi M 1993. Exine initiation and substructure in pollen of *Caesalpinia japonica* (Leguminosae: Caesalpinoidae). Am J Bot 80: 192–197

Takahashi M 1995. Three-dimensional aspects of exine initiation and development in *Lilium longiflorum* (Liliaceae). Am J Bot 82: 847–854

Takahashi M, Skvarla J 1991. Development of striate exine in *Ipomopsis rubra* (Polemoniaceae). Am J Bot 78: 1724–1733

Takahata Y, Brown DCW, Keller WA, Kiazuma N 1993. Dry artificial seeds and desiccation tolerance induction in microspore-derived embryos of broccoli. Plant Cell Tissue Org Cult 35: 121–129.

Takasaki T, Hatakeyama K, Suzuki K, Watanabe G, Isogai M, Hinata K 2000. The S receptor kinase determines self-incompatibility in *Brassica* stigma. Nature 403: 913–916

Takayama S, Isogel A, Tsukamoto C, Ueda Y, Hinata K, Okazaki K, Koseki K, Suzuki A 1986. Structure of carbohydrate chains of S-glycoproteins in *Brassica campestris* associated with self-incompatibility. Agric Biol Chem 50: 1673–1676

Takayama S, Isogel A, Tsukamoto C, Ueda Y, Hinata H, Okazaki K, Suzuki A 1987. Sequences of S-glycoproteins, products of the *Brassica campestris* self-incompatibility locus. Nature 326: 102–105

Takayama S, Shiba H, Iwano M, Shimosato H, Che F-S, Kai N, Watanabe M, Suzuki G, Hinata K, Isogai A 2000. The pollen determinants of self-incompatibility in *Brassica campestris*. Proc Natl Acad Sci USA 97: 1920–1925

Takegami MH, Yoshika M, Tanaka I, Ito M 1981. Characteristics of isolated microsporocytes from liliaceous plants

for studies of the meiotic cell cycle in vitro. Plant Cell Physiol 22: 1–10

Tamas V, Salajan G, Bodea C 1970. Effectul polenului de porumb din hrana gainilor asupra pigmentatiei galbenusului de ou. Stud Cerc biochem 13: 423–429

Tandon R 1997. Reproductive biology of two leguminous trees: *Acacia senegal* and *Butea monosperma*. Ph.D. Thesis, Univ Delhi, Delhi

Tandon R, Manohara TN, Nijalingappa BHM, Shivanna KR 1999. Polyethylene glycol enhances in vitro germination and tube growth of oil palm pollen. Indian J Exptl Biol 37: 169–172

Tandon R, Manohara TN, Nijalingappa BHM, Shivanna KR 2001. Pollination and pollen-pistil interaction in oil palm, *Elaeis guineensis*. Ann Bot 87: 831–838

Tang PH 1988. Interaction of vegetative nucleus and generative cell (then sperms). In: Sexual Reproduction in Higher Plants. M Cresti, P Gori, E Pacini (eds), Springer-Verlag, Berlin, Heidelberg, New York, pp 227–232

Tang XJ, Hepler PK, Scordilis SP 1989. Immunochemical and immunocytochemical identification of myosin heavy chain polypeptide in *Nicotiana* pollen tubes. J Cell Sci 92: 569–574

Tanksley SD, Zamir D, Rick CM 1981. Evidence for extensive overlap of sporophytic and gametophytic gene expression in *Lycopersicon esculentum*. Science 213: 453–454

Tara CP, Namboodiri AN 1974. Aberrant microsporogenesis and sterility in *Impatiens sultani* (Balsaminaceae). Am J Bot 61: 585–591

Tara CP, Namboodiri AN 1976. Association between defective stigmatic exudate and sterility in *Impatiens*: chromatographic evidences. Indian J Exp Biol 14: 354–355

Tao G, Yang R 1986. Use of CO_2 and salt solutions to overcome self-incompatibility of Chinese cabbage (*B. campestris* ssp. *pekinensis*). Cruciferea Newslett 11: 73–74

Taylor JP 1982. Carbon dioxide treatment as an effective aid to the production of selfed seed in kale and Brussels sprouts. Euphytica 31: 957–964

Taylor LP, Jorgensen R 1992. Conditional male fertility in chalcone synthase-deficient *Petunia*. J. Hered 83: 11–17

Taylor LP, Hepler PK 1997. Pollen germination and tube growth. Ann Rev Plant Physiol Plant Mol Biol 48: 461–491

Taylor PE, Singh MB, Knox RB 1993. Strategies for the immunocytochemical localization of rapidly diffusible proteins in pollen. J Computer-Assisted Microscopy 5: 53–56.

Taylor PE, Staff IA, Singh MB, Knox RB 1994. Localization of the two major allergens in rye-grass pollen using specific monoclonal antibodies and quantitative analysis of immunogold labelling. Histochem J 26: 392–401

Tejaswini, Ganeshaiah KN, Shaankar UR 2001. Sexual selection in plants: the process, components and significance. PINSA B 67: 423–432

Telmer CA, Newcomb W, Simmonds DH 1993. Microspore development in *Brassica napus* and the effect of high temperature on division in vivo and in vitro. Protoplasma 172: 154–165.

Tepfer SS, Greyson RI, Craig WR, Hindman JL 1963. In vitro culture of floral buds of *Aquilegia*. Am J Bot 50: 1035–1045

Thanikaimoni G 1978. Pollen morphological terms: proposed definitions—I. In: Proc IV Intl Palynol Conf, Lucknow, UP, India, vol 1, pp 228–239

Theerakulpisut P, Xu H, Singh MB, Pettitt JM, Knox RB 1991. Isolation and developmental expression of *Bcp* 1, an anther specific cDNA clone in *Brassica campestris*. Plant Cell 3: 1073–1084

Theunis CH, McConchie CA, Knox RB 1985. Three dimensional reconstruction of the generative cell and its wall connections in mature bicellular pollen of *Rhododendron*. Micron and Microscopica Acta 1: 225–231

Theunis CH, Pierson ES, Cresti M 1991. Isolation of male and female gametes in higher plants. Sex Plant Reprod 4: 145–154

Thimmappaiah 1982. Breakdown of self-incompatibility in *Dendrobium parardii* Roxb. (Orchidaceae) following ultraviolet irradiation of pollinia. Curr Sci 51: 372–373

Thomas SM, Murray BG 1975. A new site for the self-incompatibility reaction in the Gramineae. Incompatibility Newslett 6: 22–23

Thompson KF 1978. Application of recessive self-incompatibility to production of hybrid rapeseed. In: Proc 5th Intl Rapeseed Conf, Malmo, Sweden, pp 56–59

Thomson WW, Platt-Aloia K 1982. Ultrastructure and membrane permeability in cow-pea seeds. Plant Cell Environ 5: 367–373

Tiara T, Larter EN 1977a. Effects of E-amino-n-caproic acid and L-lysine on the development of hybrid embryo of *Triticale* (x-*Triticale-Secale*). Can J Bot 55: 2330–2334

Tiara T, Larter EN 1977b. The effects of variation in ambient temperature alone and in combination with E-amino-n-caproic acid on development of embryos from wheat-rye crosses (*Triticum turgidum* var *durum* cv gora x *Secale cereale*). Can J Bot 55:2335–2337

Tiezzi A, Moscotelli A, Cai G, Bartalesi A, Cresti M 1992. An immunoreactive homolog of mammalian kinesin in *Nicotiana tabacum* pollen tubes. Cell Motility Cytoskel 21: 132–137

Tilton VR 1981a. Ovule development in *Ornithogalum caudatum* (Liliaceae) with a review of selected papers on angiosperm reproduction. II. Megasporogenesis. New Phytol 88: 459–476

Tilton VR 1981b. Ovule development in *Ornithogalum caudatum* (Liliaceae) with a review of selected papers on angiosperm reproduction. IV. Egg apparatus structure and function. New Phytol 88: 505–532

Tilton VR, Horner HJ Jr 1980. Stigma, style, and obturator of *Ornithogalum caudatum* (Liliaceae) and their function in the reproductive process. Am J Bot 67: 1113–1131

Tilton VR, Lersten NR 1981a. Ovule development in *Ornithogalum caudatum* (Liliaceae) with a review of selected papers on angiosperm reproduction. I. Integument funiculus and vascular tissue. New Phytol 88: 439–458

Tilton VR, Lersten NR 1981b. Ovule development in *Ornithogalum caudatum* (Liliaceae) with a review of selected papers on angiosperm reproduction. III. Nucellus and megagametophyte. New Phytol 88: 477–504

Tirlapur UK, Shiggaon SV 1988. Distribution of calcium and calmodulin in the papillar cells of stigma surface visualized using chlorotetracycline and fluorescing calmodulin binding phenothiazines. Ann Biol 4: 49–53

Tirlapur UK, Willemse MTM 1992. Changes in calcium and calmodulin levels during microsporogenesis, pollen development and germination in *Gasteria verrucosa* (Mill.) H. Duval. Sex Plant Reprod 5: 214–223

Tiwari SC 1994. An intermediate voltage electron microscopic study of freeze-substituted generative cell in pear (*Pyrus communis* L.): features with relevance to cell-cell communication between the two cells of germinating pollen. Sex Plant Reprod 7: 177–186

Tiwari SC, Gunning BES 1986a. Development of an invasive but non-syncytial tapetum. Ann Bot 57: 557–563

Tiwari SC, Gunning BES 1986b. Cytoskeleton, cell surface and development of a non-syncytial invasive tapetum in *Canna*: Ultrstructural, freeze-substitution, cytochemical and immunofluorescence study. Protoplasma 134: 1–16

Tiwari SC, Gunning BES 1986c. Cytochemical, cell surface and development of invasive plasmodial tapetum in *Tradescantia virginiata* L. Protoplasma 133: 89–99

Tiwari SC, Polito VS 1988. Organization of the cytoskeleton in pollen tubes of *Pyrus communis*: a study employing conventional and freeze-substitution electron microscopy, immunofluorescence and rhodamine-phalloidin. Protoplasma 147: 100–112

Tobias CM, Howlett BJ, Nasrallah JB 1992. An *Arabidopsis thaliana* gene with sequence similarity to the S-locus receptor kinase of *Brassica oleracea*. Plant Physiol 99: 284–290.

Torchio PF 1987. Use of non-honeybee species as pollinators of crops. Proc Entomol Soc, Ontario 118: 39–44

Torchio PF 1990. Diversification of pollination strategies for US crops. Environ Entomol 19: 1649–1656

Toriyama K, Thorsness MK, Nasrallah JB, Nasrallah ME 1991. A *Brassica* S-locus gene promoter directs sporophytic expression in the anther tapetum of transgenic *Arabidopsis*. Dev Biol 143: 427–431

Touraev A, Fink CS, Stoger E, Heberle-Bors E 1995. Pollen selection: a transgenic reconstruction approach. Proc Natl Acad Sci USA 92: 12165–12169

Towbin H, Stachelin T, Gordon J 1979. Electrophoretic transfer of proteins from polyacrylamide gels to nitrocellulose sheets. Proc Natl Acad Sci USA 76: 4350–4354

Towill LE 1985. Low temperature and freeze- (vacuum-) drying preservation of pollen. In: Cryopreservation of Plant Cells and Organs. KK Kartha (ed). CRC Press, Boca Raton, FL, pp 171–198

Towill LE 1991. Cryopreservation. In: In vitro Method for Conservation of Plant Genetic Resources. JH Dodds (ed), Chapman & Hall, London, pp 41–70

Townsend CE 1966. Self-compatibility response to temperature and the inheritance of the response in tetraploid alsike clover, *Trifolium hybridum* L. Crop Sci 6: 409–414

Trick M 1990. Genome sequence of a *Brassica* S-locus related gene. Plant Mol Biol 15: 203–205

Trick M, Flavell RB 1989. A homozygous S-genotype of *Brassica oleracea* expressed two S-like genes. Mol Gen Genet 218: 112–117

Tschabold EE, Heim DR, Beck JR, Wright FL, Rainey DP, Terando NH, Schwer JF 1988. LY195259, new chemical hybridizing agent for wheat. Crop Sci 28: 583–588

Tsuchia K 1993. Development of new rice varieties Hirohikari and Hirohonami, using haploid method of breeding by anther culture. In: Plant Tissue Culture and Gene Manipulation for Breeding and Formation of Phytochemicals. K Oono (ed) Sato Press, Mito, Japan pp 145–176

Tupy J 1961. Investigation of free amino acids in cross-, self- and non-pollinated pistils of *Nicotiana alata*. Biol Plant 3: 47–64

Tupy J, Hrabetova E, Balatkova V 1974. The release of RNA from pollen tubes. In: Fertilization in Higher Plants. HF Linskens (ed). North-Holland Publ, Amsterdam, The Netherlands, pp 145–150

Tupy J, Hrabetova E, Balatkova V 1977. Evidence for ribosomal RNA synthesis in pollen tubes in culture. Biol Plant 19: 226–230

Tupy J, Suss J, Rihova L 1983. Developmental changes in gene expression during pollen differentiation and maturation in *Nicotiana tabacum* L. Biol Plant 25: 231–237

Turcotte EL, Feaster CV 1974. Semigametic production of cotton haploids. In: Haploids in Higher Plants: Advances and Potential. KJ Kasha (ed). Univ Guelph, Guelph, pp 53–63

Turkeltaub PC 1989. Biological standardization of allergenic extracts. Allergy Immunol Pathol 17: 53–65

Turner J, Facciotti D 1990. High oleic acid *Brassica napus* from mutagenized microspores. In: Proc 6th Crucifer Genetics Workshop. JR McFerson, S Kresovich, SG Dwyer (eds). USDA-ARS Geneva, NY, USA, p 24

Turpen T, Garger SJ, Grill LK 1988. On the mechanism of cytoplasmic male sterility in the 447 line of *Vicia faba*. Plant Mol Biol 10: 489–497

Twell D 1994. The diversity and regulation of gene expression in the pathway of male gametophyte development. In: Molecular and Cellular Aspects of Plant Reproduction. RJ Scott, AD Stead (eds). Cambridge Univ Press, Cambridge, England, pp 83–135

Twell D, Klein TM, McCormick S 1991. Transformation of pollen by particle bombardment. In: Plant Tissue Culture Manual. K Lindsey (ed). Kluwer Academic Publ, Dordrecht, The Netherlands, pp D2: 1–14.

Twell D, Klein TM, Fromm ME, McCormick S 1989. Transient expression of chimeric genes delivered into pollen by microprojectile bombardment. Plant Physiol 91: 1270–1274

Ueda K, Tanaka I 1994. The basic proteins of male generative nuclei isolated from pollen grains of *Lilium longiflorum*. Planta 192: 446–452

Ueda K, Tanaka I 1995. Male gametic nucleus-specific H2B and H3 histones designated gH2B and gH3 in *Lilium longiflorum*. Planta 197: 289–295

Uemura S, Ohkawara K, Kudo G, Wada N, Higashi S 1993. Heat-production and cross-pollination of the Asian skunk cabbage *Symplocarpus renifolius* (Araceae). Am J Bot 80: 635–640

Vain P, McMullen MD, Finer JJ 1993. Osmotic treatment enhances particle bombardment-mediated transient and stable transformation of maize. Plant Cell Rep 12: 84–88

Vaknin Y, Gan-mor S, Bechar A, Ronen B, Bisikowitch D 2001. Are flowers morphologically adapted to take advantage of electrostatic forces in pollination. New Phytologist 152: 301–306

Valenta R, Vrtala S, Ebner C, Kraft D, Scheiner O 1992. Diagnosis of grass pollen allergy with recombinant timothy grass (*Phlem pratens*) pollen allergens. Intl Arch Allergy Appl Immunol 97: 287–294

van der Donk JAWM 1974. Differential synthesis of RNA in self- and cross-pollinated styles of *Petunia hybrida* L. Mol Gen Genet 131: 1–8

van der Donk JAWM 1975. Recognition and gene expression during the incompatibility reaction in *Petunia hybrida* L. Mol Gen Genet 141: 305–316

van der Meer, Stam ME, Tunen J, Mol MN, Stultje AR 1992. Antisense inhibition of flavonoid biosynthesis in *Petunia* anthers results in male sterility. Plant Cell 4: 253–262

van der Woude WJ, Morre DJ 1968. Endoplasmic reticulum-dictyosome-secretory vesicle associations in pollen tubes of *Lilium longiflorum* Thunb. Proc Indian Acad Sci 77: 164–170

van der Woude WJ, Morre DJ, Bracker CE 1971. Isolation and characterization of secretory vesicles in germinated pollen of *Lilium longiflorum*. J Cell Sci 8: 331–351 van Eijk JP, Raamsdonk LWD, Eikelboom W, Bino RJ 1991. Interspecific crosses between *Tulipa gesneriana* cultivars and wild *Tulipa* species: a survey. Sex Plant Reprod 4: 1–5

van Gastel AJG, de Nettancourt D 1975. The effects of different mutagens on self-incompatibility in *Nicotiana alata* Link and Otto. II. Acute irradiation of X-rays and fast neutrons. Heridity 34: 381–392

van Heemert C, de Ruijter A, van den Eijnde J, van den Steen J 1990. Year round production of bumble bee colonies for crop production. Bee World 71: 54–57

van Lalonde BA, Nasrallah ME, Dwyer KG, Chen C-H, Barlow B, Nasrallah JB 1989. A highly conserved *Brassica* gene with homology to the S-locus specific glycoprotein structural gene. Plant Cell 1: 249–252

van Lammeren AAM, Keijzer CJ, Willemse MTM, Kieft H 1985. Structure and function of the microtubular cytoskeleton during pollen development in *Gasteria verrucosa* (Mill) H. Duval. Planta 165: 1–11

van Tuyl JM, de Jeu MJ 1997. Methods for overcoming interspecific crossing barriers. In: Pollen Biotechnology for Crop Production and Improvement. KR Shivanna, VK Sawhney (eds). Cambridge University Press, New York, NY, pp 273–292

van Tuyl JM, van Dien MP, van Creij MGM, van Kleinwee TCM, Franken J, Bino RJ 1991. Application of in vitro pollination, ovary culture, ovule culture and embryo rescue for overcoming incongruity barriers in interspecific *Lilium* crosses. Plant Sci 74: 115–126

van Went JL 1970. The ultrastructure of the fertilized embryo sac of *Petunia*. Acta Bot Neerl 19: 468–480

Vasil IK 1980. Androgenic haploids. Intl Rev Cytol Suppl 11A: 195–223

Vasil IK, 1987. Physiology and culture of pollen. Intl Rev Cytol 107: 127–174

Vasil IK, Aldrich HC 1970. A histochemical and ultrastructural study of ontogeny and differentiation of pollen in *Podocarpus macrophylla*. Protoplasma 71: 1–37

Vazart B 1970. Morphogenese du sporoderm et participation des mitochondries à la mise en place de la primexine dans le pollen de *Linum usitatissimum* L. C R Acad Sci Paris 270: 3210–3212

Veldhuis J 1968. Methods of assisted pollination for oil palms. In: Oil Palm Developments in Malaysia. PD Turner (ed). Incorp Soc Planters, Kuala Lumpur, Malaysia, pp 72–81

Verma SC, Malik R, Dhir I 1977. Genetics of the incompatibility system in the crucifer *Eruca sativa* L. Proc R Soc London Ser B 196: 131–159

Vesselina I, Alipievr M 1996. Breaking the cabbage self-incompatibility by means of pollen's laser treatment. Cruciferae Newslett 18: 60–61

Villar M, Gaget-Faurobert M 1997. Mentor effects in pistil mediated pollen-pollen interaction. In: Pollen Biotechnology for Crop Production and Improvement. KR

Shivanna, VK Sawhney (eds). Cambridge Univ Press, New York, NY, pp 315–332

Visser T 1955. Germination and storage of pollen. Meded Landb-Hoogesch 55: 1–68

Vitanova G 1984. A simple method for seed production from self-incompatible lines of cabbage (*Brassica oleracea* var. *capitata* F. *alba*). Cruciferae Newslett. 9: 41

Vogt T, Pollak N, Tarlyn N, Taylor 1994. Pollination- or wound-induced kaempferol accumulation in *Petunia* stigmas enhances seed production. Plant Cell 6: 11–23

von Schmidt H, Schmidt V 1981. Untersuchungen in pollensteriten, stamen-less-ahnlichen Mutanten von *Lycopersicon esculentum* Mill. II. Normalisievung von *ms-15* and *ms-33* mit Gibberellinsaure (GA₃). Biol Zhl 100: 691–696

Voorrips RE, Visser DL 1990. Doubled haploid lines with clubroot resistance in *Brassica oleracea*. In: Proc 6th Crucifer Genetics Workshop, JR McFerson, S Kresovich, SG Dwyer (eds). USDA-ARS, Geneva, NY, USA, p 40

Vuilleumier BS 1967. The origin and evolutionary development of heterostyly in the angiosperms. Evolution 21: 210–226

Vyas P. 1993. Studies on wide hybridization between crop brassicas and species of *Diplotaxis*. Ph.D. thesis, Univ Delhi, Delhi

Vyas P, Prakash S, Shivanna KR 1995. Production of wide hybrids and backcross progenies between *Diplotaxis erucoides* and crop Brassicas. Theor Appl Genet. 90: 549–553

Wagner G, Hess D 1973. In vitro-Befruchtungen bei *Petunia hybrida*. Z Pflanzenphysiol 69: 262–269

Wagner VT, Mogensen HL 1988. The male germ unit in the pollen and pollen tubes of *Petunia hybrida*: Ultrastructural, quantitative and three-dimensional features. Protoplasma 143: 101–110.

Walker JC, Zhang R 1990. Relationship of a putative receptor protein kinase from maize to the S-locus glycoprotein of *Brassica*. Nature 345: 743–746

Walker JW 1971. Unique type of angiosperm pollen from the family Annonaceae. Science 172: 565–567

Walker JW, Doyle JA 1975. The bases of angiosperm phylogeny: palynology. Ann Missouri Bot Garden 62: 664–723

Wallace DH 1979. Procedures for identifying S-allele genotypes of *Brassica*. Theor Appl Genet 54: 249–265

Waller GD 1980. A modification of the O A C pollen trap. Am Bee J 120: 119–121

Walsh P, Matthews JA, Denmeade R, Maxwell P, Davidson M, Walker MR 1990. Monoclonal antibodies to proteins from cock's foot grass (*Dactylis glomerata*) pollen: Isolation and N-terminal sequence of major allergen. Intl Arch Allergy Appl Immunol 91: 419–425

Wan Y, Lemaux PG 1994. Generation of large number of independently transformed fertile barley plants. Plant Physiol 104: 37–48

Wang H, Wu H-M, Cheung AY 1993. Development and pollination regulated accumulation and glycosylation of a stylar transmitting tissue-specific proline-rich protein. Plant Cell 5: 1639–1650

Wang W 1989. The development and utilization of the resources of bee pollen in China. Proc Intl Cong Apiculture (Apimondia). 32: 239

Wang W, Hu J, Cheng J 1984. Biological effect of honey bee pollen: I Radioprotective activity on hematopoietic tissues of irradiated mice. J Hangzhou Univ 11: 231–240

Warmke HE, Lee SJ 1977. Mitochondrial degeneration in Texas cytoplasmic male-sterile corn anthers. J Hered 68: 213–222

Warmke HE, Lee SJ 1978. Pollen aboration in T cytoplasmic male sterile corn (*Zea mays*): a suggestive mechanism. Science 200: 561–562

Warmke HE, Overman MA 1972. Cytoplasmic male sterility in *Sorghum*. 1. Callose behaviour in fertile and sterile anthers. J Hered 68: 103–112

Waterkeyn I, Beinfait A 1970. On a possible function of callose special wall in *Ipomoea purpuria* (L.) Roth. Grana 10: 13–20

Webb CJ, Lloyd DG 1988. The avoidance of interference between the presentation of pollen and stigmas in angiosperms: Herkogamy. New Zealand J Bot 24: 163–178

Wee YC, Rao AN 1979. *Ananas* pollen germination. Grana 18: 33–39

Weiss̀ MR 1991. Floral colour changes as cues for pollinators. Nature 354: 227–229

Welsh JR, Klatt AR 1971. Effects of temperature and photoperiod on spring wheat pollen viability. Crop Sci 11: 864–866

Wenzel G 1980. Recent progress in microspore culture of crop plants. In: The Plant Genome. DR Davies, DA Opwood (eds). The John Innes Charity, Norwich (UK), pp 185–213

Wernsman EA, Davis RL, Keim WFm 1965. Interspecific fertility of two *Lotus* species and their F₁ hybrids. Crop Sci 5: 452–454

Westphal W, Becker WM, Schlaak M 1988. Analysis of rye pollen (*Secale cerele*) allergins using patients' IgE, immunoprints, western blot and monoclonal antibodies. Int Arch Allergy Appl Immunol 86: 69–75

Wheeler MJ, Franklin-Tong VE, Franklin FCH 2001. The molecular and genetic basis of pollen-pistil interactions. New Phytologist 151: 565–584

Whelan EDP 1974. Discontinuities in the callose wall, intermeiocyte connections, and cytomixis in angiosperm meiocytes. Can J Bot 52: 1219–1224

Whitehead CS, Halevy CS, Reid MS 1984. Role of ethylene and ACC in pollination and wound-induced senescence of *Petunia hybrida* flowers. Physiol Plant 61: 643–648

Whitehead DR 1983. Wind pollination. Some ecological and evolutionary perspectives. In: Pollination Biology. L Real (ed), Acad Press, Ornaldo, FL, pp 97–108

Whitehouse HLK 1950. Multiple allelomorph incompatibility of pollen and style in the evolution of angiosperms. Ann Bot 1: 199–216

Wiberg E, Rahlen L, Hellman M, Tillberg E, Glimelius K, Stymne S 1991. The microspore-derived embryo of *Brassica napus* L. as a tool for studying embryo-specific lipid biogenesis and regulation of oil quality. Theor Appl Genet 82: 515–520

Wide L, Bennich H, Johansson SGO 1967. Diagnosis of allergy by an in vitro test for allergen antibodies. Lancet 2: 1105–1107

Widholm JM 1972. The use of fluorescein diacetate and phenosafranin for determining viability of cultured plant cells. Stain Technol 47: 189–194

Wiebe GA 1960. A proposal for hybrid barley. Agron J 52: 181–182

Wiermann R, Vieth K 1983. Outer pollen wall, an important accumulation site for flavonoids. Protoplasma 118: 230–233

Wiermann R, Gubatz S 1992. Pollen wall and sporopollenin. Intl Rev Cytol 140: 35–72

Wietsma WA, De Jong KY, van Tuyl JM 1994. Overcoming prefertilization barriers in interspecific crosses of *Frittillaria imperialis* and *F. raddeana*. Plant Cell Incomp Newslett 26: 89–92

Willemse MTM 1971. Morphological and quantitative changes in the population of cell organelles during microsporogenesis of *Pinus sylvestris* L. 1. Morphological changes from zygotene until prometaphase I. Acta Bot Neerl 20: 261–274

Willemse MTM, Franssen-Verheijen MAW 1988. Pollen tube growth and its pathway in *Gasteria verrucosa* (Mill.) H. Duval. Phytomorphology 38: 127–132

Willemse MTM, Plyusheh TA, Reinders MC 1995. In vitro micropylar penetration of the pollen tube of *Gasteria verrucosa* (Mill.) H. Duval. and *Lilium longiflorum* Thunb, conditions, attraction and application. Plant Sci 108: 201–208

Williams EG, de Lautour G 1980. The use of embryo culture with transplanted nurse endosperm for the production of interspecific hybrids in pasture legumes. Bot Gaz 141: 252–257

Williams EG, Maheshwaran G, Hutchinson JF 1987. Embryo and ovule culture and crop improvement. Plant Breed Rev 5: 181–236

Williams EG, Clarke AE, Knox RB (eds) 1994. Genetic Control of Self-incompatibility and Reproductive Development in Flowering Plants. Kluwer Publ, Dordrecht, The Netherlands

Williams ME, Leemans J, Michiels F 1997. Male sterility through recombinant DNA technology. In: Pollen

Biotechnology for Crop Production and Improvement. KR Shivanna, VK Sawhney (eds). Cambridge University Press, New York, NY, pp 237–257

Williams RR, Legge AP 1979. Pollen application by mechanical dusting in English apple orchards. J Hort Sci 54: 67–74

Williams RR, Church RM, Wood DES, Flook VA 1979. Use of an anthocyanin progeny marker to determine the value of hive pollen dispensers in apple orchards. J Hort Sci 54: 75–78

Willing RP, Bashe D, Mascarenhas JP 1988. An analysis of quantity and diversity of messenger RNAs from pollen and shoots of *Zea mays*. Theor Appl Genet 75: 751–753

Willing RR, Pryor LK 1976. Interspecific hybridization in poplar. Theor Appl Genet 47: 141–151

Willingle J, Mantle PG 1985. Stigma constriction in pearl millet, a factor influencing reproduction and disease. Ann Bot 56 109–113

Wilson AT, Vickers M, Mann LRB 1979. Metabolism in dry pollen—a novel technique for studying anhydrobiosis. Naturwissenschaften 66: 53–54

Winston ML, Slessor KN 1993. Application of queen honey bee mandibular pheromone to beekeeping and crop pollination. Bee World 74: 111–128

Wirth M, Withner CL 1959. Embryology and Development in the Orchidaceae. In: The Orchis: A Scientific Survey. CL Withner (ed), The Ronald Press Co. New York, pp 155–188

Witherspoon WD Jr, Wernsman EA, Gooding GV, Rufty RC 1991. Characterization of a gametoclonal variant controlling virus resistance in tobacco. Theor Appl Genet 81: 1–5

Withner CL, Nelson PK, Wejksnora PJ 1974. The anatomy of orchids. In: The Orchids: Scientific Studies. John Wiley & Sons, New York, NY, pp 267–347

Woittiez RD, Willemse MTM 1979. Sticking of pollen on stigmas: The factors and a model. Phytomorphology 29: 57–63

Wolters JHB, Martens JM 1987. Effects of air pollution on pollen. Bot Rev 53: 372–414

Wolters-Arts M, Lush WM, Marium C 1998. Lipids are required for directional pollen tube growth. Nature 392: 818–821

Wong KC, Watanabe M, Hinata K 1994. Fluorescence and scanning electron microscopic study on self-incompatibility in distylous *Averrhoa carambola* L. Sex Plant Reprod 7: 116–127

Wong YK, Hardon JJ 1971. A comparison of different methods of assisted pollination in the oil palm. Chemara Communication (Agronomie), No. 9, Chemara Res Stn, Seremban, Malaysia

Worrall D, Hird DL, Hodge R, Paul W, Draper J, Scott R 1992. Premature dissolution of the microsporocyte callose wall causes male sterility in transgenic tobacco. Plant Cell 4: 759–771

Wu H-M, Wang H, Cheung AY 1995. A pollen tube growth stimulating glycoprotein is deglycosylated by pollen tubes and displays a glycosylation gradient in the flower. Cell 82: 393–403

Wyatt R 1983. Pollinator-plant interactions and the evolution of breeding systems. In: Pollination Biology. L Real (ed). Academic Press, Orlando, FL, pp 51–95

Xu Y Carpenter R, Dickinson H, Coen ES 1996. Origin of allelic diversity in *Antirrhinum* S-locus RNases. Plant Cell 8: 805–814

Xu B, Grun P, Kheyr-Pour A, Kao T-h 1990a. Identification of pistil-specific proteins associated with three self-incompatibility alleles in *Solanum chacoense*. Sex Plant Reprod 3: 54–60.

Xu B, Mu J, Nevins DL, Grun P, Kao T-h 1990b. Cloning and sequencing of cDNAs encoding two self-incompatibility associated proteins in *Solanum chacoense*. Mol Gen Genet 224: 341–346.

Xu H, Knox RB, Taylor PE, Singh MB 1995a. *Bcp 1*, a gene required for male fertility in *Arabidopsis*. Proc Natl Acad Sci USA 92: 2106–2110

Xu H, Theerakulpisut P, Taylor PE, Knox RB, Singh MB, Bhalla PL 1995b. Isolation of a gene preferentially expressed in mature anthers of rice (*Oryza sativa* L.). Protoplasma 187: 127–131

Yan H, Yang H-Y, Jensen WA 1991. Ultrastructure of the micropyle and its relationship to pollen tube growth and synergid degeneration in sunflower. Sex Plant Reprod 4: 166–175

Yang H-Y, Zhou C 1982. In vitro induction of haploid plants from unpollinated ovaries and ovules. Theor Appl Genet 63: 97–104

Yang H-Y, Zhou C 1990. In vitro gynogenesis. In: Plant Tissue Culture: Applications and Limitations. SS Bhojwani (ed). Elsevier, Amsterdam, The Netherlands, pp 242–258

Yang H-Y, Zhou C 1992. Experimental plant reproductive biology and reproductive cell manipulation in higher plants. Am J Bot 79: 354–363

Ye JM, Kao KN, Harvey BL, Rossangal BG 1987. Screening of salt-tolerant barley genotypes via anther culture in salt stress media. Theor Appl Genet 74: 426–429

Yokota E, Shimmen T 1994. Isolation and characterization of plant myosin from pollen tubes of lily. Protoplasma 177: 153–162

Yokota E, Muto S, Shimmen T 1999. Inhibitory regulation of higher-plant myosin by Ca^{2+} ions. Plant Physiology 119: 231–239.

Yu H-S, Hu S-Y, Shu C 1989. Ultrastructure of sperm cells and the male germ unit of *Nicotiana tabacum*. Protoplasma 152: 29–36

Yu H-S, Hu S-Y, Russell SD 1992. Sperm cells in pollen tubes of *Nicotiana tabacum*: Three-dimensional reconstruction, cytoplasmic diminution and quantitative cytology. Protoplasma 168: 172–183

Zamir D, Gadish I 1987. Pollen selection for low temperature adaptation in tomato. Theor Appl Genet 74: 545–548

Zamir D, Vellejos EC 1983. Temperature effects on haploid selection in tomato microspores and pollen grains. In: Pollen Biology and Implications for Plant Breeding. DL Mulcahy, E Ottaviano (eds). Elsevier, New York, NY, pp 335–342

Zamir D, Tanksley S, Jones J 1982. Haploid selection for low temperature tolerance of tomato pollen. Genetics 101: 129–137

Zenkteler M 1967. Test-tube fertilization of ovules in *Melandrium album* Mill. with pollen grains of several species of the Caryophyllaceae family. Experientia 23: 775–777

Zenkteler M 1970. Test-tube fertilization of ovules in *Melandrium album* Mill. with pollen grains of *Datura stramonium* L. Experientia 26: 661–662

Zenkteler M 1980. Intra-ovarian and in vitro pollination. In: Perspectives in Plant Cell and Tissue Culture. Vasil IK (ed). Intl Rev Cytol (Suppl) 11B: 137–156

Zenkteler M 1990. In vitro fertilization and wide hybridization in higher plants. Critical Rev Plant Sci 9: 267–279

Zenkteler M 1992. In vitro fertilization: A method facilitating the production of hybrid embryos and plants. In: Angiosperm Pollen and Ovules. E Ottaviano, DL Mulcahy, M Sari-Gorla, GB Mulcahy (eds). Springer-Verlag, New York, NY, pp 331–335

Zenkteler M, Bagniewska-Zadworna A 2001. Distant in vitro pollination of ovules. Phytomorphology Golden Jubilee Issue: Trends in Plant Sciences. NS Rangaswamy (ed). Intl Soc Plant Morphologists, New Delhi, pp 225–235

Zeven AC, Keijzer CJ 1980. The effect of the number of chromosome in the rye on its crossability with wheat. Cereal Res Commun 8: 491–494

Zhang C, Qifeng C 1993. Genetic studies of rice (*Oryza sativa* L.) anther culture response. Plant Cell Tissue Organ Cult 34: 177–182

Zhang H-Q, Croes AF 1982. A new medium for pollen germination in vitro. Acta Bot Neerl 31: 113–119

Zhang H-Q, Croes AF 1983a. Proline metabolism in pollen: Degradation of proline during germination and early tube growth. Planta 159: 46–49

Zhang H-Q, Croes AF 1983b. Protection of pollen germination from adverse temperature: a possible role for proline. Plant Cell Environ 6: 471–476

Zhang H-Q, Croes, AF, Linskens HF 1982. Protein synthesis in germinating pollen of *Petunia*: role of proline. Planta 154: 199–203

Zhang XS, O'Neill SD 1993. Ovary and gametophyte development are coordinately regulated by auxin and ethylene following pollination. Plant Cell 5: 403–418

Zhou WM, Yoshida K, Shintaku Y, Takeda G 1991. The use of IAA to overcome interspecific hybrid inviability in reciprocal crosses between *Nicotiana tabacum* L. and *N. repanda* Wild. Theor Appl Genet 82: 657–661

Zhu T, Mogensen HL, Smith SE 1992. Heritable paternal cytoplasmic organelles in alfalfa sperm cells: ultrastructural reconstruction and qualitative cytology. Eur J Cell Biol 59: 211–218

Zinkl GM, Preuss D 2000. Dissecting *Arabidopsis* pollen-stigma interactions reveals novel mechanisms that confer mating specificity. Ann Bot 85: 15–21

Zinkl GM, Zweibel BI, Grier DG, Preuss D 1999. Pollen-stigma adhesion in *Arabidopsis*: a species-specific interaction mediated by hydrophobic molecules in the pollen exine. Development 126: 5431–5440

Zou J-T, Zhan X-Y, Wu H-M, Wang H, Cheung AY 1994. Characterization of a rice pollen-specific gene and its expression. Am J Bot 81: 552–561

Zuberi MI, Lewis D 1988, Gametophytic-sporophytic incompatibility in Cruciferae, *Brassica campestris*. Heredity 61: 367–377.

Index

Printed and bound by CPI Group (UK) Ltd, Croydon, CR0 4YY

23/10/2024

01777685-0006